Polymer Nanomaterials for Specialty Applications

Related Titles

Biopolymers Based Advanced Materials
ISBN: 978-0-6482205-4-1 (e-book)
ISBN: 978-0-6482205-5-8 (hardcover)

Functional Polymer Blends and Nanocomposites
ISBN: 978-0-6482205-6-5 (e-book)
ISBN: 978-0-6482205-7-2 (hardcover)

Polymers in Oil and Gas Industry
ISBN: 978-0-6482205-0-3 (e-book)
ISBN: 978-0-6482205-1-0 (softcover)

Functional Nanomaterials and Nanotechnologies: Applications for Energy & Environment
ISBN: 978-0-6482205-2-7 (e-book)
ISBN: 978-0-6482205-3-4 (softcover)

Technology in Business
ISBN: 978-1-925823-02-8 (softcover)

Advances in Polymer Technology: Material Development, Properties and Performance Evaluation
ISBN: 978-1-925823-00-4 (e-book)
ISBN: 978-1-925823-01-1 (hardcover)

Polymer Nanomaterials for Specialty Applications

Dr. Vikas Mittal
Editor and Lead Author

CWP
Central West Publishing

This edition has been published by Central West Publishing, Australia

For more information about the books published by Central West Publishing, please visit https://centralwestpublishing.com

Disclaimer
Every effort has been made by the publisher, editor and authors while preparing this book, however, no warranties are made regarding the accuracy and completeness of the content. The publisher, editor and authors disclaim without any limitation all warranties as well as any implied warranties about sales, along with fitness of the content for a particular purpose. Citation of any website and other information sources does not mean any endorsement from the publisher and authors. For ascertaining the suitability of the contents contained herein for a particular lab or commercial use, consultation with the subject expert is needed. In addition, while using the information and methods contained herein, the practitioners and researchers need to be mindful for their own safety, along with the safety of others, including the professional parties and premises for whom they have professional responsibility. To the fullest extent of law, the publisher, editor and authors are not liable in all circumstances (special, incidental, and consequential) for any injury and/or damage to persons and property, along with any potential loss of profit and other commercial damages due to the use of any methods, products, guidelines, procedures contained in the material herein.

NATIONAL LIBRARY OF AUSTRALIA

A catalogue record for this book is available from the National Library of Australia

ISBN (print): 978-1-925823-04-2
ISBN (e-book): 978-1-925823-03-5

Contents

Preface

Nanomaterials based on polymers and polymer nanocomposites have found increasing use in many specialty applications. Structure, morphology and property profiles of these materials have been effectively understood in the recent years, which has resulted in tuning the performance of these materials for specific use. Some of these materials and applications include anti-corrosion coatings for long term protection of metal substrates, active layers in solar cells, nanoparticles or polymer modified nanoparticles for biomedical applications, thermally responsive polymers for drug delivery and tissue engineering, polymer nanofoams for insulation and EMI shielding, polymer sensors for sensing gas and humidity, magnetic polymer nanocomposites for oil spill clean up and contaminant removal, etc. In addition, the use of biopolymers in many of these applications further adds the functionality of eco-friendliness in the developed nanomaterials, which also helps to overcome the challenges posed by the use of conventional polymers owing to difficulty in waste disposal as well as depleting fossil fuels. This book aims to review the recent advances in the development of a variety of polymer nanomaterials for these advanced applications. Experimental studies have also been presented in order to provide readers in depth know how on the development of these nanomaterials for specific applications.

Chapter 1 presents an overview of some of the useful polymeric coating systems suitable for application in oil and gas sectors. A detailed discussion on the methods adopted to enhance their corrosion resistive properties has also been presented. Various polymeric systems useful for gas sensing applications are reported in Chapter 2. With developments continuing at a fast pace in this direction, commercial applications of such sensors at a large scale can be envisaged in the coming years. Recent developments in the generation of many thermally responsive polymeric nanomaterials have been discussed in Chapter 3, along with their applications in drug delivery, tissue engineering, gene delivery, etc. Chapter 4 highlights various techniques used to prepare polymer nanoparticles (or polymer modified nanoparticles) and to further develop the key features for biomedical applications. It also covers the role of different polymeric nanoparticles employed in this field. Chapter 5 presents thorough compilation of recent studies reporting the design, synthesis and property evaluation of various polymers and copolymers for utilization in photovoltaic applications. Chapter 6 reviews the literature studies describing polymer nanofoams as well as polymer nanocomposite nanofoams for use in a wide variety of advanced applications such as insulation and EMI shielding. Chapter 7 focuses on the advances in the development of magnetic polymer nanocomposites, along with their environmental applications. In Chapter 8, a systematic experimental study has been presented to confirm the potential of graphene reinforced polymeric coatings as effective anti-corrosion alternatives for long-term protection of metals. Further exploring the development of eco-friendly coatings, Chapter 9 reports the potential of polylactic acid as a strong coating material for the corrosion protection of metal substrates.

The support of chapter contributors for the successful completion of the book is highly appreciated. The book is dedicated to my family for constant motivation and encouragement.

Vikas MITTAL
Australia

1

Polymer Coatings for Corrosion Protection at Oil and Gas Facilities

1.1 Introduction

The metallic structures in oil and gas facilities undergo destructive processes in the presence of aggressive environments like chemicals, gases, humidity, etc., which lead to a significant loss to the economy of the country. According to a recent report, the total estimated annual cost of corrosion in the oil and gas production industry is $1.372 billion. Application of polymeric coatings is the most widely used method for the corrosion protection of the metallic structures. Polymeric coatings protect the metal substrates from external corrosive agents by acting as an effective barrier. Besides that, depending on the morphology and structure of polymers, the coatings possess the properties like thermal, chemical and mechanical stability. These properties combined with the ability to strongly adhere to metal surfaces yield durable coatings with longer service life time compared with other metallic and inorganic coatings. Multilayers of polyethylene (PE)/polypropylene (PP), fusion bonded epoxy (FBE) and polyurethane (PU) applied in few hundred nanometers thickness are generally used for the external anti-corrosion coatings. Material and application costs make the PE/PP more attractive candidates for the coating applications. However, limited heat resistance, lack of dimensional stability and limited resistance to oxygen and other oxidants limit their applications in the severe environmental conditions. The fusion bonded epoxy coatings have good chemical and abrasion resistance and excellent adhesion, though, the material cost is high. Polyurethane displays excellent color and gloss retention, good abrasion resistance and flexibility. The poor chemical resistance, severe temperature limitations and dimensional instability limit its application in the coating field.

Coating application methods and curing processes also lead to the

Gisha E. Luckachan and Vikas Mittal, The Petroleum Institute (part of Khalifa University of Science and Technology), Abu Dhabi, UAE*
**Current address: Bletchington, Wellington County, Australia*

creation of cavities, free volume and defects inside the coating matrix which make organic coatings permeable to oxygen, water and chloride ions. The penetration of these corrosive species through the microscopic pathways degrades gradually the coating resistance and the species diffuse further through the matrix, thus, reaching the coating/metal interface and initiating corrosion processes. The hydroxide ions formed by the cathodic reaction increase the pH which lead to reduction in adhesion bond as well as coating degradation and slowly accelerate the coating blistering. It is clear that obtaining a long-term corrosion protection using polymeric coatings without reinforcing these with other inhibiting materials is challenging.

In this review, an overview of some of the useful polymeric coating systems suitable for application in oil and gas sectors has been presented, along with discussion on the methods adopted to enhance their corrosion resistive properties.

1.2 High Performance Polymer Coatings

1.2.1 Epoxy Resins

Epoxy resins (polyepoxide) are thermoset polymeric materials where primary crosslinking (curing) occurs through reactions of the epoxide groups. The chemical inertness of cured epoxy resins is attributed to the stable carbon-carbon and ether bonds in their molecular structure. Epoxy combines the properties such as chemical resistance, electrical insulation, strong adhesion/affinity to heterogeneous materials and good processability for a coating material and is, therefore, widely used in paint and coating industries [1-5]. Epoxy coatings provide a physical barrier which protects metal from corrosion by separating it from the outside environment. In addition to this, these also serve as corrosion inhibitor reservoir that helps the steel surface to resist the attack from corrosive species [6]. However, epoxy coatings are susceptible to surface abrasion and wear damage which hamper down the successful application of epoxy in the barrier coatings [7-9]. The coatings are more prone to initiation and propagation of surface cracks which diminish the mechanical strength and glossy appearance [10]. The hydrophilic nature imparts a large volume shrinkage to epoxy coatings on curing, which enables the coatings to absorb water from the surroundings. The absorbed water and other corrosive species reach metal surface by migrating through the pores formed in the cured epoxy coating, thus, leading to corrosion processes on the metal surface and delamination of the coating [11,12]. Application of multilayer coatings effectively blocks the pores and enhances the coating protection. Massiani *et al.* [13] reported protection of a double coated sample with 320 μm coating thickness over 1 year compared to a single coated sample. The observed efficiency was attributed to the blockage of pores. The defects existing in the two successive layers have poor probability of connecting with one another, which, thus, reduces the chances of aggressive species to reach the metal surface for corrosion.

Several studies have reported the incorporation of a second phase into the epoxy matrix to improve the barrier performance of coatings [14-17]. One class of materials in this category are the inorganic fillers or nanoparticles, as their incorporation into the polymeric matrices significantly enhances the durability and integrity of coatings. These nanometer sized fine particles can disperse well in the coatings and fill the cavities which leads to crack deflection, bowing and bridging [14-17]. These particles occupying the holes formed during curing processes interconnect polymer chains and, thus, reduce the total free volume, increase the cross-linking density and enhance significantly the barrier properties of epoxy coatings [18-21]. Clay [22-26], titanium oxide (TiO_2) [27-29], silica (SiO_2) [30-33], silver oxide (Ag_2O) [34-36], iron oxide (Fe_2O_3) [37,38], zinc oxide (ZnO) [39], cerium oxide (Cr_2O_3) [40], alumina (Al_2O_3) [41] and zirconia (ZrO_2) [42,43] are commonly employed inorganic filler nanoparticles in polymeric coatings. Shi *et al.* [1] studied the effect of size and shape of nanoparticles on the corrosion resistive properties of epoxy coatings by using Zn, SiO_2, Fe_2O_3 and halloysite clay at a concentration of 1 wt%. A significantly enhanced Young's modulus of epoxy coating was observed with SiO_2 nanoparticles. The improvements were attributed to the bridging or interconnection of nearby chains with silica particles occupying the pinholes and voids of the coating. The enhancement in stiffness was also attributed to the chain segmental motion which was reduced by the curing of nanocomposite. Additionally, the silica nanoparticles reduced the viscosity of epoxy solution, which resulted in a thin layer of coating on the steel surface and reduced the internal stresses of the coating while curing [44,45]. Nano-Zn modification increased Young's modulus by ~30%, whereas ~25% to ~30% decrease occurred for Zn and Fe_2O_3 nanoparticles. The authors correlated the observed behavior to the specific surface area which was low for Zn and Fe_2O_3 nanoparticles (30-50 m^2/g and ≥30 m^2/g, respectively) as compared to SiO_2 particles. The agglomeration of these nanoparticles during curing created micro-cracks which diminished the strength of coating and, thus, resulting in a lower Young's modulus among other nanoparticles. The findings also emphasized the importance of uniform dispersion of nanoparticles in the formulation to obtain coatings with desirable mechanical properties. Shape and structure of the nanoparticles were also reported to play a significant role in generating epoxy coatings with desirable mechanical properties. The addition of phyllosilicate montmorillonite nanoclay ($(Na,Ca)_{0.33}(Al,Mg)_2(Si_4O_{10})(OH)_2{\cdot}nH_2O$) with layered structure also improved the Young's modulus of epoxy coatings [46-48]. On the other hand, halloysite clay nanoparticles with hollow cylindrical structure diminished the modulus. The incorporation of silica nanoparticles significantly enhanced the coating resistance (R_1) and the charge transfer resistance (R_2) of the epoxy coatings. As mentioned earlier, the observed behavior resulted from the nanoparticles depositing in the voids and holes, thus, diminishing the porosity of the cured epoxy coatings. The deleterious species underwent a tortuous diffusion path through the coatings which improved further the barrier performance of the coatings. It was also confirmed from the reduced coating capacitance (C_1) and the double layer capacitance (C_2). The strong adherence of cured epoxy coating with steel surface due to the presence of silica particles changed the physiochemical properties of the interface between the coating and metal [1].

Care is also needed during the mixing of polymer matrix with inorganic nanoparticles due to the weak interfacial interactions between the two. Inefficient mixing can lead to the formation of agglomerates of nanoparticles in the polymer coatings which could act as defects and, thus, reduce the barrier performance of the coatings [49,50]. Surface modifiers prevent the agglomeration of nanoparticles by increasing their compatibility with polymer matrices in composite coatings [40-52]. It, thus, also affects the corrosion protection and mechanical properties of the coatings significantly. Organosilanes ($(RO)_3SiX$) are effective surface modifiers for inorganic nanoparticles [51,53]. These consist of hydrolysable alkoxy (RO) such as ethoxy and methoxy as well as organofunctional group (X) such as vinyl, methyl and methacrylate groups. Normally sol-gel method of hydrolysis and condensation is used for the silanization of inorganic fillers. An improved dispersability of zirconium oxide, iron oxide and chromium oxide in the polymeric matrix was reported by the surface modification with organosilanes which contributed to an enhancement of the mechanical and anti-corrosion properties of the coatings

[54]. Gharagozlou *et al.* [55] used silica matrix incorporated with cobalt ferrite nanoparticles ($CoFe_2O_4$-SiO_2) to introduce corrosion inhibitive nature to epoxy coating. The dispersed $CoFe_2O_4$ particles filled and blocked the cavities, free volume and all electrolyte pathways in the coating. Corrosion inhibition by the released inhibitive species from the nano-pigment at the scratched areas of the coating was also reported. It was clearly observed from lesser extent of blisters and corrosion products appearing on the metal surface near the scratches of $CoFe_2O_4$ and $CoFe_2O_4$-SiO_2 nano-pigmented coatings exposed in salt spray chamber over 400 h compared with blank epoxy coating. The corrosion inhibition of $CoFe_2O_4$-SiO_2 nano-pigment coatings was attributed to the formation of a protective passive layer on the metal surface by the interaction of metal ions with Co^{2+}cations released from the nano-pigment. The corrosion inhibitive and barrier performance of $CoFe_2O_4$ nano-pigment coatings was comparatively poorer than the $CoFe_2O_4$-SiO_2 nano-pigmented coatings exhibiting the advantage of surface modification of nano-pigments by SiO_2 particles. The surface modification of Fe_3O_4 nanoparticles with 3-aminopropyltrimethoxysilane (APTMS) was also conducted by Javidparvar *et al.* [56] to study the changes in mechanical as well as corrosion inhibition properties of epoxy coatings. More effective dispersion of the nano-pigments in epoxy matrix was obtained by the silane functionalization. The interfacial interaction through covalent bonding between silane functionalized Fe_3O_4 nanoparticles with epoxy coating resulted in well dispersed nanoparticles which reduced the diffusion pathways and blocked the coating porosity. Mechanical properties of epoxy coating also increased significantly with the modification of nano-pigment with silane.

Apart from above mentioned nanoparticles, the use of graphene and its analogs has also become widespread as nano-reinforcement for anti-corrosion coatings. In a recent study, the addition of reduced graphite oxide (RGO) was observed to significantly enhance the anti-corrosion performance of the epoxy coating (Figure 1.1) [57]. The addition of 1% RGO filler to the epoxy thermoset resin yielded an initial impedance modulus of $10^{8.21}$ Ω cm^2 at the beginning of the 1 day EIS (electrochemical impedance spectroscopy) analysis, which indicated an increase of almost 2.5 orders of magnitude over neat epoxy. The impedance in the low frequency region of the Bode plot was $10^{6.83}$ Ω cm^2 at the end of the 1 day analysis which was an improvement of 2 orders of magnitude over the neat epoxy system. The enhancement in the anti-corrosion performance with the addition of RGO was maintained over the whole immersion cycle. Further addition of RGO to the epoxy coating did not enhance the anti-corrosion performance further.

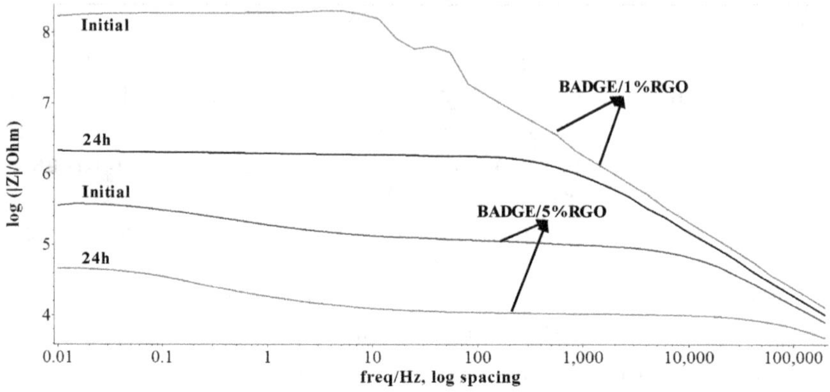

Figure 1.1 Bode plots of BADGE/RGO composite coatings in 3.5% sodium chloride solution. Reproduced from Reference 57.

Luo and Mather [58] also reported an advanced system based on shape memory assisted self-healing coatings. This was achieved by distribution of electrospun thermoplastic polycaprolactone in a shape memory epoxy matrix. As shown in Figure 1.2, the damage to the coatings could be healed automatically on the application of heat.

1.2.2 Polyurethane Resin

Polyurethane (PU) is a versatile material used widely in reaction injection molding, foams, coatings, fibers, adhesives, thermoplastic composites and elastomers. PU possesses excellent surface mechanical properties, like hardness, wear, scratch and abrasion resistance, effective weather and chemical resistance and high temperature stability which make it a suitable material for coating applications [59]. However, pure polyurethane coatings exhibited a diminished protective ability in aggressive environments of high salt spray, sewage, high heat, chemicals and other harsh conditions [60]. The modification of PU either by changing its microstructure or by the incorporation of organic or inorganic fillers within the matrix is an effective method to adjust its properties based on the application. PU commonly consists of two phases, soft and hard segments. Thus, the performance of PU as a coating material along with physical properties

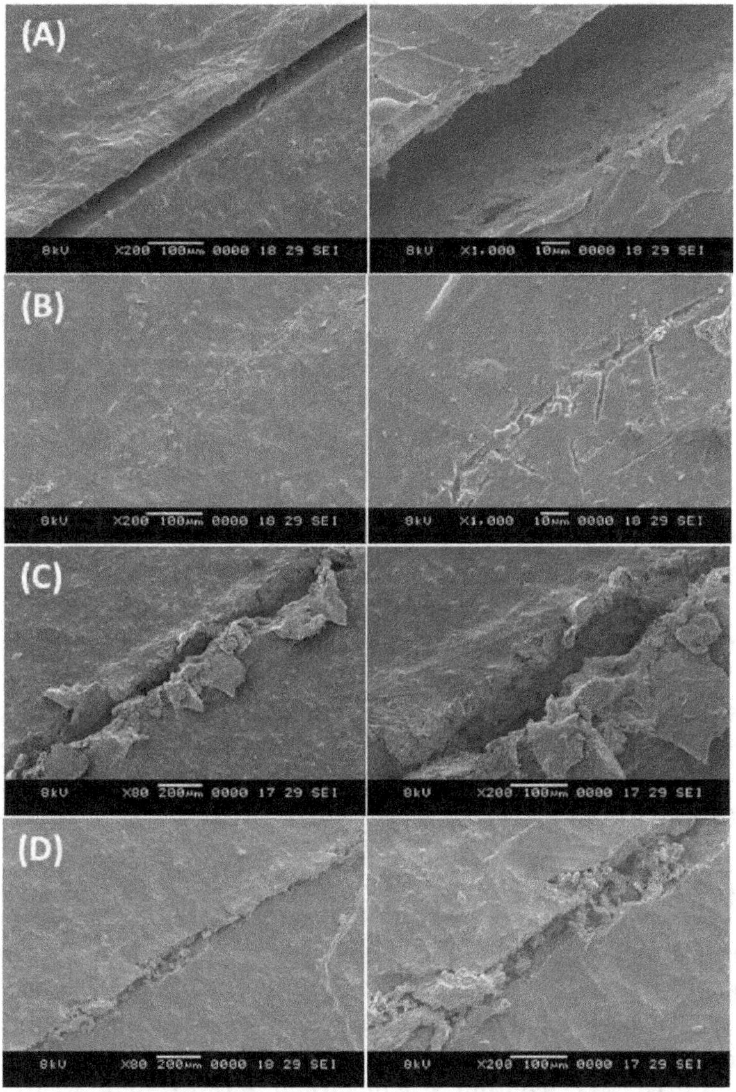

Figure 1.2 SEM micrographs of (A) cracked coating, (B) crack coating after self-healing, (C) scribed coating, and (D) scribed coating after self-healing. Reproduced from Reference 58 with permission from American Chemical Society.

in the solid-state are dependent on the nature of segments. The length of segments, the distribution sequences of each segment, crosslinking and branching of polymer chains are some of the important factors determining the properties [61-63].

Like epoxy, the incorporation of nanoparticles in the polymer matrix at very low volume fractions is generally employed to enhance the properties of PU for various applications. Organoclay in PU matrix reinforces the cohesive strength, fracture toughness, specific heat, hardness, electrical resistivity, barrier, thermal, flame-retardance and mechanical properties of the composite [64-75]. Based on the dispersion level of nanoparticles in the polymer matrix, two types of nanocomposites have been reported: intercalated and exfoliated nanocomposites. These two types of polymer/clay nanocomposites exhibit different physical properties because of the difference in their micro-phase morphology. In intercalated nanocomposites, clay expands by entering of polymer chains into the interlayer spacing of unit cells, keeping the periodicity of clay crystal structure. In exfoliated nanocomposites, the clay layers disperse individually into the polymer matrix which delaminate the crystal structure completely [76-79].

Ghosal *et al.* [80] reported soya polyurethane based silica hybrid composite coatings for anti-corrosion performance. In-situ generation of silica in the polymer led to enhanced thermal, physico-mechanical and corrosion resistance of the hybrid coatings. Figure 1.3 shows the potentiodynamic polarization (PDP) studies of the composites in comparison with pure polymer and pure substrate. The corrosion protection ability of the coatings enhanced with increasing the amount of silica in the composites. The enhanced corrosion protection efficiency of the composite coatings was attributed to the blocking effect and good adhesion of coating with the substrate surface.

Heidarian *et al.* [81] prepared PU/clay nanocomposite by dispersing organically modified montmorillonite (OMMT) nanolayers into castor oil based polyurethane matrix using ultrasonication assisted process. Corrosion protective nature of PU increased significantly by the incorporation of OMMT clay particles. The corrosion potential (E_{corr}) of PU/OMMT coating was observed in the Tafel plots to move to positive direction compared to pure PU coating. Furthermore, it kept traversing in the positive direction further with addition of OMMT clay particles up to 3 wt%. Also, smaller values of I_{corr} and higher values of R_p were noted for substrates coated with PU/OMMT nanocomposites. The observed results exhibited the barrier nature of PU/OMMT nanocomposite coatings. It was attributed to the dispersed silicate nano-layers in the PU matrix which increased the diffusion path of water and oxygen molecules through the coating, thus, leading to enhanced performance.

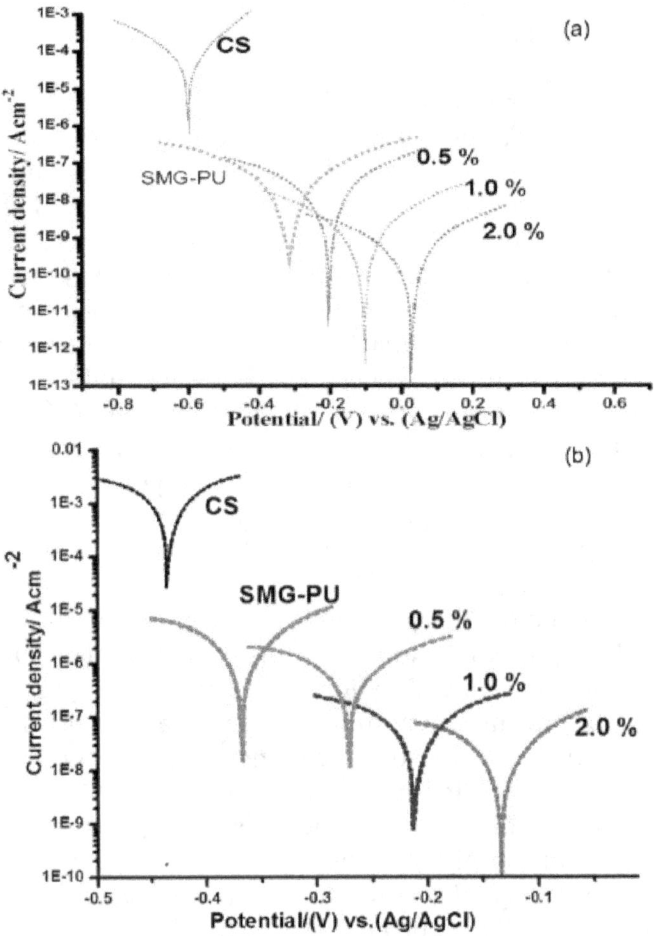

Figure 1.3 PDP curves of (a) CS, (b) SMG-PU, and composites with (c) 0.5%, (d) 1% and (e) 2% silica content in 3.5 wt % NaCl medium (SMG: soy oil monoglyceride, CS: carbon steel). Reproduced from Reference 80 with permission from American Chemical Society.

In another work, Yeh *et al.* [82] analyzed the effect of Na^+-montmorillonite (Na^+-MMT) clay particles in a waterborne polyurethane (WPU) composite coating [82]. WPU/Na^+-MMT nanocomposite coating with 3 wt% filler loading on cold-rolled steel (CRS) coupons exhibited superior corrosion protection than that of neat WPU based coatings. UV-cured polyurethane coating incorporated with organic phosphorous compounds was reported by Truc *et al.* [83] for the protection of carbon steel. The protection mechanism was attributed to the deposition of an insoluble layer of Fe(TDP)*n* on the metal surface by the interaction of metal ions with tridecylphosphate (TDP) in the composite coating [84].

Due to the increasing demand of polymer nanocomposites in coating industry, a variety of nanoparticles like ZnO, CeO_2, Al_2O_3, $CaCO_3$, TiO_2, etc., have also been incorporated in the PU formulations to modify their properties for coating applications. Since the size of nanoparticles is less than 100 nm, their incorporation into the transparent polymeric matrix didn't affect the optical properties. In certain cases, an increased transparency and light fastness of the coatings was also observed on addition of nanoparticles [85]. As mentioned earlier, the high surface area of nanoparticles makes them suitable for various surface modifications which result in uniform dispersion of particles in the polymeric matrices. The concentration of nanoparticles is also important in enhancing the properties of polymer composites. A small variation in the nanoparticle concentration can remarkably affect the coating characteristics such as optical, chemical and mechanical properties [86].

The effect of nano-silica particles on the corrosion protective behavior of PU coating was studied by Mills *et al.* [87]. The authors incorporated nano-silica into acrylic polyol/HDI polyisocyanate polyurethane matrix and cured it at different temperatures of 20 °C, 70 °C and 110 °C. Ionic resistance of PU coating increased significantly by the addition of 5 wt% silica. It was attributed to the reinforcement of polymer network which reduced the possible areas for water accumulation and, thus, minimized the water affinity of PU coatings. In addition, the presence of silica improved the abrasion resistance of PU coatings. It was also reported that effective bonding of nanoparticles with functional groups on polymer chain was essential to obtain well dispersed morphology and, thus, abrasion resistant coating. The high temperature curing of PU/nano-silica coating produced a highly crosslinked structure which was attributed to higher deformation resistance. Nano-silica incorporated PU coatings, thus, represent systems which can be used as barrier coatings with an aim of reducing the number of coating layers [88]. In another study, Rashvand *et al.* [89] fabricated nano-ZnO particles on polyurethane based waterborne coatings. The authors used nano-ZnO as anti-migrating additive and cathodically electrodeposited 3 wt% of nanoparticles onto the polyurethane films. The nano-ZnO filled coatings provided protection in 3.5 wt% sodium chloride solution over 120 days compared to neat PU coatings.

Similar to epoxy based materials, the corrosion protection as well as barrier properties of PU coatings were also further improved by the use of surface modified nanoparticles [90-93]. The organosilane modified silica nanoparticles of <50 nm size were reported to be effectively dispersed in the polyurethane matrix. Surface modification improved the effective crosslink density and the nanosized silica particles occupied the superfine pores in the coating matrix. The combined action of both properties reduced the diffusion rate of corrosive ions through the polyurethane coating. The interaction between silica particles and polyurethane matrix was improved when hydrocarbon groups containing long chains or higher numbers of short chains existed on the nanoparticle surface. Such highly crosslinked strong coating on the metal surface effectively slowed down the penetration of corrosive species through the coating. Figure 1.4 demonstrates an example of core shell Fe-silica nanoparticles for reinforcing polyurethane coatings [90].

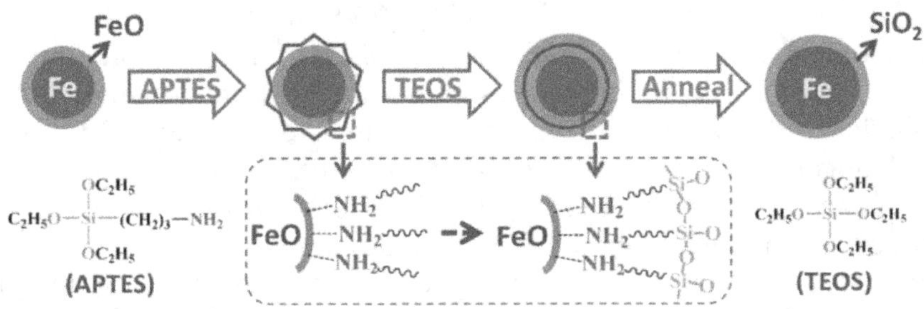

Figure 1.4 Synthesis of core shell Fe-silica nanoparticles. Reproduced from Reference 90 with permission from American Chemical Society.

Palimi *et al.* [40,49] modified Cr_2O_3 nanoparticles using 3-aminopropyltrimethoxysilane. The silane modified Cr_2O_3 nanoparticles improved the compatibility with polyurethane and, thus, resulted in a dispersed coating matrix. The size of nanoparticles decreased to 38 nm after modification. The average particle size of unmodified particles was observed to be 102 nm which exhibited a tendency to agglomerate on the coating surface. The significantly enhanced corrosion protection of polyurethane coating with the incorporation of surface modified Cr_2O_3 nanoparticle was confirmed through EIS analysis. The shift of maximum phase angle to higher frequencies at 2 wt% nanoparticle fraction exhibited that the barrier properties of polyurethane were enhanced significantly. It occurred due to the combined action of small size and greater surface area of modified nanoparticles. In addition, the surface modification enabled the particles to involve in strong bonding with the coating matrix. Such coatings are interesting as these reduce the adhesion bonds destruction by lowering the OH^- ions creation at metal-coating interface. Furthermore, Cr^{3+}ions released from Cr_2O_3 nanoparticles could form protective layers on the metal surface by reaction with OH^-ions at cathodic sites, which further prevented the cathodic reactions [94]. Additionally, the amount of particles able to provide protection is needed to be optimized for coating conditions. At high loadings, the chances of particle agglomeration are high which negatively affects the coating performance.

1.2.3 Polyetherimide

Among the different polymers used as protective coatings, polyetherimide (PEI) has drawn special attention because of its hydrophobicity, thermal and mechanical stability [95-99]. Polyetherimide is a thermoplastic amorphous polymer consisting of ether and imide groups, used mainly for high performance applications. The melt flow behavior and flexibility of PEI are attributed to ether units and the mechanical properties and thermal resistance are due to the presence of aromatic imide units [98,100]. The adhesion of PEI with metal surfaces is enhanced by the polar aromatic imide rings in the backbone of the polymer structure. PEI is stable in extreme environmental conditions. Its high glass transition temperature T_g (216.7 °C) and a low dielectric constant enable the polymer to form thermally stable coatings at environmental conditions [101-103]. At high temperatures of 320 °C to 380 °C, PEI films undergo crosslinking which results in coatings with higher resistance towards chemical attack, along with reduction in the permeability and pore size in the coatings [104]. Another interesting characteristic of PEI is its film forming property and flexibility to conform to the metal surface. Therefore, with the precise choice of solvent and heating conditions, it can form corrosion resistant, transparent and defect-free coatings. Conceicao *et al.* [97,105] studied the potential use of PEI in corrosion protective coatings. PEI coating of 3 μm thickness was prepared by dip-coating PEI/DMAc (20/80) solutions on hydrofluoric acid (HF) treated substrates. The coating maintained impedance in the order of $10^7\ \Omega\ cm^2$ even after 3312 h of exposure to 3.5 wt% sodium chloride solution [105]. The authors reported that the choice of solvent was significant in producing high quality defect-free polymer coatings. Solvents have been reported to affect gloss, appearance, porosity and adhesion of coatings [106-109]. Also, solvent retention in the coatings causes plasticization of the polymer, reduction in the T_g of the coatings and enhancement of influx of water and oxygen which subsequently reduce the protective behavior of the coatings.

Metal oxide nanoparticles (NPs) are potential fillers in enhancing the properties of polymer coatings, however, viscosity, appearance and rheological properties of coatings are strongly influenced by their tendency of agglomeration after the grinding and mixing processes. As a solution to these challenges, Rout *et al.* [104] generated metal oxide/PEI nanocomposite coatings by in-situ method. Steel substrate coated with TiO_2, MgO and Al_2O_3 nanoparticles in polyamic acid (PAA) solution were cured at 250 °C for 2-5 min to form a PEI@metal oxide composite coatings. The T_g of PEI was increased to a value ranging from 250 °C to 300 °C by the incorporation of metal oxide nanoparticles showing the improved cross-linking density of PEI matrix [110]. It was attributed to an enhanced rubbery as well as glassy region and increased storage modulus of PEI. It also underlined that the uniform distribution of NPs in the polymer matrix restricted the free molecular motion of the PEI chains. The relaxation of imide segments of PEI chains at high temperature contributed to a higher tan δ value to PEI@Al_2O_3 coating. The Tafel plots of PEI, PEI@MgO and PEI@Al_2O_3 didn't show anodic and cathodic current responses indicating a barrier/insulating nature of protection of these coatings. Compared with zinc (10^{-5} A/cm^2) coating and bare steel (10^{-6} A/cm^2), a reduced corrosion current density (10^{-8} A/cm^2) was observed for PEI@TiO_2 composite coating. The surface passivation of TiO_2 composite coating was emerged in the Tafel plot with a significant passivation band between -0.8 mV and 1.5 mV. Such a behavior of passivation band and corrosion potential (E_{corr}) was not observed in the PEI, PEI@MgO and PEI@Al_2O_3 coatings. Thus, each oxide NPs exhibited different protective nature in PEI matrix. Magnesia and alumina nanocomposites exhibited a barrier nature of protection, whereas titania nanocomposite had an active-passive behavior.

In a recent study, Akhtar *et al.* [111] reported fabrication of high performance polymer coatings of PEI on mild carbon steel and evaluated the protection offered by these coatings against corrosion in aggressive environments. Different solvent formulations of dichloromethane (DCM), n-methyl pyrolidinone (NMP) and n,n-dimethylacetamide (DMAc) with PEI were used. The effect of solvent choice, heating conditions and coating adhesion on corrosion performance was studied. An optimized heating cycle starting from room conditions to high temperature (270 °C) produced protective, stable high impedance (10^{11} Ωcm^2) PEI-DMAc coatings of thickness 7-8 μm, which resisted corrosion over 10 weeks of testing in 3.5 wt% sodium chloride solution. The performance of PEI-DMAc coating was attributed to removal of a large amount of residual solvent from coatings. It reduced plasticization and increased the T_g or rigidity of coatings, thus, making them less vulnerable to water permeation and resulting in the development of enhanced barrier properties and better wet adhesion (Figure 1.5).

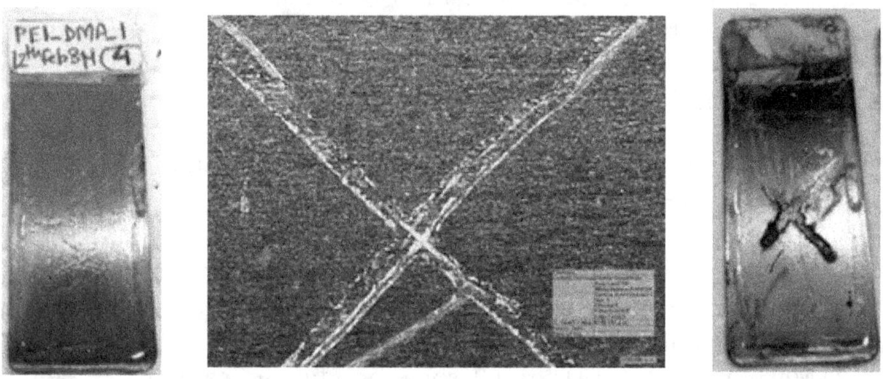

Figure 1.5 Appearance of PEI-DMAc coatings after 10 weeks of immersion in 3.5 wt.% NaCl solution (left) digital image (middle) optical image of scribed area and (right) digital image of the scribed area after 6 weeks of immersion in 3.5 wt% NaCl solution. Reproduced from Reference 111.

1.2.4 Conducting Polymer Composite Coatings

Intrinsically conducting polymers (ICPs) have recently emerged as potential candidates to replace the hexavalent chromium from the coatings used in oil and gas sectors because of their self-healing and environment friendly nature [112-113]. Polypyrrole (PPy), polyaniline (PANI), poly (*N*-ethyl aniline), polyindole and polythiophene are the commonly used ICPs for coating applications [114-125]. Anti-corrosive coatings can be prepared by directly polymerizing the ICPs on the metal substrate [126]. The solubility of chemically synthesized ICPs in common organic solvents is very low, therefore, these can be used for coating applications by blending with other polymers. Such blends provide coatings with desired mechanical properties especially at low concentrations by the right choice of host polymer matrix.

Yeh *et al.* [73] reported enhancement of corrosion protection effect by the generation of polyaniline-clay nanocomposites. Polyaniline-clay nanocomposite coatings with low filler fraction (0.75 wt%) on cold-rolled steel (CRS) were observed to have superior corrosion protection over pure polyaniline coatings, as confirmed from electrochemical measurements of corrosion potential, polarization resistance and corrosion current in 5 wt% aqueous NaCl electrolyte. Figure 1.6 also shows the reduction in the gas permeability of the composites as a function of filler content.

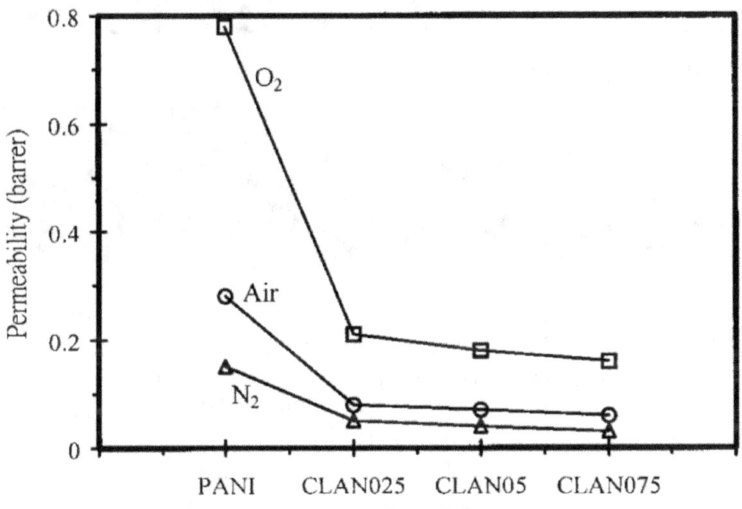

Figure 1.6 Permeability of O_2, N_2, and air as a function of the MMT clay content in the polyaniline-clay nanocomposite materials. Reproduced from Reference 73 with permission from American Chemical Society.

In a recent study, Chaudhry and Mittal [127] generated polyaniline/graphene composite nanoparticle pigments with different compositions by in-situ polymerization of aniline in the presence of graphene nano-platelets. Poly(vinyl butyral) (PVB) was used as the coating matrix. Figure 1.7 demonstrates the surface of the substrate after the corrosion test as well as removal of the coating. Substrates coated with pure PVB (Figure 1.7(a)) and pure PANI in PVB (Figure 1.7(b)) formulations exhibited extensive surface corrosion. It indicated that the corrosion media could penetrate faster through the coatings, thus, initiating the corrosion process. The substrate coated with PANI/10r-GO (Figure 1.7(c)) in PVB exhibited improved anti-corrosion performance as the extent of the surface corrosion was markedly reduced in comparison with Figure 1.7(a) and (b). Further increase in the graphene content enhanced the performance further. PANI/20r-GO, PANI/30r-GO and PANI/40r-GO coated substrates optically exhibited high level of stability to the immersion conditions.

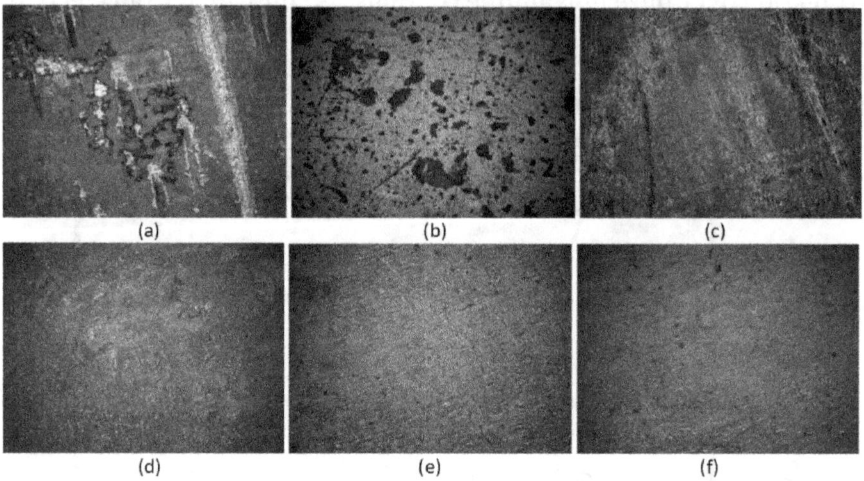

Figure 1.7 Surface of the substrates after corrosion testing and coating removal; (a) blank (pure PVB), (b) PANI, (c) PANI/10r-GO, (d) PANI/20r-GO, (e) PANI/30r-GO and (f) PANI/40r-GO. The width of the images equals 200 μm. Reproduced from Reference 127.

Several possible mechanisms have been reported on the corrosion inhibition of ICPs. Mengoli *et al.* [128] and DeBerry [129] reported for the first time the ability of ICPs in preventing metal corrosion [128,129]. The authors studied this behavior for electrodeposited polyaniline coatings on steel surface. Since then, a large number of studies have been reported on ICPs' performance focusing mainly on polyaniline and polypyrrole based coatings. Surface ennobling and anodic passivation of steel induced by the redox behavior of ICPs is the most cited corrosion protection mechanism [130-132]. As polypyrrole or polyaniline based coatings are applied on the metal surface in the oxidized state, the coatings behave as an oxidizer and improve the nature of oxide layer formed at the interface of polymer matrix and metal. The coatings even provide an anodic protection by keeping the metal in passive domain at small defects. The oxygen present in the coating system oxidizes the reduced ICPs to the initial state. These continuous redox reactions of ICPs passivate the metal surface and block the possible defects and pinholes in the coatings. The doping anions released from ICPs react with the metal cations and create a second complex barrier layer on the metal surface [133-136]. Another proposed corrosion protection mechanism is that ICPs shift the oxygen reduction site from polymer/metal interface into the polymer coatings. It reduces significantly the concentration of radicals over the metal for oxygen reduction and, hence, prevents the corrosion processes and coating disbandment [137-140]. Michalik [133] conducted a detailed study on the possible protection of ICPs through the ennobling mechanism. Using a special model sample set up, it was shown that electrochemically deposited polypyrrole (PPy) films could indeed protect small pinhole like defects from corrosion. Polypyrrole was able to passivate the metal surface at defects during immersion in chloride free solution for a few hours, but this passivity started to break down in chloride containing solution after few seconds of immersion. The size of the defect significantly determined the degree of possible protection. Polypyrrole exhibited the ability to recharge itself after disconnection from the defect, so it could provide a real protection under cyclic climatic conditions. In the presence of a larger defect, however, the positive effect of the conducting polymer turned into a negative one and fast delamination was observed. It was observed that instead of electrodeposition, if the conductive polymer was added as an additive to an insulating matrix, a fast reduction of the whole coating should be inhibited if the mean percolation path of conductive particles is not too high. In this case, the insulating matrix polymer prevents the disastrous reduction of the whole coating, while a capability for small pinhole passivation should still exist. Also, another positive effect might apply: the high potential of the additives might inhibit oxygen reduction at the interface, therefore, preventing fast delamination. The study confirmed the advantage of using conducting polymers as additive pigments in the barrier polymer coatings.

Several other studies have been reported on the use of ICPs as additive component in barrier coatings. Jafarzadeh *et al.* [134] reported active protection of UV-cured polyester acrylate (PEA) coatings by the incorporation of phosphoric acid doped polyaniline (PANI-PA). PEA provided a typical barrier protection to carbon steel. However, a significantly enhanced protective performance of PEA was reported by the incorporation of PANI-PA particles into the insulating PEA coating matrix. It was confirmed in the open circuit potential (OCP) of PEA coatings containing different amounts of PANI-PA particles immersed in 3 wt% sodium chloride solution. 1 wt% PANI-PA particles exhibited longest active protection and the dispersion of more than 3 wt% particles significantly reduced the protection because of the hydrophilicity of PANI which enhanced the water uptake.

The main factors affecting the protective performance of polyaniline is the form of polyaniline and pH of the environment [132,140]. Talo *et al.* [115] analyzed the effect of different forms of polyaniline in providing steel corrosion protection. The authors used emeraldine base (EB), phenyl sulfonic acid (PANI*PSA) and polyaniline protonated with camphor sulfonic acid (PANI*CSA) in epoxy blended coatings on mild carbon steel at neutral, alkaline and acidic solutions. Emeraldine base was reported to be the most effective form of polyaniline in providing corrosion protection to mild steel in saline solution. Fahlman *et al.* [141] suggested an anodic passivation mechanism of EB form of polyaniline. The form was reported to effectively remove the charge from metal and form strong protective oxide layer which passivated the defects on the coating. EB also formed low-porosity, dense and adherent film which maintained a basic environment on the metal surface [141]. The protective iron oxide layer was analyzed as consisting of a thin layer of Fe_2O_3 (~15 A°) and a thick layer of Fe_3O_4 at the steel/coating interface [141,142]. On the other hand, the protonated polyaniline exhibited a different protective mechanism. An effective protection of PANI*CSA was reported in acidic solution than in neutral solution. PANI*PSA coatings didn't provide steel protection in sodium chloride solution. An enhanced corrosion protection was reported for PANI*PSA/epoxy coatings compared to pure epoxy coatings. Thus, based on the requirement or the environment, the properties of protective coatings containing polyaniline can be tailored.

Another proposed mechanism of ICP blended coatings is the self-healing, where ICP is considered as corrosion inhibitor reservoir if it can be doped with effective anionic inhibitors. As corrosion starts the oxidation of metal, the surface changes the oxidation state of inhibitor doped ICPs which triggers the release of the doping agent at the defected areas [143-147]. Such smart coatings can even prevent corrosion on the bare steel surfaces which get exposed to aggressive environment through the defects or scratches on the coating [143,146]. In certain cases, ICPs show a duel protection mechanism by releasing dopant anion which form a passive complex with metal ions on the defects and stop the corrosion processes. Subsequently, this complex acts as a physical barrier and avoids the penetration of corrosive species further down to the metal surface which improves the protective nature of coatings. Such a mechanism was reported in the acrylic blends formed by poly(methyl methacrylate) and sulfonate-doped polyaniline [135,148,149].

As mentioned earlier, the blending of ICPs together with other pigments and anti-corrosive agents in paints can also significantly impart an efficient protection to the underlying metal surfaces. Zinc oxide is the most commonly used additive and several studies have reported the corrosion inhibitive nature of primers containing zinc pigments and PANI [150,151]. Patil *et al.* [142] reported the anti-corrosion behavior of PVAc coating matrix enriched with PANI hybrid with ZnO nanoparticles. Corrosion protection was achieved by the combined effect of both PANI and ZnO nanoparticles. The redox nature of PANI and the semiconducting behavior of ZnO improved the barrier properties of PVAc coating by creating a p-n junction at the metal-coating interface. The concertation of nanometer sized ZnO particles and PANI in the coating was above the percolation threshold. It provided a strong network of ZnO and PANI in the resultant composite coatings. ZnO being n-type semiconductor and PANI behaving as a p-type semiconductor formed a p-n junction which allowed the transport of electrons only in one direction and, thus, prevented the flow of ions or electrons across the film. The observed behavior of ZnO-PANI composite significantly enhanced the barrier properties of the PVAc composite coating. In addition, the metal scribes were protected by the dopant anions, which were released from PANI by redox reactions.

PVB coating on steel surface incorporated with PANI nano-particulate TiO_2 hybrid composite was also reported to exhibit a similar protective mechanism [152]. Armelin *et al.* [153] also proposed a slightly different protective mechanism for PANI-EB/zinc rich epoxy coatings on steel surfaces. Here, zinc oxide (ZnO) formed a passive protective layer on the metal surface which hindered the reach of dissolved oxygen onto the metal surface and prevented the receiving of electrons from the metal. At the same time, the chlorine anions and neutral or slightly acidic pH (pH ~ 6-7) converted the undoped PANI to the doped form. The doped PANI captured the electrons formed due to oxidation reactions and got reduced again to neutral form. These autocatalytic redox processes stabilized the potential of iron in the passive region in the ferrous form Fe (II) and zinc in the active region. The results suggested that incorporation of conducting polymers would be a promising approach to reinforce the protective ability of epoxy primer containing conventional anti-corrosive pigments. It was also confirmed that the conducting polymers can be used as a good replacement for the existing corrosion inhibitors like hexavalent chromate. However, it should be noted that the performance of the conducting polymers depends on the corrosive environment as well as the way these polymers are applied on the metal surface. Depending on the conditions, a conducting polymer may lead to an enhanced corrosive attack or may have excellent protection capability.

A new method was proposed by Kalendova *et al.* [151,154,155] to combine the use of PANI and inorganic pigments. It also addressed the problems associated with PANI in resin blended coatings, that is, inefficient distribution, lack of excellent polymer-polymer contact, poor adhesion and change in volume of PANI due to redox reaction. To overcome these problems, the authors modified the surfaces of four pigments goethite FeO(OH), talc ($Mg_3(OH)_2$- (Si_4O_{10}), specularite, graphite and Fe_2O_3 with PANI and subsequently mixed with epoxy binder. In all cases, the PANI modified pigments exhibited a better corrosion resistance than unmodified pigments. A similar behavior of enhanced corrosion protection of steel was reported by Sathiyanarayanan *et. al.* [156] for the coatings containing PANI modified TiO_2 and Fe_2O_3. Several other reports are also available on the use of PANI with other additives, most of which were used in the macro-particulate form [157-160]. As the size of particles play a significant role in enhancing the corrosion protection of polymer coatings, thus, compared with the conventional micron size additives, nano-particulate fillers have been reported to provide much better barrier properties even at low concentrations [161,162].

Limited number of studies have been reported on the use of PPy as corrosion inhibitive pigment. Yan *et al.* [163] used oxyanion (molybdate, phosphate or vanadate) doped PPy deposited on aluminium flakes as inhibitive pigments in epoxy primer for AA 2024-T3 aluminium [163]. Among the different oxyanion pigments, vanadate doped coatings exhibited the best protection performance. It was observed from the scanning vibrating electrode technique (SVET) current density maps collected from the areas near the defect on oxyanion doped PPy Al flake coatings immersed in dilute Harrison's solution (DHS) over 24 h [164]. Well-defined cathodic as well as anodic current peaks attributed respectively to the reduction of PPy and/or O_2 and active dissolution of the substrate wereobserved in the SVET current maps of PSAF (PPy–sulfate Al flake) and PPAF (PPy–phosphate Al flake) coatings. In the case of PVAF (PPy–vanadate Al flake) and PMAF (PPy–molybdate Al flake) coatings, such behavior of current flow was not detected which indicated that the vanadate and molybdate anions suppressed the active dissolution of AA 2024-T3 exposed in the defects. In addition, due to the reduced anodic dissolution by the inhibitive effect of oxyanion dopants, the PPy component also provided protection through oxygen-scavenger effect.

In another study, Niratiwongkorn *et al.* [165] reported a similar protective mechanism of PPy pigment in a polypyrrole-carbon black (PPyCB) composite coating with PVB. The incorporation of PPyCB pigment at 20 wt% didn't affect the adhesion and barrier properties of PVB coatings. During the immersion in 4 wt% sodium chloride solution, PPy imparted a self-healing protection to the PVB/PPyCB composite coatings. It occurred by the redox behavior as well as diffusion barrier nature of PPy (demonstrated in Figure 1.8). The passive layer formed on the metal surface by the interaction of released dopant from the organic sulfonic acid doped PPy with metal oxide enhanced the barrier protection of coatings. As a result, this physical barrier

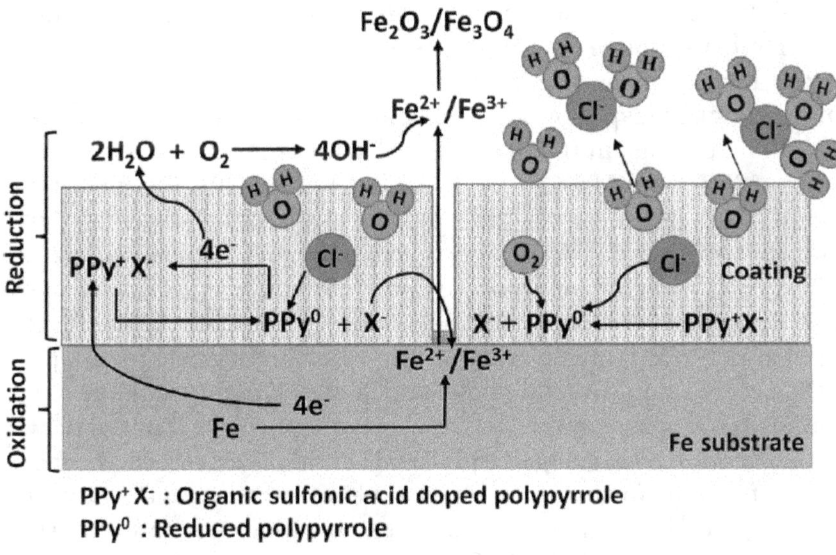

Figure 1.8 Schematic representation of corrosion protection imparted by PPy pigments in PVB/PPyCB composite coatings. Reproduced from Reference 165 with permission from Royal Society of Chemistry.

along with reduced PPy prevented the oxygen and water diffusion down to the metal surface, which reduced the metal corrosion processes further.

1.2.5 Polymer/Graphene Nanocomposite Coatings

The role of reinforcing fillers in enhancing the barrier protection of polymeric coatings has been clearly established in the previous sections. The presence of fillers in the polymer matrix increases the tortuous path for oxygen and water diffusion through the coating and improves the adhesion strength of polymer with metal surface, which subsequently enhances the corrosion protective nature of polymer coatings. As also mentioned earlier, size, shape, orientation of the nanofiller as well as the nature of interactions between the matrix and nanofiller determine the end use properties of polymer nanocomposites (PNCs) [166].

Graphene reinforced polymers have been extensively reported in the recent years as ideal formulations for corrosion resistant coatings on metal substrates. Graphene has a lamellar structure with one-atom thick two dimensional layers of sp^2 carbon atoms bonded together in a honey comb lattice. It is considered as "thinnest material in the universe" and has a wide variety of potential applications [167,168]. Graphene has high mechanical strength, surface area, chemical resistance, aspect ratio, flexibility, electrical and thermal conductivity, transparency, EMI shielding ability and low coefficient of thermal expansion (CTE) compared with other conventional nanofillers (CNT, Na-MMT, CNF, LDH, EG, etc.) [169,170]. The incorporation of graphene enhances the mechanical, electrical, thermal, flame retardant and gas barrier properties of polymers in nanocomposites [171-173]. Graphene incorporated polymer nanocomposites also exhibit improved electrical and mechanical properties, which otherwise cannot be attained by clay or other carbon fillers [174-176].

The corrosion protection mechanism of polymer/graphene nanocomposites is based on the "active-passive" approach. It prevents water permeation by serving as a physical barrier and protects the metal surface from the formation of ion channels. It passivates the interface between metal and metal oxide by establishing a Schottky barrier through electron depletion processes [95,177]. The barrier effect of graphene in preventing the penetration of water and corrosive species occurs due to its high aspect ratio and platelet structure which enhances the diffusion pathways in the coating. The electrons in the p orbitals of graphene create a repelling field which repel the negatively charged ions (like chloride ions) from entering the coating, thus, improving the anti-corrosion performance [177-179]. Graphene eliminates the potential differences between the cathodic and anodic sites on the metal surface and, thus, prevents the redox reactions of corrosion processes [178]. A Schottky barrier of p-n junction is formed at the interface of metal surface and graphene platelets when the p-type metal comes in contact with n-type graphene particles. This potential barrier disrupts the easy flow of electrons from anode to cathode through the coating, thereby, preventing the metal corrosion by an active-passive approach [95,177]. In addition to the in-plane alignment and aspect ratio, graphene dispersion in a polymeric matrix is another important factor in determining the properties of composites and their coatings.

Several studies have been reported on the corrosion inhibitive nature of graphene in polymer composite protective coatings [180-188]. Most of these studies demonstrated the diffusion barrier nature of graphene, thus, preventing direct interaction between underlying metal and corrosion promoting species. Chang *et al.* [185] studied the corrosion protective behavior of polyaniline/graphene composites (PAGCs) in comparison with polyaniline/clay composites (PACCs) and neat polyaniline. In order to enhance the dispersion of graphene in polyaniline matrix, graphene was functionalized with 4-aminobenzoic acid. PAGCs exhibited improved gas barrier properties than PACCs and led to a 68% and 66% reduction in water and oxygen permeability respectively. The observed performance was attributed to high aspect ratio of graphene compared with clay and its uniform dispersion in polyaniline matrix due to the functionalization. The authors concluded that PAGCs are ideal materials for providing a gas barrier and synthesizing anti-corrosion coatings. In addition, the barrier properties of PAGCs were observed to enhance with graphene loading. These results suggested that polymer/graphene composite coatings have potential advantage for application in oil and gas field and can replace the traditional polymer/clay composite coatings. Similar mechanisms were reported for epoxy/graphene [180,181], polyimide/graphene [184], polyurethane/graphene [182,183] and poly (sodium styrene sulfonate)/graphene [186] composites based protective coatings.

In addition to the diffusion barrier protection, the ability of graphene in passivating the metal surface through electron depletion was reported in polyetherimide/graphene composite coatings

[95,177,189]. This electron depletion creates a Schottky barrier at metal/graphene junction, which requires more bias to facilitate electron transfer. Such potential barrier at metal/coating interface restricts the flow of electrons to the metal surface and, thus, impedes the oxidation reactions and further corrosion processes. In addition, a band gap will open in the semi-metallic graphene by the graphene hybridization with metal ions, which leads to a semiconducting metal/graphene interface. Such a passivating semiconducting layer leads to charge depletion and, thus, suppresses the corrosion processes on the metal surface further. Such strong protection was reported for carbon nanomaterial dispersed in PEI coatings [95,177,189]. It was attributed to the uniform dispersion of carbon nanomaterials in the PEI matrix due to the interaction of aromatic moieties on the polymer backbone with the π-conjugated graphene basal planes. Thus, it can be suggested that the delamination of composite coatings from the metal could be prevented by the incorporation of well dispersed conductive fillers which aided the formation of strong passive layers at the metal surface. It would be a potentially efficient method to enhance the corrosion resistive behavior of existing barrier coatings used in oil and gas sectors.

In a recent study, Chaudhry *et al.* [190] compared the barrier properties of nanocomposites prepared from Haydale's graphene and PVB. In specific, four kinds of graphene, i.e., U-GP and Haydale's processed graphene nanoplatelets produced using three types of process gases, i.e., fluorine (F-GP), oxygen (O-GP), and argon (A-GP) were used. As shown in Figure 1.9, phase angle exhibited distinct behavior for PVB and composites coatings. The range of frequency corresponds to the various regions in the coating, i.e., starting from coating surface to substrate surface. After 1 h of immersion, the PVB coating at higher frequency ~100 kHz exhibited a mixed capacitive and resistive behavior (~-43°), which indicated the barrier and diffusive nature of the coating. The same behavior continued over intermediate frequency range with a variation of theta around ~-45°±-5°, afterwards it changed to resistive behavior in the low-frequency spectrum. At frequency ~25 kHz, the slight increase in the angle ~-37° indicated dominant resistive behavior. A little bump could be observed at 12.4 Hz corresponding to the phase angle ~-46° between two minima of theta ~-49° at 100 Hz and ~-48.8° at 2.5 Hz. The continuous resistive and capacitive behavior ranging from higher to intermediate frequency could be an indication of constrained diffusion of electrolyte through the coating corresponding to Porous Bounded

Warburg (PB-W) as mentioned in Gamry Echem Analyst™ manual. The major transition from mixed resistive-capacitive behavior to resistive behavior occurs at ~0.9 Hz corresponding to the metal/interface and metal characteristics.

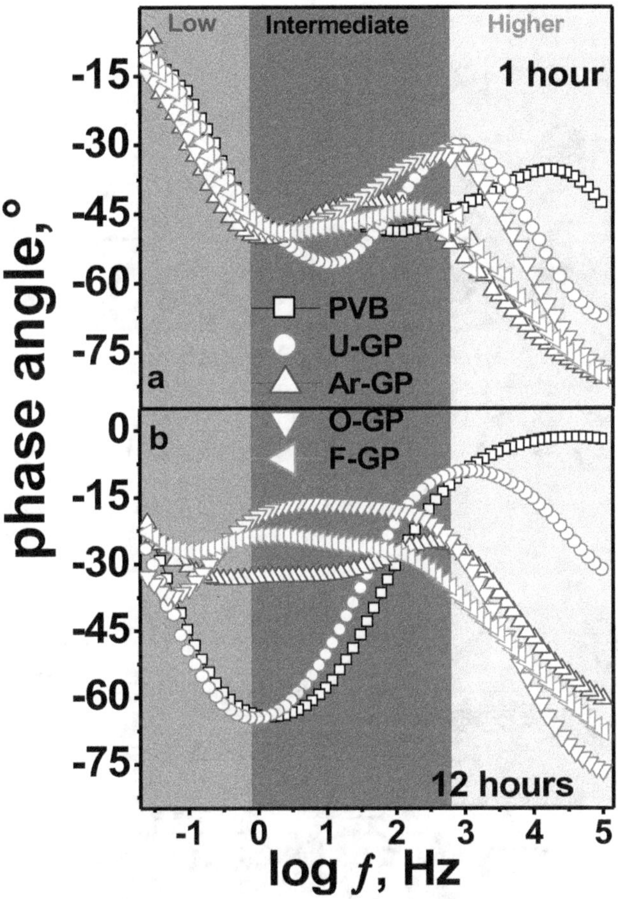

Figure 1.9 Phase angle behavior for PVB and GP-PVB nanocomposites coating on carbon steel recorded after 1 h and 12 h of immersion in 4 wt% NaCl. Reproduced from Reference 190 with permission from Elsevier.

Chaudhry *et al.* [191] also studied the anti-corrosion performance of unmodified graphene platelets (UGP) dispersed in PVB. As shown by the schematic in Figure 1.10, the barrier nature of the coatings could be explained by the low diffusivity of corrosion reactants due to sheet-like structure of graphene platelets. The corrosive solution has to adopt longer diffusion and tortuous paths to reach at the metal surface which, in result, slows down the corrosion rate. In case, the defects are created by the incorporation of filler due to the poor dispersion and agglomerations of fillers, this phenomenon may cause inverse results.

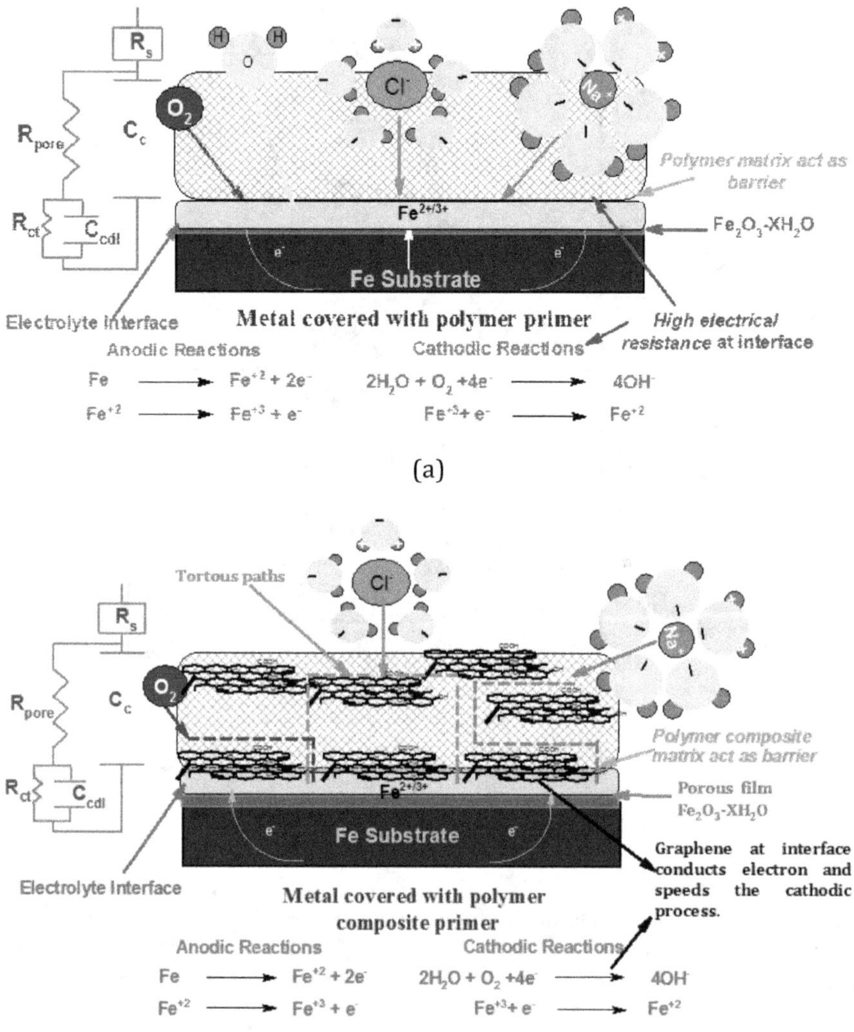

Figure 1.10 Scheme of corrosion protection phenomena in the (a) absence and (b) presence of UGP in PVB with electrochemical corrosion models for 26 h immersion in 0.1 NaCl. Reproduced from Reference 191 with permission from Royal Society of Chemistry.

Another study by Luckachan and Mittal [192] reported that graphene incorporation (5 wt%) decreased water permeability of the chitosan-SiO_2 (Ch-SiO_2) hybrid coatings, as observed from the lower value of C_c in the order of 10^{-9} Fcm^2 compared with the value in the order of 10^{-5} Fcm^2 for 20Ch-SiO_2/PVB hybrid coatings (chitosan corresponding to 20 wt% of total silica content). The behavior was attributed to the increased diffusion pathways for water molecules inside the coating, thus, reducing the chances of water molecules to reach the metal surface as free electrolyte. Enhanced corrosion resistance of 20Ch-5Gr-SiO_2/PVB hybrid coatings was further evidenced from the decreased I_{corr} as compared with the native 20Ch-SiO_2/PVB hybrid coatings. Therefore, it could be concluded that graphene incorporation would be promising to obtain excellent corrosion protection of Ch-SiO_2 hybrid coatings. A schematic representation of the tortuous movement of water through 20Ch-5Gr-SiO_2/PVB coating is also presented in Figure 1.11.

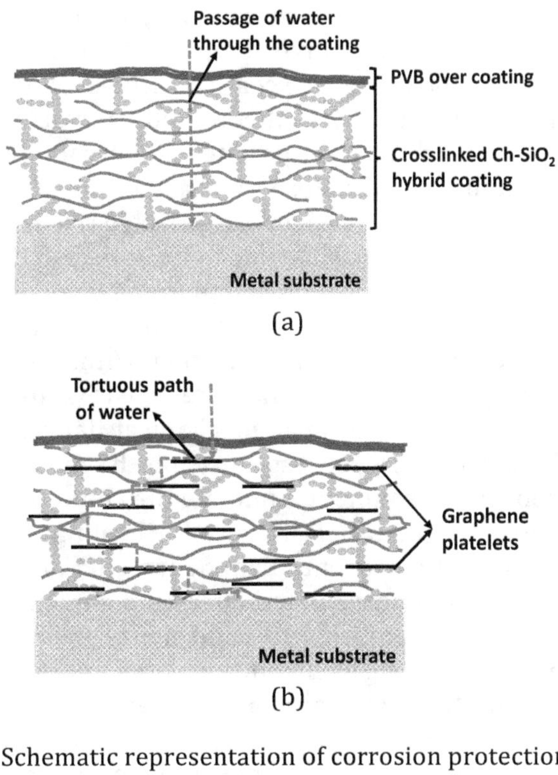

Figure 1.11 Schematic representation of corrosion protection provided by Ch-SiO_2 hybrid coatings in the absence (a) and presence (b) of graphene platelets. Reproduced from Reference 192.

1.2.6 Smart Polymer Coatings with Microcapsules and Nano-containers

As mentioned earlier, polymer coatings protect metals from aggressive environments by creating a passive barrier on the metal surface. During service life, the mechanical properties of the coatings change and lead to the formation of micro-pores or cracks on the coating film, which subsequently propagate and create paths for atmospheric oxygen and moisture to reach the metal surface. This leads to initiation of corrosion and formation of flakes on the metal coating interface leading to the disbanding of the coatings. These drawbacks of polymer barrier coatings have motivated the development of new generation of protective coatings, where the mechanical properties of the coating matrix provide a passive protection and the changes of local environment or the integrity of the passive matrix stimulate the active responses of the coating. Active corrosion protection is aimed to decrease the corrosion rate when the barrier protection is damaged and the corrosive species come in contact with metal [193]. An established approach for the active corrosion protection is the addition of inhibitors directly to the protective coatings. However, this approach is not environmentally friendly as the inhibitor molecules can leach out from the coating matrix and pollute the surrounding environment. Corrosion inhibitors also lose their activity by chemical interaction with coating matrix. Moreover, a controlled release of inhibitor at the damaged sites is not possible due to its bulk distribution in the coating matrix. These issues have resulted in the development of advanced coating systems based on the "passive" matrix-"active" container structure. Active containers include nano/microscale particles loaded with corrosion inhibitor which are sensitive to the internal (e.g., pH changes) or external (e.g., mechanical damage) corrosion triggers [193]. As corrosion process starts on the metal surface, the changes in the local environment stimulate the containers to release the encapsulated active material (inhibitor or healing agent) directly at the damaged sites and prevent further corrosion processes. Since the active materials are encapsulated inside the nanoparticles, these would not affect the barrier properties and, hence, the integrity of the coating. Recently, the healing agent encapsulated microcapsules have attracted much attention in the area of polymeric composites for developing active/smart protective coatings [194]. Such microcapsules or nano-container systems can significantly enhance the performance of polymeric barrier coatings and, thus, provide coatings with long service life time which subsequently reduces the maintenance cost of metal structures.

Microcapsules are nanometer to micrometer sized tiny particles consisting of a core material and a covering shell [195]. Encapsulated active material can be in solid, liquid or gaseous form. The encapsulating shell can be natural or synthetic polymeric material based on the desired characteristics of the microcapsules. Microcapsules, thus, act as a reservoir of healing agent. These must possess certain properties, like sufficient strength to withstand the application of stress, ability of healing agent to diffuse out from the shell, high bonding capacity with the host polymer matrix and impermeable shell wall to prevent the leakage of encapsulated healing agent. Siva *et al.* [196] used mercaptobenzothiazole (MBT) and linseed oil as core materials and urea formaldehyde (UF) as shell material for the preparation of 70-200 nm sized microcapsules [196]. An enhanced corrosion protection was reported for the UF microcapsules incorporated epoxy coating as compared to pure epoxy coating. It was confirmed from the SVET results that the initial anodic current flow at the affected area got suppressed due to the self-healing nature of the microcapsules. Anodic current flow was clearly visible during the initial immersion time in 3% sodium chloride solution. With time of immersion, it exhibited a decreasing trend and a fully suppressed anodic current was noticed after one day of exposure. The rupture of microcapsule by cutting the coating released the linseed oil which sealed the crack with a precipitate formed by oxidation with atmospheric oxygen [197].

Huang *et al.* [198] prepared polyurethane (PU) microcapsules encapsulated with hexamethylene diisocyanate (HDI) as core material *via* interfacial polymerization reaction of 1,4-butanediol and commercial methylene diphenyldiisocyanate (MDI) prepolymer in an oil-in-water emulsion. The diameter of the resulted PU microcapsules was in the range 5-350 μm and shell thickness from 1-15 μm, both related directly to the agitation rate in the double logarithm coordinates. The microcapsules embedded epoxy coatings exhibited a self-healing nature. The HDI release from the ruptured microcapsules formed a passive product on the metal surface by reaction with water, which healed completely the scratches during an immersion period of 48 h in sodium chloride solution. In a recent study, Tatiya *et al.* [199] also developed novel polyurea microcapsules containing linseed oil as the active healing agent by interfacial polymerization of commercial methylene diphenyl diisocyanate (MDI) and dendritic polyamidoamine (PAMAM) (Figure 1.12). The amino functional groups on microcapsules improved the adhesion performance with both metal substrate and polyurethane matrix. Microcapsules exhibited an excellent healing ability with a 2 to 5% linseed oil loading and a better anti-corrosion property at 5% capsule loading.

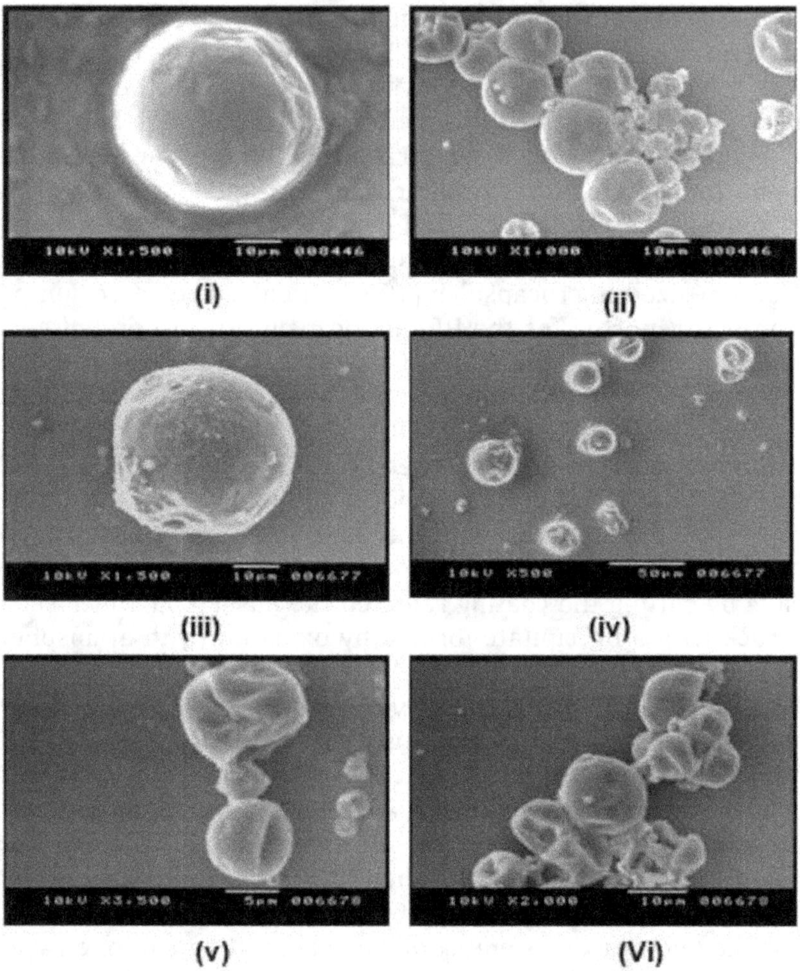

Figure 1.12 SEM Micrographs of polyurea microcapsules obtained using different rotational speeds. Reproduced from Reference 199 with permission from American Chemical Society.

The microcapsule assisted self-healing processes generally operate by the conversion of released healing agents into hard materials in the vicinity of damaged sites. However, these solidified healing agents are prone to secondary damage either on the surface or at the boundary between matrix and healing agent under the mechanical stress. To address this issue, Yang *et al.* [200] developed advanced microcapsule type organogel based self-healing systems in which the healing agent prevented the chances of secondary damage. A combination of tetramethyl tetraphenyl trisiloxane (MPS), hexatriacontane (C36) and cyclohexane was used to prepare the healing agent. Urea formaldehyde (UF) polymer was used to microencapsulate the viscoelastic organogels formed by the evaporation of cyclohexane. The synthesized microcapsules were incorporated into the epoxy polymer coating. Scanning electron microscopy (SEM) analysis confirmed the self-healing ability of organogels which healed the scratches on the coating (Figure 1.13). Secondary damage of healed coatings was not observed after subjecting them to vigorous vibrations. The ability of organogels to prevent the secondary damage is based on their property of viscoelasticity. Such "permanent" self-healing coating systems are potential alternatives to commercially used chromate based protective coatings in oil and gas industries. These would result in enhanced service life and prevent the environmental pollution caused by leaching of inhibitors used in commercial coatings.

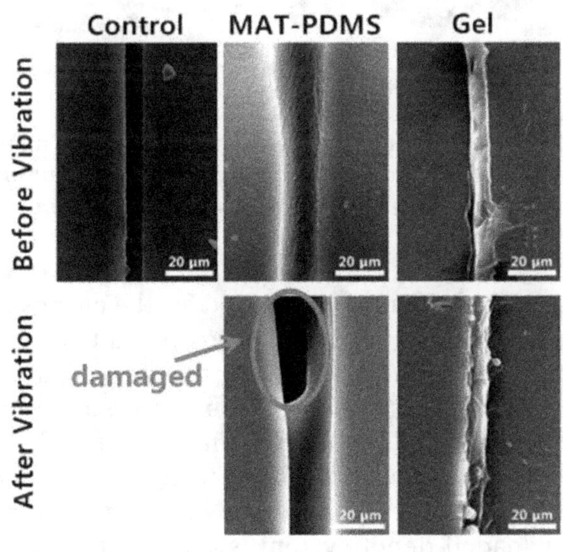

Figure 1.13 SEM images of scratched control and self-healing coating samples. Reproduced from Reference 200 with permission from American Chemical Society.

Fu *et al.* [201] reported acid and alkaline dual stimuli-responsive mechanized hollow mesoporous silica nanoparticles as smart nano-containers for anti-corrosion coatings. As shown in Figure 1.14, the smart nano-containers delayed the penetration of the corrosive species in the coating. In addition, the nano-containers were also observed to repair the damaged aluminum oxide layer, thus, exhibiting self-healing behavior to provide the long term anti-corrosion behavior. The nano-containers were observed to encapsulate the corrosion inhibitor at neutral pH, whereas the release was observed at both acidic and alkaline conditions, thus, confirming the functional nature of the nano-containers.

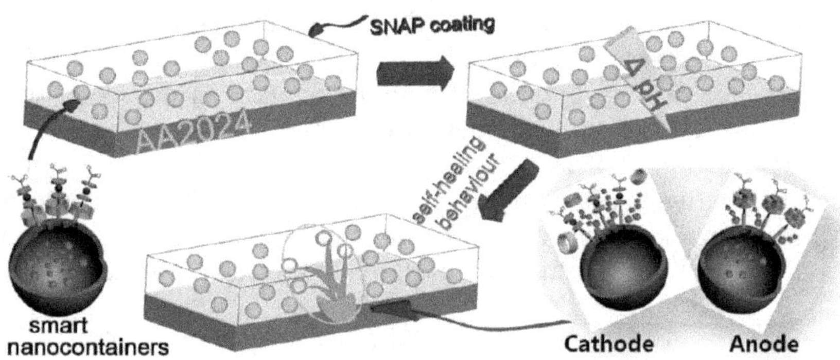

Figure 1.14 Schematic of the self-healing behavior of the nano-containers generated using mechanized hollow mesoporous silica nanoparticles. Reproduced from Reference 201 with permission from American Chemical Society.

In recent years, the synthesis method for the encapsulation of corrosion inhibitor by the use of polyelectrolyte multilayers has received increasing research attention. Polyelectrolytes are macromolecules with large number of functional groups possessing some charges or can become charged under suitable conditions [202]. Active species can incorporate in between the polyelectrolyte network or can be deposited as a dopant which will release on demand at required sites. The inhibitor loaded depot system is coated onto the substrate by mixing with the standard anti-corrosion coating formulations. Corrosion is accompanied by significant changes in local pH, formation of corrosive species, flux of the corrosion products and changes in the electrochemical potential. Polyelectrolyte based nano-containers are sensitive to these local changes occurring during the corrosion processes and, thus, initiate the release of entrapped corrosion inhibitor [203,204]. Till date, only a few studies have explored the utilization of polyelectrolyte nano-containers in self-healing anti-corrosion coatings. In one of these studies, Grigoriev *et al.* [203] prepared silica based nano-containers by incorporating benzotriazole (inhibitor) in between the oppositely charged polyelectrolytes by layer-by-layer addition on the mesoporous silica particles. The authors used a sol-gel process to generate a nano-containers based coating on metal surface. A significantly reduced corrosion due to the effect of released corrosion inhibitor from the nano-containers was observed. Attempts have also been made to use inexpensive and economically viable nano-containers for the encapsulation of corrosion inhibitors. Shchukin *et al.* [205] used low cost cylindrical shaped halloysite nano-containers filled with 2-mercaptobenzothiazole as corrosion inhibitor (Figure 1.15). The nano-containers were covered with polyelectrolyte multilayers to avoid the undesirable leakage and interaction of inhibitor with the coating. The release of the inhibitor was triggered by changes in local pH due to the sensitivity of polyelectrolyte layers.

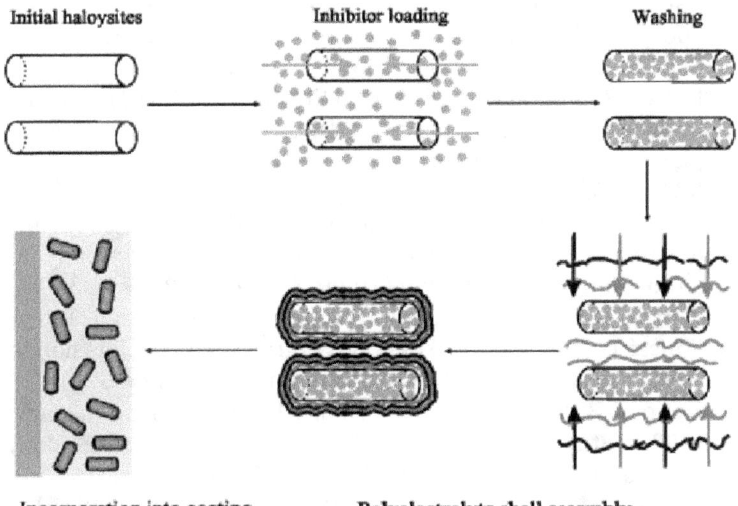

Figure 1.15 Fabrication of halloysite nano-containers as self-healing filler. Reprinted from Reference 205 with permission from American Chemical Society.

Angele *et al.* [206] used a melt extrusion process at 140 °C to incorporate nano-containers in the polymer blend. The obtained nano-container/polymer granules were further processed with solvents, pigments, plasticizers, surfactants, coalescing agents, etc., to generate the coatings. In another study, Kopec *et al.* [207] also prepared a system of water based epoxy coating embedded with corrosion inhibitor (2-methylbenzothiazole (BT) and 2-mercaptobenzothiazole (MBT)) encapsulated polyelectrolyte nano-capsules for self-healing ability.

1.3 Conclusions

The studies reviewed in the chapter confirm the immense potential of polymer nanomaterials for corrosion protection applications, especially under severe conditions at oil and gas facilities. The addition of fillers, inhibitors, nano-containers with inhibiting agents, cross-linking agents, etc., constitute functional strategies to reinforce the corrosion protection abilities of a wide variety of polymers.

References

1. Shi, X., Nguyen, T. A., Suo, Z., Liu, Y., and Avci, R. (2009) Effect of nanoparticles on the anticorrosion and mechanical properties of epoxy coating. *Surface Coating and Technology*, **204**, 237-245.
2. Chatterjee, S., Wang, J. W., Kuo, W. S., Tai, N. H., Salzmann, C., Li, W. L., Hollertz, R., Nuesch, F. A., and Chu, B. T. T. (2012) Mechanical reinforcement and thermal conductivity in expanded graphene nanoplatelets reinforced epoxy composites. *Chemical Physics Letters*, **531**, 6-10.
3. Yu, J., Huo, R., Wu, C., Wu, X., Wang, G., and Jiang, P. (2012) Influence of interface structure on dielectric properties of epoxy/alumina nanocomposites. *Macromolecular Research*, **20**, 816-826.
4. Zaman, I., Phan, T. T., Kuan, H. C., Meng, Q., Bao La, L. T., Luong, L., Youssf, O., and Ma, J. (2011) Epoxy/graphene platelets nanocomposites with two levels of interface strength. *Polymer*, **52**, 1603-1611.
5. Smaoui, H., Mir, L. E. L., Guermazi, H., Agnel, S., and Toureille, A. (2009) Study of dielectric relaxations in zinc oxide-epoxy resin nanocomposites. *Journal of Alloys and Compounds*, **477**, 316-321.
6. Conradi, M., Kocijan, A., Zorko, M., and Verpoest, I. (2015) Damage resistance and anticorrosion properties of nanosilica epoxy-resin composite coatings. *Progress in Organic Coatings*, **80**, 20-26.
7. Wetzel, B., Haupert, F., and Zhang, M. Q. (2003) Epoxy nanocomposites with high mechanical and tribological performance. *Composite*

Science and Technology, **63**, 2055-2067.
8. Zhang, M. Q., Rong, M. Z., Yu, S. L., Wetzel, B., and Friedrich, K. (2002) Improvement of tribological performance of epoxy by the addition of irradiation grafted nano-inorganic particles. *Macromolecular Materials and Engineering*, **287**, 111-115.
9. Yamini, S., and Young, R. J. (1977) Stability of crack propagation in epoxy resins. *Polymer*, **18**, 1075-1080.
10. Perreux, D., and Suri, C. (1997) Study of the coupling between the phenomena of water absorption and damage in glass/epoxy composite pipes. *Composite Science and Technology*, **57**, 1403-1413.
11. Loos, C., and Springer, G. S. (1979) Moisture absorption of graphene-epoxy composites immersed in liquids and in humid air. *Journal of Composite Materials*, **13**, 131-147.
12. Spanoudakis, J., and Young, R. J. (1984) Crack propagation in a glass particle filled resin. *Journal of Material Science*, **19**, 487-496.
13. Massiani, Y., Crousier, J. P., and Lazzeri, A. (1993) Studies of the corrosion protection of immersed metallic structures by sub-aquatic applications of epoxy coatings. *Corrosion Science*, **34**, 1685-1696.
14. Lam, K., and Lau, K. T. (2006) Localized elastic modulus distribution of nanoclay/epoxy composites by using nanoindentation. *Composite Structures*, **75**, 553-558.
15. Shi, G., Zhang, M. Q., Rong, M. Z., Wetzel, B., and Friedrich, K. (2003) Friction and wear of low nanometer Si_3N_4 filled epoxy composites. *Wear*, **254**, 784-796.
16. Hartwig, A., Sebald, M., Putz, D., and Aberle, L. (2005) Preparation, caharacterization and properties of nanocompoaites based on epoxy resins - an overview. *Macromolecular Symposia*, **221**, 127-135.
17. Dietsche, F., Thomann, Y., Thomann, R., and Mulhaupt, R. (2000) Translucent acrylic nanocomposites containing anisotropic laminated nanoparticles derived from intercalated layered silicates. *Journal of Applied Polymer Science*, **75**, 396-405.
18. Huong, N. (2006) *Improvement of Bearing Strength of Laminated Composites by Nanoclay and Z-pin Reinforcement*, PhD. Dissertation, University of New South Wales, Australia.
19. Becker, O., Varley, R., and Simon, G. (2002) Morphology, thermal relaxations and mechanical properties of layered silicate nanocomposites based upon high-functionality epoxy resins. *Polymer*, **43**(16), 4365-4373.
20. Yang, L. H., Liu, F. C., and Han, E. H. (2005) Effects of P/B on the properties of anticorrosive coatings with different particle size. *Progress in Organic Coatings*, **53**, 91-98.
21. Lamaka, S. V., Zheludkevich, M. L., Yasakau, K. A., Serra, R., Poznyak, S. K., and Ferreira, M. G. S. (2007) Nanoporous titania interlayer as reservoir of corrosion inhibitors for coatings with self-healing abili-

ty. *Progress in Organic Coatings*, **58**, 127-135.

22. Yu, Y., Yeh, J., Liou, S., Chen, C., Liaw, D., and Lu, H. (2004) Preparation and properties of polyimide–clay nanocomposite materials for anticorrosion application. *Journal of Applied Polymer Science*, **92**, 3573-3582.
23. Behzadnasab, M., Mirabedini, S. M., and Esfandeh, M. (2013) Corrosion protection of steel by epoxy nanocomposite coatings containing various combinations of clay and nanoparticulate zirconia. *Corrosion Science*, **75**, 134-141.
24. Bagherzadeh, M. R., and Mahdavi, F. (2007) Preparation of epoxy-clay nanocomposite and investigation on its anti-corrosive behaviour in epoxy coating. *Progress in Organic Coatings*, **60**, 117-120.
25. Hang, T. T. X., Truc, T. A., Olivier, M. G., Vandermiers, C., Guerit, N., and Pebre, N. (2010) Corrosion protection mechanisms of carbon steel by an epoxy resin containing indole-3 butyric acid modified clay. *Progress in Organic Coatings*, **69**, 410-416.
26. Singh-Beemat, J., and Iroh, J. O. (2012) Characterization of corrosion resistant clay/epoxy ester composite coatings and thin films. *Progress in Organic Coatings*, **74**, 173-180.
27. Tatsuma, T., Saitoh, S., Ohko, Y., and Fujishima, A. (2001) TiO_2-WO_3 photoelectrochemical anticorrosion system with an energy storage ability. *Chemistry of Materials*, **13**, 2838-2842.
28. Ohko, Y., Saitoh, S., Tatsuma, T., and Fujishima, A. (2001) Photoelectrochemical anticorrosion and self-cleaning effects of a TiO_2 coating for type 304 stainless steel. *Journal of Electrochemical Society*, **148**, B24-B28.
29. Matinlinna, J. P., Ozcan, M., Lassila, L. V. J., and Vallittu, P. K. (2004) The effect of a 3-methacryloxypropyltrimethoxysilane and vinyltriisopropoxysilane blend and tris(3-trimethoxysilylpropyl)isocyanurate on the shear bond strength of composite resin to titanium metal. *Dental Materials*, **20**, 804-813.
30. Kissounko, D. A., Deitzel, J. M., Doherty, S. P., Shah, A., and Gillespie, J. W. (2008) Understanding the role of clay silicate nanoparticles with organic modifiers in thermal curing of cyanate ester resin. *European Polymer Journal*, **44**, 2807-2819.
31. Yuan, M. R., Lu, J. T., and Kong, G. (2010) Effect of SiO_2:Na_2O molar ratio of sodium silicate on the corrosion resistance of silicate conversion coatings. *Surface and Coating Technology*, **204**, 1229-1235.
32. Lin, J. M., and Ma, C. C. M. (2000) Thermal degradation of phenolic resin/silica hybrid ceramers. *Polymer Degradation and Stability*, **69**, 229-235.
33. Borisova, D., Mohwald, H., and Shchukin, D. G. (2011) Mesoporous silica nanoparticles for active corrosion protection. *ACS Nano*, **5**, 1939-1946.
34. Suriati, G., Mariatti, M., and Azizan, A. (2013) Silver-filled epoxy co-

mposites: effect of hybrid and silane treatment on thermal properties. *Polymer Bulletin*, **70**, 311-323.

35. Tee, D. I., Mariatti, M., Azizan, A., See, C. H., and Chong, K. F. (2007) Effect of silane-based coupling agent on the properties of silver nanoparticles filled epoxy composites. *Composites Science and Technology*, **67**, 2584-2591.
36. Akbarian, M., Olya, M. E., Mahdavian, M., and Ataeefard, M. (2014) Effect of nanoparticulate silver on the corrosion protection performance of polyurethane coatings on mild steel in sodium chloride solution. *Progress in Organic Coatings*, **77**, 1233-1240.
37. Palimi, M. J., Rostami, M., Mahdavian, M., and Ramezanzadeh, B. (2014) Surface modification of Fe_2O_3 nanoparticles with 3-aminopropyltrimethoxysilane (APTMS): An attempt to investigate surface treatment on surface chemistry and mechanical properties of polyurethane/Fe_2O_3 nanocomposites. *Applied Surface Science*, **320**, 60-72.
38. Liu, X., Shao, Y., Zhang, Y., Meng, G., Zhang, T., and Wang, F. (2014) Using high-temperature mechanochemistry treatment to modify iron oxide and improve the corrosion performance of epoxy coating – I. High-temperature ball milling treatment. *Corrosion Science*, **90**, 451-462.
39. Ramezanzadeh, B., Attar, M. M., and Farzam, M. J. (2011) Effect of ZnO nanoparticles on the thermal and mechanical properties of epoxy-based nanocomposite. *Journal of Thermal Analysis and Calorimetry*, **103**, 731-739.
40. Palimi, M. J., Rostami, M., Mahdavian, M., and Ramezanzadeh, B. (2014) Application of EIS and salt spray tests for investigation of the anticorrosion properties of polyurethane-based nanocomposites containing Cr_2O_3 nanoparticles modified with 3-amino propyl trimethoxy silane. *Progress in Organic Coatings*, **77**, 1935-1945.
41. Ahmad, S., Ashraf, S. M., Hasnat, A., Yadav, S., and Jamal, A. (2001) Studies on urethane-modified alumina-filled polyesteramide anticorrosive coatings cured at ambient temperature. *Journal of Applied Polymer Science*, **82**, 1855-1865.
42. Behzadnasab, M., Mirabedini, S. M., Kabiri, K., and Jamali. S. (2011) Corrosion performance of epoxy coatings containing silane treated ZrO2 nanoparticles on mild steel in 3.5% NaCl solution. *Corrosion Science*, **53**, 89-98.
43. Haddadi, S. A., Mahdavian, M., and Karimi, E. (2015) Evaluation of the corrosion protection properties of an epoxy coating containing sol–gel surface modified nano-zirconia on mild steel. *RSC Advances*, **5**, 28769-28777.
44. Vanlandingham, M. R., McKnight, S. H., Palmese, G. R., Elings, J. R., Huang, X., Bogett, T. A., Eduljee, R., and Gillespie, J. W. (1997) Nanoscale indentation of polymer systems using the atomic force mic-

roscope. *The Journal of Adhesion*, **64**, 31-59.
45. Brindley, G. W. (1952) Geology and mineralogy of clays structural mineralogy of clays. *Clay and Clay Minerals*, **1**, 33-43.
46. Qi, B., Zhang, Q. X., Bannister, M., and Mai, Y. W. (2006) Investigation of the mechanical properties of DGEBA-based epoxy resin with nanoclay additives. *Composite Structures*, **75**, 514-519.
47. Wang, L., Wang, K., Chen, L., Zhang, Y., and He, C. (2006) Preparation, morphology and thermal/mechanical properties of epoxy/nanoclay composite. *Composites Part A: Applied Science and Manufacturing*, **37**, 1890-1896.
48. Khanbabaei, G., Aalaie, J., Rahmatpour, A., Khoshniyat, A., and Gharabadian, M. A. (2007) Preparation and properties of epoxy-clay nanocomposites. *Journal of Macromolecular Science Part B*, **46**, 975-986.
49. Palimi, M. J., Rostami, M., Mahdavian, M., and Ramezanzadeh, B. (2014) Surface modification of Cr_2O_3 nanoparticles with 3-amino propyl trimethoxy silane (APTMS). Part 1: Studying the mechanical properties of polyurethane/Cr_2O_3 nanocomposites. *Progress in Organic Coatings*, **77**, 1663-1673.
50. Akbarian, M., Olya, M. E., Ataeefard, M., and Mahdavian, M. (2012) The influence of nanosilver on thermal and antibacterial properties of a 2 K waterborne polyurethane coating. *Progress in Organic Coatings*, **75**, 344-348.
51. Mittal, K. L. (2007) *Silanes and Other Coupling Agents*, 4th volume, CRC Press, Netherlands.
52. Matinlinna, J. P., Ozcan, M., Lassila, L. V. J., and Vallittu, P. K. (2004) The effect of a 3-methacryloxypropyltrimethoxysilane and vinyltriisopropoxysilane blend and tris(3-trimethoxysilylpropyl)isocyanurate on the shear bond strength of composite resin to titanium metal. *Dental Materials*, **20**, 804-813.
53. Tracton, A. A. (2010) *Coatings Technology Handbook*, 3rd edition, CRC Press, Netherlands.
54. Ghazizadeh, A., Haddadib, S. A., and Mahdavian, M. (2016) The effect of sol–gel surface modified silver nanoparticles on the protective properties of the epoxy coating. *RSC Advances*, **6**, 18996-19006.
55. Gharagozlou, M., Ramezanzadeh, B., and Baradaran, Z. (2016) Synthesize and characterization of a novel anticorrosive cobalt ferrite nanoparticles dispersed in silica matrix ($CoFe_2O_4$-SiO_2) to improve the corrosion protection performance of epoxy coating. *Applied Surface Science*, **377**, 86-98.
56. Javidparvar, A. A., Ramezanzadeh, B., and Ghasemi, E. (2016) The effect of surface morphology and treatment of Fe3O4 nanoparticles on the corrosion resistance of epoxy coating. *Journal of Taiwan Institute of Chemical Engineers*, **61**, 356-366.
57. Luckachan, G. E., Hassan, K., and Mittal, V. (2018) Epoxy composite

coatings for enhanced corrosion resistant. In: *Polymers in Oil and Gas Industry*, Mittal, V. (ed.), Central West Publishing, Australia, pp. 293-310.

58. Luo, X., and Mather, P. T. (2013) Shape memory assisted self-healing coating. *ACS Macro Letters*, **2**, 152-156.
59. Meier-Westhues, U. (2008) *Polyurethanes Coatings, Adhesives and Sealants*, European Coating Tech Files, Germany.
60. Wu, Z. F., Wang, H., Tian, X. Y., Xue, M., Ding, X., Ye, X. Z., and Cui, Z. Y. (2014) Surface and mechanical properties of hydrophobic silica contained hybrid films of waterborne polyurethane and fluorinated polymethacrylate. *Polymer*, **55**, 187-194.
61. Paul, P. J., Nair, M. G. R., Koshy, P., and Idage, B. B. (1999) Segmented block copolymers of natural rubber and bisphenol A–toluene diisocyanate oligomers. *Journal of Applied Polymer Science*, **74**, 706-721.
62. Pandya, M. V., Desphande, D. D., and Hundiwale, D. G. (1986) Effect of diisocyanate structure on viscoelastic, thermal, mechanical and electrical properties of cast polyurethanes. *Journal of Applied Polymer Science*, **32**, 4959-4969.
63. Blackwell, J., Nagarajan, M. R., and Haitink, T. B. (1982) Structure of polyurethane elastomers: effect of chain extender length on the structure of MDI/diol hard segments. *Polymer*, **23**, 950-956.
64. Gleiter, H. (1989) Nanocrystalline materials. *Progress in Materials Science*, **33**, 223-315.
65. Baer, D. R., Burrows, P. E., and El-Azab, A. A. (2003) Enhancing coating functionality using nanoscience and nanotechnology. *Progress in Organic Coatings*, **47**, 342-356.
66. Chen, Y., Zhou, S., Yang, H., Gu, G., and Wu, L. (2004) Preparation and characterization of nanocomposite polyurethane. *Journal of Colloid and Interface Science*, **279**, 370-378.
67. Finnigan, B., Martin, D., Haiiey, P., Truss, R., and Cambell, K. (2004) Morphology and properties of thermoplastic polyurethane nanocomposites incorporating hydrophilic layered silicates. *Polymer*, **45**, 2249-2260.
68. Ahmadi, B., Kassiriha, M., Khodabakhshi, K., and Mafi, E. R. (2007) Effect of nano layered silicates on automotive polyurethane refinish clear coat. *Progress in Organic Coatings*, **60**, 99-104.
69. Chen-Yang, Y. W., Lee, Y. K., Chen, Y. T., and Wu, J. C. (2007) High improvement in the properties of exfoliated PU/clay nanocomposites by the alternative swelling process. *Polymer*, **48**, 2969-2979.
70. Ray, S. S., and Okamoto, M. (2003) Polymer/ layered silicate nanocomposites: a review from preparation to processing. *Progress in Polymer Science*, **28**, 1539-1641.
71. Rehab, A., and Salahuddin, N. (2005) Nanocomposite materials based on polyurethane intercalated into montmorillonite clay. *Material Science Engineering*, **399**, 368-376.

72. Yusoh, K., Jin, J., and Song, M. (2010) Subsurface mechanical properties of polyurethane/organoclay nanocomposite thin films studied by nanoindentation. *Progress in Organic Coatings*, **67**, 220-224.
73. Yeh, J. M., Liou, S. J., Lai, C. Y., Wu, P. C., and Tsai, T. Y. (2001) Enhancement of corrosion protection effect in polyaniline via the formation of polyaniline–clay nanocomposite materials. *Chemistry of Materials*, **13**, 1131-1136.
74. Yeh, J. M., Liou, S. J., Lin, C. Y., Cheng, C. Y., Chang, Y. W., and Lee, K. R. (2002) Anticorrosively enhanced PMMA-clay nanocomposite materials with quaternary alkylphosphonium salt as an intercalating agent. *Chemistry of Materials*, **14**, 154-161.
75. Yu, Y. H., Yeh, J. M., Liou, S. J., and Chang, Y. P. (2004) Organo-soluble polyimide (TBAPP–OPDA)/clay nanocomposite materials with advanced anticorrosive properties prepared from solution dispersion technique. *Acta Materialia*, **52**, 475-486.
76. Xia, H. S., Shaw, S. J., and Song, M. (2005) Relationship between mechanical properties and exfoliation degree of clay in polyurethane nanocomposites. *Polymer International*, **54 (10)**, 1392-1400.
77. Jin, J., Chen, L., Song, M., and Yao, K. J. (2006) An analysis on enhancement of fatigue durability of polyurethane by incorporating organoclay nanofillers. *Macromolecular Materials and Engineering*, **291**, 1414-1421.
78. Choi, W. J., Kim, S. H., Kim, Y. J., and Kim, S. C. (2004) Synthesis of chain-extended organifier and properties of polyurethane/clay nanocomposites. *Polymer*, **45**, 6045-6057.
79. Spirkova, M., Brus, J., Brozova, L., Strachota, A., Baldrian, J., Urbanova, M., Kotek, J., Strachotova, B., and Slouf, M. (2008) A view from inside onto the surface of self-assembled nanocomposite coatings. *Progress in Organic Coatings*, **61**, 145-155.
80. Ghosal, A., Rahman, O. U., and Ahmad, S. (2015) High-performance soya polyurethane networked silica hybrid nanocomposite coatings. *Industrial and Engineering Chemistry Research*, **54**, 12770-12787.
81. Heidarian, M., Shishesaz, M. R., Kassiriha, S. M., and Nematollahi, M. (2010) Characterization of structure and corrosion resistivity of polyurethane/organoclay nanocomposite coatings prepared through an ultrasonication assisted process. *Progress in Organic Coatings*, **68**, 180-188.
82. Yeh, J. M., Yao, C. T., Hsieh, C. F., Lin, L. H., Chen, P. L., Wu, J. C., Yang, H. C., and Wu, C. P. (2008) Preparation, characterization and electrochemical corrosion studies on environmentally friendly waterborne polyurethane/Na+-MMT clay nanocomposite coatings. *European Polymer Journal*, **44**, 3046-3056.
83. Truc, T. A., Pebere, N., Hang, T. T. X., Hervaud, Y., and Boutevin, B. (2004) Evaluation of corrosion performance of a UV-cured polyure-

thane coating in the presence of organic phosphorous compounds. *Progress in Organic Coatings*, **49**, 130-136.

84. Trinh, T. A., Pebere, N., To, X. H., Boutevin, B., and Hervaud, Y. (2002) Study of the synergistic effect observed for the corrosion protection of a carbon steel by an association of phosphates. *Corrosion Science*, **44**, 2055-2071.
85. Sawitowska, T. (2005) *International Congress of Nanotechnology (ICNT)*, San Francisco, USA.
86. Dashtizadeh, A., Abdouss, M., Mahdavi, H., and Khorassani, M. (2011) Acrylic coatings exhibiting improved hardness, solvent resistance and glossiness by using silica nano-composites. *Applied Surface Science*, **257**, 2118-2125.
87. Mills, D. J., Jamali, S. S., and Paprocka, K. (2012) Investigation into the effect of nano-silica on the protective properties of polyurethane coatings. *Surface and Coatings Technology*, **209**, 137-142.
88. Dastmalchian, H., Moradian, S., Jalili, M. M., and Mirabedini, S. M. (2009) Investigating changes in electrochemical properties when nano-silica is incorporated into an acrylic-based polyurethane clearcoat. *Journal of Coating Technology and Research*, doi:10.1007/s11998-009-9222-0.
89. Rashvand, M., and Ranjbar, Z. (2013) Effect of nano-ZnO particles on the corrosion resistance of polyurethane-based waterborne coatings immersed in sodium chloride solution via EIS technique. *Progress in Organic Coatings*, **76**, 1413-1417.
90. Zhu, J., Wei, S., Haldolaarachchige, N., Young, D. P., and Guo, Z. (2011) Electromagnetic field shielding polyurethane nanocomposites reinforced with core–shell Fe–silica nanoparticles. *The Journal of Physical Chemistry C*, **115**, 15304-15310.
91. Zhu, J., Wei, S., Lee, I.Y., Park, S., Willis, J., Haldolaarachchige, N., Young, D. P., Luo, Z., and Guo, Z. (2012) Silica stabilized iron particles toward anti-corrosion magnetic polyurethane nanocomposites. *RSC Advances*, **2**, 1136-1143.
92. Guo, Z., Park, S., Wei, S., Pereira, T., Moldovan, M., Karki, A. B., Young, D. P., and Hahn, H. T. (2007) Flexible high-loading particle-reinforced polyurethane magnetic nanocomposite fabrication through particle-surface-initiated polymerization. *Nanotechnology*, **18**, 335704.
93. Guo, Z., Lee, S. E., Kim, H., Park, S., Hahn, H. T., Karki, A. B., and Young, D. P. (2009) Fabrication, characterization and microwave properties of polyurethane nanocomposites reinforced with iron oxide and barium titanate nanoparticles. *Acta Materialia*, **57**, 267-277.
94. Chang, Y. T., Wen, N. T., Chen, W. K., Ger, M. D., Pan, G. T., and Yang, T. C. K. (2008) The effects of immersion time on morphology and electrochemical properties of the Cr(III)-based conversion coatings on zinc coated steel surface. *Corrosion Science*, **50**(12), 3494-3499.

95. Dennis, R. V., Viyannalage, L. T., Gaikwad, A. V., Rout, T. K., and Banerjee, S. (2013) Graphene nanocomposite coatings for protecting low-alloy steels from corrosion. *American Ceramic Society Bulletin*, **92**, 18-24.
96. Gaikwad, A. V., and Rout, T. K. (2011) In situ synthesis of silver nanoparticles in polyetherimidematrix and its application in coatings. *Journal of Materials Chemistry*, **21**, 1234-1239.
97. Conceicao, T. F., Scharnagl, N., Blawert, C., Dietzel, W., and Kainer, K. (2010) Corrosion protection of magnesium alloy AZ31 sheets by spin coating process with poly(ether imide). *Corrosion Science*, **52**, 2066-2079.
98. Chen, B. K., Su, C. T., Tseng, M. C., and Tsay, S. Y. (2006) Preparation of polyetherimide nanocomposites with improved thermal, mechanical and dielectric properties. *Polymer Bulletin*, **57**, 671-681.
99. Bijwe, J., Indumathi, J., John Rajesh, J., and Fahim, M. (2001) Friction and wear behavior of polyetherimide composites in various wear modes. *Wear*, **249**, 715-726.
100. Kostina, Y. V., Bondarenko, G., Alent'Ev, A. Y., and Yampol'Skii, Y. P. (2007) Effect of structure and conformational composition on the transport behaviour of poly(ether imide). *Polymer Science Series A*, **49**, 77-88.
101. Xian, G., and Zhang, Z. (2005) Sliding wear of polyetherimide matrix composites: I. Influence of short carbon fibre reinforcement. *Wear*, **258**, 776-782.
102. Amancio-Filho, S., Roeder, J., Nunes, S., Dos Santos, J., and Beckmann, F. (2008) Thermal degradation of polyetherimide joined by friction riveting (FricRiveting). Part I: Influence of rotation speed. *Polymer Degradation and Stability*, **93**, 1529-1538.
103. Liu, T., Tong, Y., and Zhang, W. D. (2007) Preparation and characterization of carbon nanotube/polyetherimide nanocomposite films. *Composites Science and Technology*, **67**, 406-412.
104. Rout, T. K., and Gaikwad, A. V. (2015) *In-situ* generation and application of nanocomposites on steel surface for anti-corrosion coating. *Progress in Organic Coatings*, **79**, 98-105.
105. Da Conceicao, T. F., Scharnagl, N., Dietzel, W., and Kainer, K. (2011) Corrosion protection of magnesium AZ31 alloy using poly(ether imide) [PEI] coatings prepared by the dip coating method: Influence of solvent and substrate pre-treatment. *Corrosion Science*, **53**, 338-346.
106. Wicks Jr, Z. W., Jones, F. N., Pappas, S. P., and Wicks, D. A. (2007) *Organic Coatings: Science and Technology*, 3rd edition, John Wiley & Sons, USA.
107. Ahmad, Z. (2006) *Principles of Corrosion Engineering and Corrosion Control*, 1st edition, Butterworth-Heinemann, USA.
108. Scharnagl, N., Blawert, C., and Dietzel, W. (2009) Corrosion protect-

tion of magnesium alloy. AZ31 by coating with poly(ether imides) (PEI). *Surface and Coatings Technology*, **203**, 1423-1428.

109. Menut, P., Pochat-Bohatier, C., Deratani, A., Dupuy, C., and Guilbert, S. (2002) Structure formation of poly (ether-imide) films using non-solvent vapor induced phase separation: relationship between mass transfer and relative humidity. *Desalination*, **145**, 11-16.
110. Tang, J. C., Yang, H. C., Chen, S. Y., and Chen-Yang, Y. W. (2007) Preparation and properties of polyimide/silica hybrid nanocomposites. *Polymer Composites*, **28**, 575-581.
111. Akhtar, T., Luckachan, G. E., and Mittal, V. (2018) Polyetherimide coatings for corrosion protection of mild carbon steel: Effect of residual solvent and temperature on coating performance, *in preparation.*
112. Qi, K., Qiu, Y., Chen, Z., and Guo, X. (2012) Corrosion of conductive polypyrrole: Effects of environmental factors, electrochemical stimulation, and doping anions. *Corrosion Science*, **60**, 50-58.
113. Xing, C., Zhang, Z., Yu, L., Zhang, L., Bowmaker, and G. A. (2014) Electrochemical corrosion behavior of carbon steel coated by polyaniline copolymers micro/nanostructures. *RSC Advances*, **4**, 32718-32725.
114. Tan, C. K., and Blackwood, D. J. (2003) Corrosion protection by multilayered conducting polymer coatings. *Corrosion Science*, **45**, 545-557.
115. Talo, A., Forsen, O., and Ylasaari, S. (1999) Corrosion protective polyaniline epoxy blend coatings on mild steel. *Synthetic Metals*, **102**, 1394-1395.
116. Tuken, T., Arslan, G., Yazici, B., and Erbil, M. (2004) The corrosion protection of mild steel by polypyrrole/polyphenol multilayer coating. *Corrosion Science*, **46**, 2743-2754.
117. Klimartin, P. A., Trier, L., and Wright, G. A. (2002) Corrosion inhibition of polyaniline and poly(o-methoxyaniline) on stainless steels. *Synthetic Metals*, **131**, 99-109.
118. Meneguzzi, A., Pham, M. C., Ferreira, C. A., Lacroix, J. C., Aeiyach, S., and Lacaze, P. C. (1999) Electroactive poly(aromatic amine) films deposited on mild steel. *Synthetic Metals*, **102**, 1390-1391.
119. Iroh, J. O., Zhu, Y., Shah, K., Levine, K., Rajagopalan, R., Uyar, T., Donley, M., Mantz, R., Johnson, J., Voevodin, N.N., Balbyshev, V. N., and Khramov, A. N. (2003) Electrochemical synthesis: a novel technique for processing multi-functional coatings. *Progress in Organic Coatings*, **47**, 365-375.
120. Popovic, M. M., Grgur, B. N., and Miskovic–Stankovic V. B. (2005) Corrosion studies on electrochemically deposited PANI and PANI/epoxy coatings on mild steel in acid sulfate solution. *Progress in Organic Coatings*, **52**, 359-365.
121. Hammache, H., Makhloufi, L., and Saidani, B. (2003) Corrosion prot-

ection of iron by polypyrrole modified by copper using the cementation process. *Corrosion Science*, **45**, 2031-2042.
122. Dominis, A. J., Spinks, G. M., and Wallace, G. G. (2003) Comparison of polyaniline primers prepared with different dopants for corrosion protection of steel. *Progress in Organic Coatings*, **48**, 43-49.
123. Yagan, A., Pekmez, N. O., and Yildiz, A. (2006) Electropolymerization of poly(N-ethyl aniline) on mild steel: Synthesis, characterization and corrosion protection. *Electrochimica Acta*, **51**, 2949-2955.
124. Guimard, N. K., Gomez, N., and Schmidt, C. E. (2007) Conducting polymers in biomedical engineering. *Progress in Polymer Science*, **32**, 876-921.
125. Tuken, T., Yazici, B., and Erbil, M. (2006) The use of polyindole for prevention of copper corrosion. *Surface and Coating Technology*, **200**, 4802-4809.
126. Santos, J. R., Mattoso, L. H. C., and Motheo A. J. (1998) Investigation of corrosion protection of steel by polyaniline films. *Electrochimica Acta*, **43**(3-4), 309-313.
127. Chaudhry, A. U., and Mittal, V. (2018) Polyaniline-graphene composite nanoparticle pigments for anti-corrosion coatings, *in preparation*.
128. Mengoli, G., Munari, M. T., Bianco, P., and Musiani, M. M. (1981) Anodic synthesis of polyaniline coatings onto Fe sheets. *Journal of Applied Polymer Science*, **26**(12), 4247-4257.
129. DeBerry, D. W. (1985) Modification of the electrochemical and corrosion behavior of stainless steels with an electroactive coating. *Journal of Electrochemical Society*, **132**(5), 1022-1026.
130. Yan, M., Vetter, C. A., and Gelling, V. J. (2013) Corroison inhibition performance of polypyrrole Al flake composite coatings for Al alloys. *Corrosion Science*, **70**, 37-45.
131. Wessling, B. (1994) Passivation of metals by coating with polyaniline: Corrosion potential shift and morphological changes. *Advanced Materials*, **6**, 226-228.
132. Lu, W. K., Elsenbaumer, R. L., and Wessling, B. (1995) Corrosion protection of mild steel by coatings containing polyaniline. *Synthetic Metals*, **71**, 2163-2166.
133. Michalik A. (2009) *Conducting Polymers for Corrosion Protection: A Critical Investigation*, PhD Thesis, Ruhr-Bochum University, Germany.
134. Jafarzadeh, S., Adhikari, A., Sundall, P. E., and Pan, J. (2011) Study of PANI-MeSA conducting polymer dispersed in UV-curing polyester acrylate on galvanized steel as corrosion protection coating. *Progress in Organic Coatings*, **70**(2-3), 108-115.
135. De Souza, S., Pereira da Silva, J. E., Cordoba de Torresi, S. I., Temperini, M. L. A., and Torresi, R. M. (2001) Polyaniline based acrylic blends for iron corrosion protection. *Electrochemical Solid-State*

Letters, **4**(8), B27-B30.

136. Kinlen, P. J., Silverman, D. C., and Jeffreys, C. R. (1997) Corrosion protection using polyanujne coating formulations. *Synthetic Metals*, **85**(1), 1327-1332.
137. Camalet, J. L., and Lacroix, J. C. (1996) Electrodeposition of protective polyaniline films on mild steel. *Journal of Electroanalytical Chemistry*, **416**, 179-182.
138. Camalet, J. L., and Lacroix, J. C. (1998) Electrosynthesis of adherent polyaniline films on iron and mild steel in aqueous oxalic acid medium. *Synthetic Metals*, **93**, 133-142.
139. Schauer, T., Joos, A., Dulog, L., and Eisenbach, C. D. (1998) Protection of iron against corrosion with polyaniline primers. *Progress in Organic Coatings*, **33**, 20-27.
140. Elsenbaumer, R. L., and Reynolds, J. R. (1998) *Handbook of Conducting Polymers*, 2nd edition, Marcel Dekker, USA.
141. Fahlman, M., Jasty, S., and Epstein, A. J. (1997) Corrosion protection of iron/steel by emeraldine base polyaniline: an X-ray photoelectron spectroscopy study. *Synthetic Metals*, **85**, 1323-1326.
142. Patil, R. C., and Radhakrishnan, S. (2006) Conducting polymer based hybrid nano-composites for enhanced corrosion protective coatings. *Progress in Organic Coatings*, **57**, 332-336.
143. Paliwoda-Porebska, G., Stratman, M., Rohwerder, M., Potje-Kamloth, K., Lu, Y., Pich, A. Z., and Adler, H. J. (2005) On the development of polypyrrole coatings with self-healing properties for iron corrosion protection. *Corrosion Science*, **47**, 3216-3233.
144. Iroh, J. O., and Su, W. (2000) Corrosion performance of polypyrrole coating applied to low carbon steel by an electrochemical process. *Electrochimica Acta*, **46**, 15-24.
145. Nguyen, T., Le, T., Garcia, B., Deslouis, C., and Le, X. Q. (2001) Corrosion protection and conducting polymers: polypyrrole films on iron. *Electrochimica Acta*, **46**, 4259-4272.
146. Kowalski, D., Ueda, M., and Ohtsuka, T. (2010) Self-healing ion-permselective conducting polymer coating. *Journal of Materials Chemistry*, **20**, 7630-7633.
147. Liu, A. S., Bezerra, M. C., and Cho, L. Y. (2009) Control of the anisotropy field of successively grown garnet films. *Material Research Bulletin*, **12** 503-509.
148. Cao, Y., and Heeger, A. J. (1992) Magnetic susceptibility of polyaniline in solution in non-polar organic solvents and in polyblends in poly(methyl methacrylate). *Synthetic Metals*, **52**, 193-200.
149. Yang, C. Y., Cao, Y., Smith, P., and Heeger, A. J. (1993) Morphology of conductive, solution-processed blends of polyaniline and poly(methyl methacrylate). *Synthetic Metals*, **53**, 293-301.
150. Meroufel, A., Deslouis, C., and Touzain, S. (2008) Electrochemical and anticorrosion performances of zinc-rich and polyaniline powd-

er coatings. *Electrochimica Acta*, **53**, 2331-2338.

151. Kalendova, A., Vesely, D., and Stejskal, J. (2008) Organic coatings containing polyaniline and inorganic pigments as corrosion inhibitors. *Progress in Organic Coatings*, **62**, 105-116.
152. Radhakrishnan, S., Siju, C.R., Mahanta, D., Patil, S., and Madras, G. (2009) Conducting polyaniline–nano-TiO2 composites for smart corrosion resistant coatings. *Electrochimica Acta*, **54**, 1249-1254.
153. Armelin, E., Pla, R., Liesa, F., Ramis, X., Iribarren, J. I., and Alema, C. (2008) Corrosion protection with polyaniline and polypyrrole as anticorrosive additives for epoxy paint. *Corrosion Science*, **50**, 721-728.
154. Brodinova, J., Stejskal, J., and Kalendova, A. (2007) Investigation of ferrites properties with polyaniline layer in anticorrosive coatings. *Journal of Physics and Chemistry of Solids*, **68**, 1091-1095.
155. Havlik, J., Kalendova, A., and Vesely, D. (2007) Electrochemical, chemical and barrier action of zinc dust/anticorrosive pigments containing coatings. *Journal of Physics and Chemistry of Solids*, **68**, 1101-1105.
156. Sathiyanarayanan, S., Syed Azim, S., and Venkatachari, G. (2008) Corrosion protection coating containing polyaniline glass flake composite for steel. *Electrochimica Acta*, **53**, 2087-2094.
157. Yagan, A., Pekmez, N. O., and Yildiz, A. (2007) Inhibition of corrosion of mild steel by homopolymer and bilayer coatings of polyaniline and polypyrrole. *Progress in Organic Coatings*, **59**, 297-303.
158. Rohwerder, M., Duc, L. M., and Michalik, A. (2009) In situ investigation of corrosion localised at the buried interface between metal and conducting polymer based composite coatings. *Electrochimica Acta*, **54**, 6075-6081.
159. Sathiyanarayanan, S., Muthkrishnan, S., and Venkatachari, G. (2006) Corrosion protection of steel by polyaniline blended coating. *Electrochimica Acta*, **51**, 6313-6319.
160. Grgur, B. N., Gvozdenovic, M. M., Miskovic-Stankovic, V. B., and Popovic, Z. K. (2006) Corrosion behavior and thermal stability of electrodeposited PANI/epoxy coating system on mild steel in sodium chloride solution. *Progress in Organic Coatings*, **56**, 214-219.
161. Chako, D. J., and Leyva, A. A. (2005) Thermal transitions and barrier properties of olefinic nanocomposites. *Chemistry of Materials*, **17**, 13-19.
162. Alexandre, M., and Dubois, P. (2000) Polymer-layered silicate nanocomposites: preparation, properties and uses of a new class of materials. *Materials Science and Engineering*, **28**, 1-63.
163. Yan, M., Vetter, C. A., and Gelling, V. J. (2013) Corrosion inhibition performance of polypyrrole Al flake composite coatings for Al alloys. *Corrosion Science*, **70**, 37-45.
164. He, J., Gelling, V. J., Tallman, D. E., and Bierwagen, G. P. (2000) A scan-

ning vibrating electrode study of chromated-epoxy primer on steel and aluminum. Journal of the *Electrochemical Society*, **147**, 3661-3666.

165. Niratiwongkorn, T., Luckachan, G. E., and Mittal, V. (2016) Self-healing protective coatings of polyvinyl butyral/polypyrrole-carbon black composite on carbon steel. *RSC Advances*, **6**, 43237-43247.
166. Abd El-Fattah, M., El Saeed, A. M., Dardir, M. M., and El-Sockary, M. A. (2015) Studying the effect of organo-modified nanoclay loading on the thermal stability, flame retardant, anti-corrosive and mechanical properties of polyurethane nanocomposite for surface coating. *Progress in Organic Coatings*, **89**, 212-219.
167. Geim, A. K., and MacDonald, A. H. (2007) Graphene: exploring carbon flatland. *Physics Today*, **60**(8), doi: http://dx.doi.org/10.1063/1.2774096 .
168. Si, Y., and Samulski, T. (2008) Synthesis of water soluble graphene. *Nano Letters*, **8**, 1679-1682.
169. Dreyer, R. D., Park, S., Bielawski, C. W., and Ruoff, R. S. (2010) The chemistry of graphene oxide. *Chemical Society Reviews*, **39**, 228-240.
170. Wang, G., Yang, J., Park, J., Gou, X., Wang, B., Liu, H., and Yao, J. (2008) Facile synthesis and characterization of graphene nanosheets. *The Journal of Physical Chemistry C*, **112**, 8192-8195.
171. Stankovich, S., Dikin, D.A., Dommett, G. H. B., Kohlhaas, K. M., Zimney, E. J., Stach, E. A., Piner, R. D., and Ruoff, R. S. (2006) Graphene-based composite materials. *Nature*, **442**, 282-286.
172. Ansari, S., and Giannelis, E. P. (2009) Functionalized graphene sheet-Poly(vinylidene fluoride) conductive nanocomposites. *Journal of Polymer Science, Part B: Polymer Physics*, **47**, 888-897.
173. Ramanathan, T., Abdala, A. A., Stankovich, S., Dikin, D. A., Alonso, M. H., Piner, R. D., Adamson, D. H., Schniepp, H. C., Chen, X., Ruoff, R. S., Nguyen, S. T., Aksay, I. A., Prud'Homme, R. K., and Brinson, L. C. (2008) Functionalized graphene sheets for polymer nanocomposites. *Nature Nanotechnology*, **3**, 327-331.
174. Quan, H., Zhang, B., Zhao, Q., Yuen, R. K. K., and Li, R. K. Y. (2009) Facile preparation and thermal degradation studies of graphite nanoplatelets (GNPs) filled thermoplastic polyurethane (TPU) nanocomposites. *Composites Part A: Applied Science and Manufacturing*, **40**, 1506-1513.
175. Eda, G., and Chhowalla, M. (2009) Graphene-based composite thin films for electronics. *Nano Letters*, **9**, 814-818.
176. Liang, J., Xu, Y., Huang, Y., Zhang, L., Wang, Y., and Ma, Y. (2009) Infrared- Triggered Actuators from Graphene-Based Nanocomposites. *The Journal of Physical Chemistry C*, **113**, 9921-9927.
177. Dumee, L. F., He, L., Wang, Z., Sheath, P., Xiong, J., Feng, C., Tan, M. Y., She, F., Duke, M., and Gray, S. (2015) Growth of nano-textured grap-

hene coatings across highly porous stainless steel supports towards corrosion resistant coatings. *Carbon*, **87**, 395-408.

178. Berry, V. (2013) Impermeability of graphene and its applications. *Carbon*, **62**, 1-10.
179. Prasai, D., Tuberquia, J. C., Harl, R. R., Jennings, G. K., and Bolotin, K. I. (2012) Graphene: Corrosion-Inhibiting Coating. *ACS Nano*, **6**, 1102-1108.
180. Zhang, Z., Zhang, W., Li, D., Sun, Y., Wang, Z., Hou, C., Chen, L., Cao, Y., and Liu, Y. (2015) Mechanical and anticorrosive properties of graphene/epoxy resin composites coating prepared by *in-situ* method. *International Journal of Molecular Science*, **16**, 2239-2251.
181. Ramezanzadeh, B., Niroumandrad, S., Ahmadi, A., Mahdavian, M., and Moghadam M. H. (2016) Enhancement of barrier and corrosion protection performance of an epoxy coating through wet transfer of amino functionalized graphene oxide. *Corrosion Science*, **103**, 283-304.
182. Li, Y., Yang, Z., Qiu, H., Dai, Y., Zheng, Q., Li, J., and Yang, J. (2014) Self-aligned graphene as anticorrosive barrier in waterborne polyurethane composite coatings. *Journal of Materials Chemistry A*, **2**, 14139-14145.
183. Mo, M., Zhao, W., Chen, Z., Yu, Q., Zeng, Z., Wu, X., and Xue, Q. (2015) Excellent tribological and anti-corrosion performance of polyurethane composite coatings reinforced with functionalized graphene and graphene oxide nanosheets. *RSC Advances*, **5**, 56486-56497.
184. Chang, K., Hsu, C., Lu, H., Ji, W., Chang, C., Li, W., Chuang, T., Yeh, J., Liu, W., and Tsai, M. (2014) Advanced anticorrosive coatings prepared from electroactive polyimide/graphene nanocomposites with synergistic effects of redox catalytic capability and gas barrier properties. *XPRESS Polymer Letters*, **8**, 243-255.
185. Chang, C. H., Huang, T. C., Peng, C. W., Yeh, T. C., Lu, H. I., Hung, W. I., Weng, C. J., Yang, T. I., and Yeh, J. M. (2012) Novel anticorrosion coatings prepared from polyaniline/graphene composites, *Carbon*, **50** 5044-5051.
186. Hsieh, Y. P., Hofmann, M., Chang, K. W., Jhu, J. G., Li, Y. Y., Chen, K. Y., Yang, C. C., Chang, W. S., and Chen, L. C. (2013) Complete corrosion inhibition through graphene defect passivation. *ACS Nano*, **8**, 443-448.
187. Stankovich, S., Piner, R. D., Nguyen, S. T., and Ruoff, R. S. (2006) Synthesis and exfoliation of isocyanate-treated graphene oxide nanoplatelets. *Carbon*, **44**, 3342-3347.
188. Chaudhry, A. U., Bhola, R., Mittal, V., and Mishra, B. (2014) $Ni_{0.5}Zn_{0.5}Fe_2O_4$ as a potential corrosion inhibitor for API 5L X80 steel in acidic environment. *International Journal of Electrochemical Science*, **9**, 4478-4492.
189. Sreevatsa, S., Banerjee, A., and Haim, G. (2009) Graphene as a perm-

eable ionic barrier. *ECS Transactions*, **19**, 259-264.

190. Chaudhry, A. U., Mittal, V., and Mioshra, B. (2017) Impedance response of nanocomposite coatings comprising of polyvinyl butyral and Haydale's plasma processed graphene. *Progress in Organic Coatings*, **110**, 97-103.
191. Chaudhry, A., Mittal, V., and Mishra, B. (2015) Inhibition and promotion of electrochemical reactions by graphene in organic coatings. *RSC Advances*, **5**, 80365-80368.
192. Luckachan, G. E., and Mittal, V. (2018) Stable corrosion-resistant chitosan coatings: Effect of crosslinking with silica, poly(vinyl butyral) over-coating and graphene on the coating performance. In: *Biopolymers Based Advanced Materials*, Mittal, V. (ed.), Central West Publishing, Australia, pp. 207-237.
193. Zheludkevich, M. L., Shchukin, D. G., Yasakau, K. A., Mohwald, H., and Ferreira, M. G. S. (2007) Anticorrosion coatings with self-healing effect based on nanocontainers impregnated with corrosion inhibitor. *Chemistry of Materials*, **19**, 402-411.
194. Plawecka, M., Snihirova, D., Martins, B., Szczepanowicz, K., Warszynski, P., and Montemor, M. F. (2014) Selfhealing ability of inhibitor-containing nanocapsules loaded in epoxy coatings applied on aluminium 5083 and galvanneal substrates. *Electrochimica Acta*, **140**, 282-293.
195. Benita, S. (2006) *Microencapsulation Methods and Industrial Applications*, 2nd edition, CRC Press, USA.
196. Siva, T., and Sathiyanarayanan, S. (2015) Self healing coatings containing dual active agent loaded urea formaldehyde (UF) microcapsules. *Progress in Organic Coatings*, **82**, 57-67.
197. Mallegol. J., Lemaire, J., and Gardette, J. L. (2000) Deier influence on the curing of linseed oil. *Progress in Organic Coatings*, **39**, 107-113.
198. Huang, M., and Yang, J. (2011) Facile microencapsulation of HDI for self-healing anticorrosion coatings. *Journal of Materials Chemistry*, **21**, 11123-11130.
199. Tatiya, P. D., Hedaoo, R. K., Mahulikar, P. P., and Gite, V. V. (2013) Novel polyurea microcapsules using dendritic functional monomer: synthesis, characterization, and its use in self-healing and anticorrosive polyurethane coatings. *Industrial and Engineering Chemistry Research*, **52**, 1562-1570.
200. Yang, H. I., Kim, D. M., Yu, H. C., and Chung, C. M. (2016) Microcapsule- type organogel-based self-healing system having secondary damage preventing capability. *ACS Applied Materials and Interfaces*, **8**, 11070-11075.
201. Fu, J.-J., Chen, T., Wang, M.-D., Yang, N.-W., Li, S.-N., and Liu, X.-D. (2013) Acid and alkaline dual stimuli-responsive mechanized hollow mesoporous silica nanoparticles as smart nanocontainers for intelligent anticorrosion coatings. *ACS Nano*, **7**(12), 11397-11408.

202. Dobrynin, A. V., and Rubinstein, M. (2005) Theory of polyelectrolytes in solutions and at surfaces. *Progress in Polymer Science*, **30**, 1049-1118.
203. Grigoriev, D. O., Kohler, K., Skorb, E., Shchukin, D. G., and Mohwald, H. (2009) Polyelectrolyte complexes as a "smart" depot for self-healing anticorrosion coatings. *Soft Matter*, **5**, 1426-1432.
204. Rubner, M. F. (2003) In: *Multilayer Thin Films*, Decher, G., and Schlenoff, J. B. (eds.), Wiley, Germany, pp. 133.
205. Shchukin, D. G., Lamaka, S. V., Yasakau, K. A., Zheludkevich, M. L., Ferreira, M. G. S., and Mohwald, H. (2008) Active anticorrosion coatings with halloysite nanocontainers. *The Journal of Physical Chemistry C*, **112**(4), 958-964.
206. Angele, S. (2014) Self-healing Coatings from Recycled Polymer Blends, US Patent WO 2014/099070 A1.
207. Kopec, M., Szczepanowicz, K., Mordarski, G., Podgorna, K., Socha, R. P., Nowak, P., Warszynski, P., and Hack T. (2015) Self-healing epoxy coatings loaded with inhibitor-containing polyelectrolyte nanocapsules. *Progress in Organic Coatings*, **84**, 97-106.

2

Polymers for Gas Sensing

2.1 Introduction

Gas Sensing is one of the functional areas where polymers are finding widespread use and have proven to be very useful candidates owing to the tunable and remarkable structural, chemical and electrical performance profiles [1]. With a multitude of different analytical technologies present, last two decades have seen significant development of chemical sensors. The speedily rising applications of chemical sensors reveal the vast extent to which analytical researchers need these sensors for low-cost, precise, appropriate and fast examination of numerous analytes. Generally, in this process, initially the analyte gets recognized, which is followed by the generation of electrical signal from the chemical information obtained in the first step. The benefit of using polymers for chemical sensors is the easier fabrication of sensory devices on electrodes using electrochemical polymerization. Depending on the mechanism of sensing, different structures of the electrodes of the sensors are needed. Amperometry and potentiometry are two methods commonly utilized for identifying analytes present in liquids, particularly in polymer-based biosensors [2-6] and polymer-based pH sensors [7-10]. On the other hand, conductimetry is the most commonly chosen method for detection of gases [1]. For instance, Gu *et al.* [11] reported a versatile nano-sensor generated from polymeric single nanowires. Depending on the optical behavior of the wave guiding polymer single nanowires after being revealed to the samples, the polymer nanowires were utilized in humidity sensors having a time of response of 30 ms and for NH_3 and NO_2 sensing below parts-per-million scale. The authors suggested that the flexible and compact sensing structure having higher sensitivity and smaller footprint could have immense potential for rapid identification in biological, physical and chemical applications [11]. Figure 2.1 shows the schematic and other characteristics of the sensor.

Swati Singh and Vikas Mittal, The Petroleum Institute (part of Khalifa University of Science and Technology), Abu Dhabi, UAE*
**Current address: Bletchinton, Wellington County, Australia*

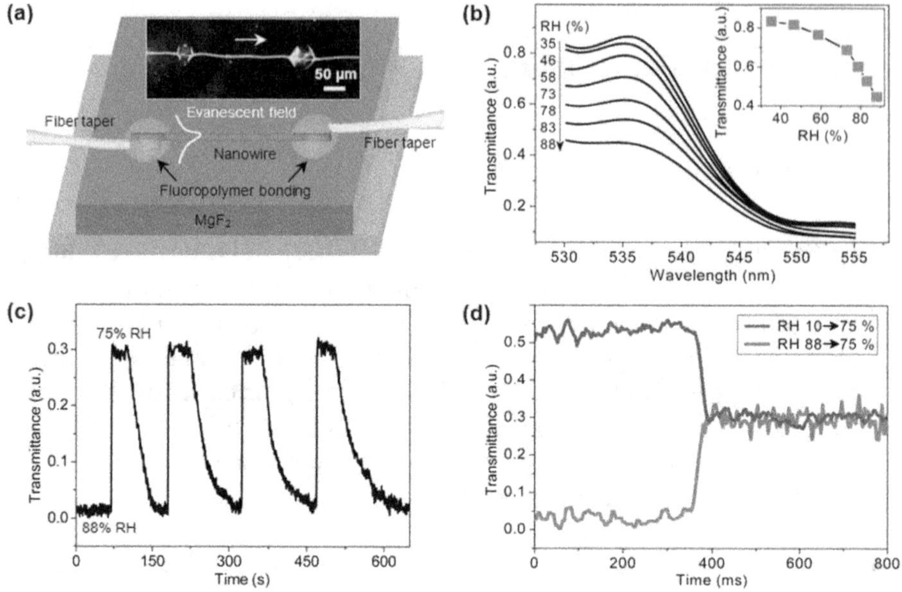

Figure 2.1 Polyacrylamide single-nanowire humidity sensors: (a) schematic illustration of the sensor, (b) transmittance of a MgF_2-supported 410-nm-diameter PAM nanowire exposed to atmosphere of RH from 35% to 88%, (c) reversible response of the nanowire tested by alternately cycling 75% and 88% RH air and (d) typical time-dependent transmittance of the sensor. Reproduced from Reference 11 with permission from American Chemical Society.

Sensors generated from metal oxide semiconductors exhibit sensitivity towards inorganic gases and volatile organic compounds like nitrogen and alcohol, respectively. However, the volatile organic compounds (VOCs) present in concentrations more than their threshold values are occasionally not detected by these sensors. These volatile organic compounds may be inhaled by humans, especially in cases where vaporization takes place at ambient temperature. Thus, it makes the monitoring of these vapors an important concern from health and safety perspectives. In addition, monitoring atmospheric releases is vital to prevent environmental hazards. This is becoming increasingly possible by utilizing polymers as sensing materials. For instance, there have been several studies where sensing materials based on polymers have been reported for detecting inorganic gases [12]. These functional materials can also be useful in detecting other classes of volatile organic compounds as well as solvent vapors like aromatic or halogenated compounds, alcohols, etc. VOCs chiefly interact with polymeric sensors through several physisorption mechanisms ranging from induced dipole/induced dipole interactions, dipole/induced dipole interactions to dipole/dipole interactions and hydrogen bonds [13]. Subsequent to the fact that mechanisms of these polymer-based sensors differ from semiconductors, detection processes in polymeric sensors usually occur at ambient temperatures. Classification of polymers used in gas-sensing processes, essentially (1) conducting polymers and (2) non-conducting polymers, is based upon the varying changes in physical properties after interactions with the vapors or gases [14].

2.2 Conducting Polymers

Nylander *et al.* [15] reported in 1983 that the conducting polymers like polyaniline, polythiophene, polypyrrole and their derivatives can be used to generate the active layer of gas sensors. There are many remarkable characteristics of conducting polymers which make them suitable for chemical sensing like fast response and high sensitivity at room temperature. In addition, different synthesis processes (chemical or electrochemical) can be used to prepare conducting polymers. One of the common preparation methods is the chemical oxidation process in which the monomer and oxidants are mixed in a solution. Nevertheless, it has also been observed that electrochemical process is generally better as compared to the chemical method as it provides regulation of doping level and polymerization. Also, this process can be carried out on different substrates so as to directly construct the devices. Electrochemical method consists of three types of electrodes known as reference, counter and working electrodes. In this method, polymers are synthesized through anodic oxidation of the analogous monomers in appropriate electrolytes at constant current, at constant potential or via cycling the potential. Furthermore, the size and nature of the counter-ions have a significant impact on the physical behavior of the polymer such as porosity, thermal stability, mechanical properties and morphology. In order to expand the performance domains, electrodeposited polymers are repeatedly assorted with other functional materials like metal oxide nanoparticles [16], organic compounds [17], carbon nanotubes [18] or metal nanoparticles [19], thus, resulting in composites having effective sensing behavior. In addition, use of structural derivations or copolymerization also leads to further modification of molecular chain structure.

For instance, Figure 2.2 demonstrates the scanning electron micrographs (SEM) of gold nanoparticles deposited on polypyrrole films [19].

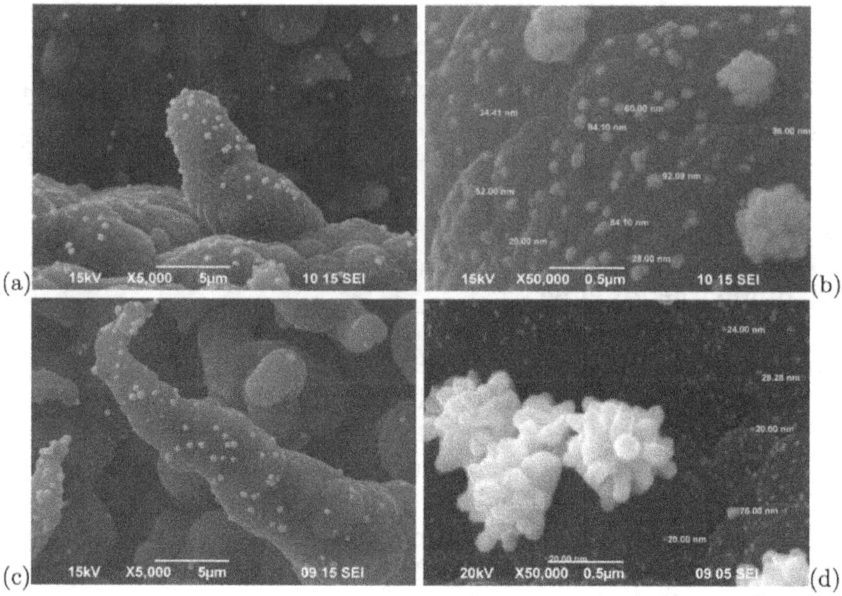

Figure 2.2 SEM images of gold deposited on polypyrrole films. Reproduced from Reference 19 with permission from American Chemical Society.

2.2.1 Synthesis of Conducting Polymers and Their Thin Films

As mentioned earlier, the two commonly used methods for synthesis of conducting polymers are chemical and electrochemical approaches [20]. Preparation of conducting polymers using chemical oxidation method involves the combination of oxidant and monomer in solution. Different oxidants used commonly for this purpose are cerium sulfate, potassium dichromate, hydrogen peroxide, ferrum chloride, ammonium persulfate, etc. Different methods like cyclic voltammetry, potentiostatic and potentiodynamic techniques can be used for electrochemical production. In these methods, three different electrodes mentioned previously as counter, working and reference electrodes are used to complete the synthesis. The working electrode, used effectively for the deposition of conducting polymers, is fabricated using a variety of materials like stainless steel, platinum, indium, tin oxide, glass, gold, etc. However, in several cases, the deposited polymers can be easily removed and turned into free-standing form. Some conducting polymers like poly (phenylene ethynylene) and poly (phenylene vinylene) along with their derivative are generally prepared using other reaction schemes such as Gilch polymerization [21], Wittig and Heck reaction [22,23], etc. Similarly, coupling reaction is used for the synthesis of polythiophene and their derivatives [24]. These methods for chemically synthesizing conducting polymers are capable of producing polymers having regular structures. For instance, Figure 2.3 shows the structures of some poly(phenylene vinylene) copolymers [22].

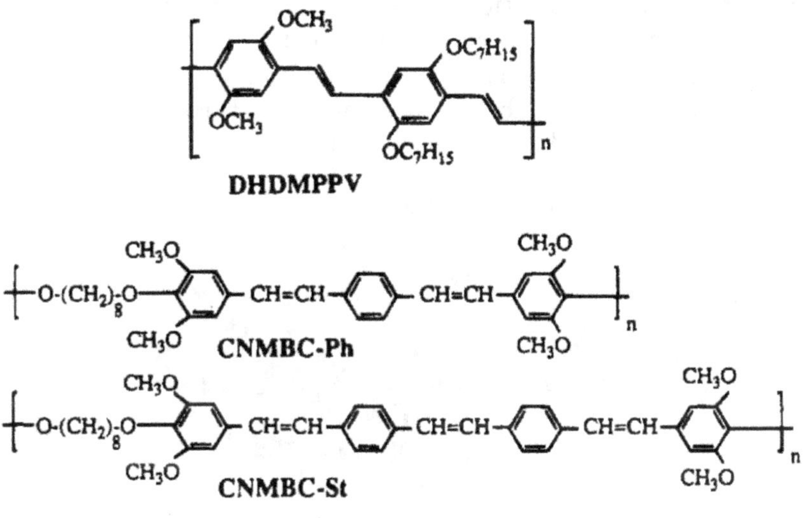

Figure 2.3 Molecular structure of poly(phenylene vinylene) copolymers. Reproduced from Reference 22 with permission from American Chemical Society.

In a sensor, the most important component is its active layer. Several methods have been developed to fabricate conducting polymer films so as to adjust with the requirements of effective sensing and diverse configurations of sensors [25]. Among all the methods, the most expedient way to deposit the layer of conducting polymers films is electrochemical deposition. By this method, it is possible to monitor the film thickness by controlling the amount of total charge passing inside the electrochemical cell during the film growing process. In addition, it is also possible to effectively layer the films on patterned microelectrodes [26]. It is necessary that such deposition takes place on a substrate which is conducting in nature. On the other hand, if there is a close insulating gap of several micrometers among the electrodes, then the growing film can cover the connect electrodes and insulated gap [27]. It is also a crucial requirement for the fabrication of chemiresistors.

Another method for the preparation of conducting polymer films is known as dip-coating. In this process, the substrate is dipped into a solution of chemical species so that some of the polymer will form a layer on the surface of substrate after polymerization [28,29]. It can be achieved on various substrates and the films thickness can be generally monitored by controlling the time of dipping. Figure 2.4 demonstrates the schematic of polyaniline growth through electrochemical deposition. A very similar method to this involves the alternate submerging of the substrate into a solution of oxidant and monomer. The monomer which gets adsorbed goes through polymerization on top surface of the substrate [30,31].

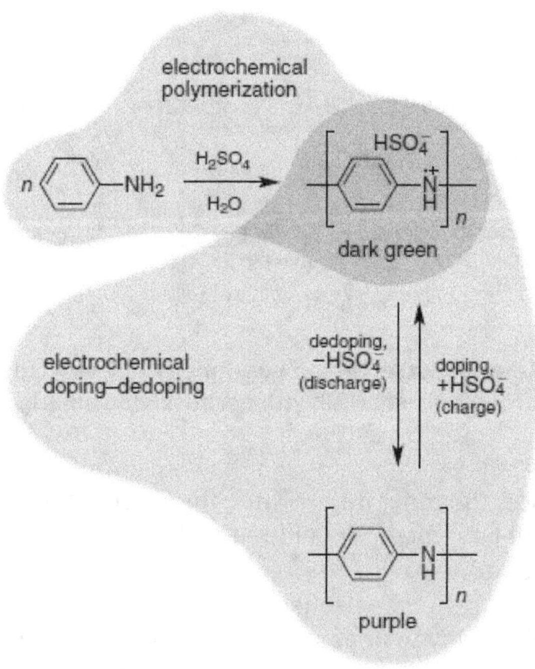

Figure 2.4 Electrochemical polymerization and doping-dedoping of polyaniline. Reproduced from Reference 28 with permission from American Chemical Society.

A simpler method to prepare films from soluble conducting polymers is known as spin-coating. Here, the solution of conducting polymer is scattered on the rotating substrate [32,33]. Once the solvent is evaporated, a thin film is left deposited on the substrate. This process can be repeated and it is possible to monitor the film thickness. The crucial factors for the adjustment of formed film thickness are rotating rate of the substrate and concentration of the solution. The method is capable to deposit the conducting polymers on insulating and conducting substrates. Also, Langmuir-Blodgett technique is a well-known method to generate thin films of surfactants and polymers. Two different approaches for deposition of polymer films are commonly reported for this method. The first approach involves the direct deposition of polymer on the substrate, whereas in the second approach, the monomer is first deposited on the substrate which is then polymerized to obtain the polymer film on the substrate surface. In this method, the films formed on the surface are ultrathin, but it is still possible to generate thicker films through the repetition of deposition steps.

Layer-by-layer (lbl) self-assembly technique is another method in which the substrate is immersed in a polymeric anionic and cationic solutions alternatively to obtain a layer by layer deposition of polymer films [34,35]. Conducting polymers which are doped (like polyaniline) have positive charges on their backbone, thus, these allow the polymeric anions to deposit on them. The number of times the process has been repeated governs the thickness of layer by layer film. In thermal evaporation technique, the conducting polymer is heated inside a vacuum environment and the evaporated polymer, in turn, deposits on the surface of the substrate. The duration of evaporation provides the thickness of as prepared film [36]. The vapor deposition polymerization method involves two steps. In the first step, a film of oxidant is prepared. The second step involves the placement of the film into monomer vapor [37]. The polymerization occurs by the diffusion of monomer on to the film. This method can be used for fabricating conducting films of pure polymers as well as composite films comprising of various conducting polymers.

In the process known as drop-coating, the solution of polymer or few drops of oxidant and monomer solutions are placed on the substrate for reaction [38,39]. The technique is though simple, however, the resultant films on the substrate are generally not uniform. There are other processes for generating films such as the fabrication of patterned conducting film of polymers using an electric field prompted electrochemical polymerization. Pyrrole can be polymerized by applying a voltage with an electrode gap of 5 mm to the saturated pyrrole vapor [40]. The directional deposit of polyaniline at a controlled voltage on the microelectrode has also been achieved by controlling the colloidal suspension of polyaniline [41]. Figure 2.5 demonstrates the preparation of polyaniline by this method [41].

In case of soluble conducting polymers, thin films can also be fabricated by using the inkjet-print method [42]. Moreover, the conducting polymers have also been packed into pellets in several research studies to prepare the active layers for the purpose of gas sensing [43,44].

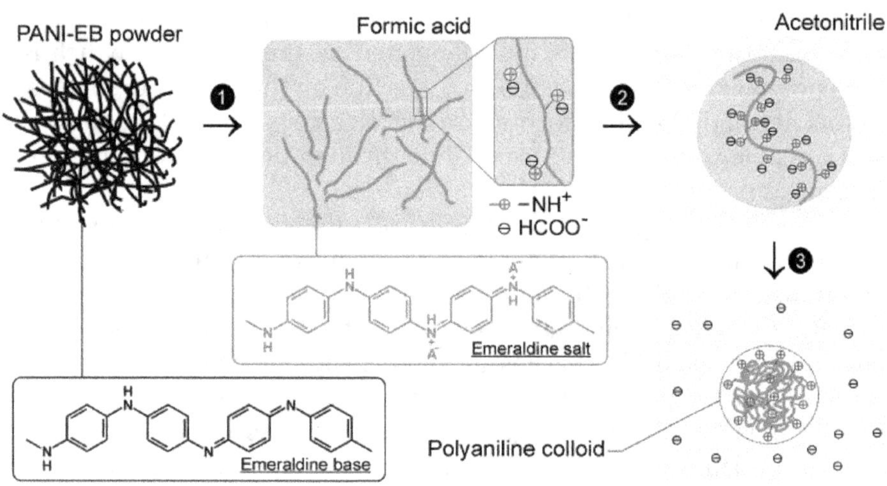

Figure 2.5 Formation of polyaniline colloids by dispersing polyaniline/formic acid solution into acetonitrile. Reproduced from Reference 41 with permission from American Chemical Society.

2.2.2 Electronic Properties of Conducting Polymers

Conjugated polymers have optical and electrical properties comparable to metals and semiconductors due to their specific molecular structure [45]. Delocalized π-electrons over the large segments in the repeating units of the polymer provide such properties to these polymers. Most of the well-established conjugated polymers have straight chain units, with five or six rings, along with their combinations. A larger variety can also be synthesized by adding different types of heteroatoms and side chains of conjugated polymers. The synthesis method employed for generating such polymers has also a significant effect on the electronic properties of the material. The electronic conductivity of such polymers generally ranges over a value of 10^{-12} S.cm^{-1} to 10^{5} S.cm^{-1}.

Doping is also carried out on the polymer to improve (or reinforce) the properties. Doping is basically the process of injecting charged holes or electrons into the conjugated spine of the polymer. Owing to chain deformation around the charge, there is self-localized electronic state formation in the formerly forbidden semiconducting band gap. Thus, the charge of the polymer is not due to a free electron or a hole, but because of the polaron or bipolaron which depends on the degree of doping. Polaron or bipolarons are localized particles connected within a chain deformation. If the doping results in a polymeric cation rich material, it called p doped material. On the other hand, an anion rich material is called n doped material. To maintain charge neutrality, the insertion or repulsion of counter ions is also carried out in addition to charge injection. Due to the long chain of monomer molecules, which can act as individual redox sites, there is the possibility of doping of high density of charge carriers on the conjugated polymer chains. Compared with common inorganic semiconductors, this charge density can range up to five times the normal charge density of a typical semiconductor.

There are a number of ways by which doping can be realized. These are electrochemical, chemical, photochemical and charge insertion on the interface of metal-insulator-semiconductor. The chemical and electrochemical doping is achieved by oxidation or reduction of neutral polymer chain to form p or n doping respectively, which later forms polaron or bipolaron states. There is a permanent induced electrical conductivity due to this process which can only be removed by reversing the electrochemical reaction by undoping the electrons at the redox sites of the polymer chain. The charge transport within a single chain is realized by creating polarons or bipolarons. However, the electronic charge carriers can't participate in electron transport other than hopping because of the special delocalization of high density charge carriers. Structural disorder acts as the major source for the localization of charge carriers in the conducting polymers. This, in turn, reduces the conductivity of the polymer by reducing the charge carrier mobility. It is also seen that hopping transport is majorly affected by the dopant compounds in the system [46].

2.3 Non-Conducting Polymers

Polymers with non-conductive nature have a wide range of applications as sorptive coatings in various sensors in which the device and coating can be considered as a shared transducer. Thus, it is possible to coat such polymers with different physisorption mechanisms and properties on particular transducers. For example, the layers of polymers which cause change in the dielectric constant and enthalpy upon desorption/absorption along with resonance frequency of analytes can be individually coated on surface transverse wave, mass-sensitive (e.g., quartz crystal microbalance), capacitive (dielectric), surface acoustic wave and calorimetric sensor devices. Thus, the properties of polymer can be converted into an electric signal as an output [47]. It has to be noted that the performance of non-conducting polymers is not straight forward to analyze, especially when these are coated on the sensor devices. For example, when a thin sensitive polymer layer is used to coat the surface transverse wave resonant device, a significant advantage is noticed in terms of complete electrical behavior and low noise of the sensor oscillator as well as relative gas probing sensitivity in contrast to their surface acoustic wave equivalents [48]. Furthermore, non-conducting polymers like polyamides can be utilized in gas sensors based on metal oxide semiconductor as molecular sieves which can improve the general selectivity [49].

2.4 Gas Sensors

As mentioned earlier, the use of conducting polymers is promising in the field of chemical sensors to detect and identify several volatile organic compounds. These can also be applied for the sensing of various combined vapors and multifarious odours. Therefore, such sensors aid in the improvement of traditional apparatuses having utilities in various fields such as food and chemical sectors, biomedical analysis and ecological checking. The basic mechanism of chemical sensor based on polymers is the change in the electrical conductivity of polymers when these are kept in the presence of vapors for analysis. Such change in electrical conductivity of polymer can be associated quantitatively with the concentration of the vapor. Persuad *et al.* [50-52] reported the improvements for conducting polymers so as to use them for gas sensing devices, especially for developing various materials for transduction of odours [50-52]. For sensing various odours, the conducting polymers are selected on the basis of the below stated reasons:

- The elements of sensor consume very less power in the range of microwatts and that's why there is no need of heating element.
- Such sensors can resist themselves from the poisonous compounds that are capable of inactivating the sensors based on inorganic semiconductor.
- At room temperature, desorption kinetics and quick adsorption is displayed by the sensors.
- It is possible to correlate the structure of polymers to specific kinds of chemical combinations [53].

2.4.1 Performance Indicators and Gas Sensor's Stability

In order to analyze the behavior of gas sensors and various approaches of gas sensing, a few factors are required to be taken into consideration. First indicator is sensitivity which is the least amount of the concentration of target gases which can be sensed. Second important indicator is selectivity which is the capability of a gas sensor to detect a particular gas from a mixture of gases [54]. Next indicator is the time of response which gives the amount of time taken by the gas sensors to reach a particular value where the sensor triggers a warning signal. Reversibility is also a factor which indicates if the sensor can go back to its original form after the process of detection. Several other factors include the amount of energy consumed by the sensors, capacity of adsorption which also impacts the selectivity and sensitivity of sensors and cost of fabrication. Moreover, the gas sensors fabricated for their commercial application should assure the operating stability which means that the sensor should display a reproducible and stable signal for a long term. There are various factors which lead to instability in gas sensors such as design errors, structural changes like deviations of grain network or grain size, phase shifts which generally means the separation of doped additives with materials used in sensing, poisoning activated through chemical reactions, change in the environment, etc. With the aim of solving these issues, the following approaches should be implemented: (1) choosing materials which are chemically and thermally stable; (2) improving grain size and elemental composition of sensing materials; (3) employing efficient technology.

2.5 Polymers for Gas Sensing and Gas Sensors

As mentioned earlier, polymers based gas sensors display significant advantages like fast response and high sensitivity. In addition, these sensors do not need high temperature while operation and can work at room temperature, unlike metal oxide semiconductor based sensors which have a very demanding operation temperature. Thus, lower consumption of energy makes polymers based gas sensors suitable for application in detection units driven by batteries. Other benefits of using polymers based sensors are cost effective fabrication, movability and simple structure [55], along with the possibility of reproduction. However, there are some disadvantages of these sensors like low selectivity, irretrievability and instability for long-term operations. The performance of these sensors can also be affected by the environment. Moreover, it is important that the working mechanism of polymer based sensors be known more clearly and convincingly.

Dan *et al.* [56] reported the gas sensing reaction and electronic behaviors of nanowires containing PEDOT/PSS (poly(3,4-ethylenedioxythiophene)/poly(styrenesulfonate)), which were prepared by using anodic aluminum oxide membranes. The as prepared nanowires had diameter of 220 nm and were almost 8 μm long. In order to place the single nanowires on the pre-fabricated gold electrodes, dielectrophoretic assembly was used. A contact resistance of 27.6±4 kΩ and a polymer conductivity of 11.5±0.7 S/cm were measured. The nanowires as a gas sensor displayed a change in resistance of 9%, 4% and 10.5%, at the saturation vapor pressure of methanol, ethanol and acetone, respectively. The response and recovery time of the sensor were in seconds with extraordinary time reproducibility. Thus, the PEDOT/PSS based conducting nanowires represented an effective candidate for nanoscale gas sensors due to their sensitivity, fast recovery and response as well as outstanding reproducibility of analyte at ambient conditions. In another study, Ji *et al.* [57] prepared a supramolecular cross-linked network and analyzed its application as fluorescent sensor. Figure 2.6 shows the fabrication of the polymer network with supramolecular cross-linked conjugation, along with the disassembly prompted by various signals. It was synthesized using a fluorescent conjugated polymer and a bisammonium salt based cross-linker driven by dibenzo[24]crown-8/secondary ammonium salt host-guest interactions. Due to the aggregation of polymer chains in the network, it had weak fluorescence as compared to the conjugated polymer. The authors observed that due to the host-guest connections having stimuli-responsiveness, the fluorescence intensity could be enhanced by four kinds of signals, such as inclusion of chloride anion, potassium cation, along with heating and pH increase. Therefore, the network had the capability to function as an anion sensor, a cation sensor, a temperature sensor as well as a pH sensor. It could be applied both in thin film and solution. Additionally, exposing the film generated from supramolecular cross-linked network to ammonia led to an improvement in fluorescence, thus, confirming its potential for gas sensing [57].

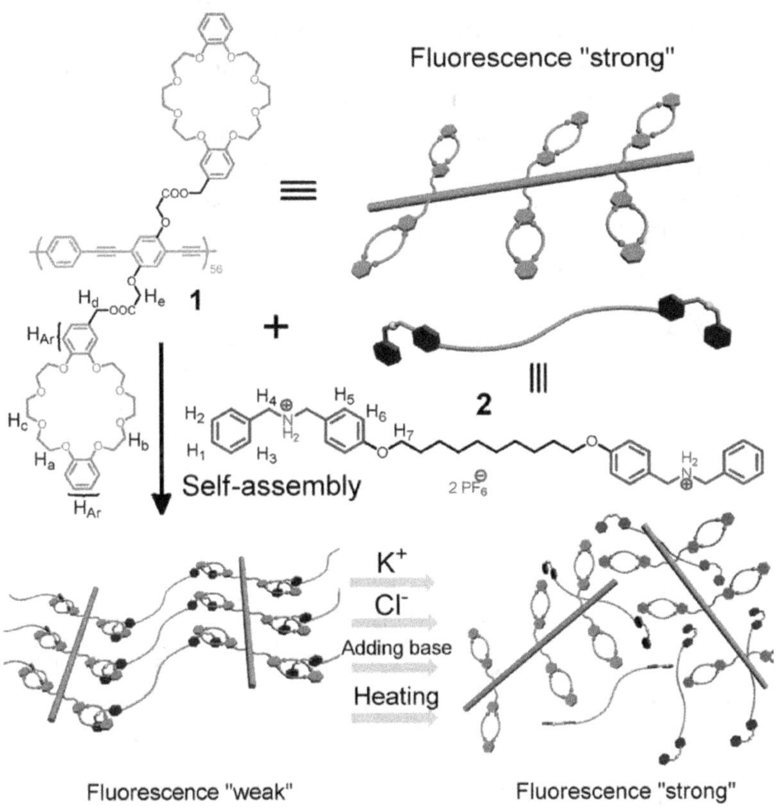

Figure 2.6 Representation of the formation of a supramolecular cross-linked conjugated polymer network and its disassembly induced by different signals. Reproduced from Reference 57 with permission from American Chemical Society.

In another study, Ding *et al.* [58] produced a polymer with well-defined structure having linear chains and higher molecular weight using the polycondensation of hexafluorobisphenol A (6F-BPA) with 2,5-Bis(pentafluorophenyl)-1,3,4-oxadiazole (FPOx) at ambient temperature in DMAc in the presence of potassium fluoride (Figure 2.7). Potassium fluoride acted as a catalyst and a base during the reaction as observed for other bis-(pentafluorophenyl) monomers. UV studies also confirmed the interaction and identified a new absorption spectrum after the mixing of FPOx solution and potassium fluoride. In order to investigate the application of the FPOx as a substance for sensing and detecting the fluoride solution, the variation in the UV spectrum after the addition of potassium fluoride was examined additionally. It was observed that such molecules exhibited high selectivity and sensitivity for the sensing of fluoride anion having a detection limit of around 0.1 ppm in both fluorescence and UV detection. The FPAEOx polymer containing FPOx displayed an equivalent selectivity and sensitivity for fluoride sensing in both acetonitrile and DMAc solutions, though a tendency of aggregation in acetonitrile was observed due to lower solubility. This led to lower sensitivity of the solution where the concentrations of fluoride were low [58].

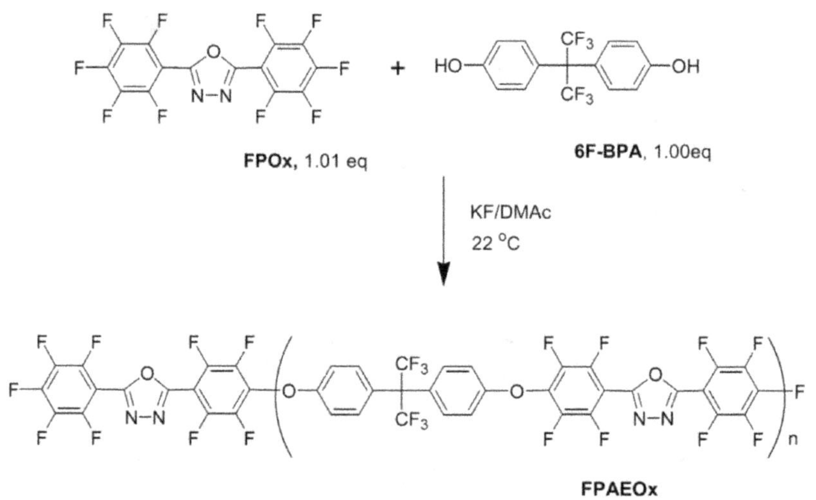

Figure 2.7 Schematic representation of synthesis of the fluorinated polymer. Reproduced from Reference 58 with permission from American Chemical Society.

In a recent study, Xue *et al.* [59] fabricated biomimetic polymer nanofibers with sensing capabilities, thus, having immense potential for applications in biomedical and biological fields. The as prepared fibers comprised of either a polysulfone (PSU) or a poly(ether sulfone) (PES) core which was layered by a biocompatible polycaprolactone (PCL) shell. Using single step coaxial electrospinning method, the luminescent probes with oxygen sensitivity were introduced in the core such as Pd(II) meso-tetra(pentafluorophenyl)porphine (PdTFPP), or Pt(II) meso-tetra(pentafluorophenyl)porphine (PtTFPP) facilitating linear response, greater sensitivity, exceptional stability and high brightness (Figure 2.8). Both PSU-PCL and PES-PCL fibers resulted in uniform distribution of probe as compared to polydimethylsiloxane (PDMS). Owing to their comparatively high permeability of oxygen, sensing fibers based on PSU had optimum sensitivity. With PtTFPP as the indicator, a I_0/I_{100} value of 6.7 was displayed

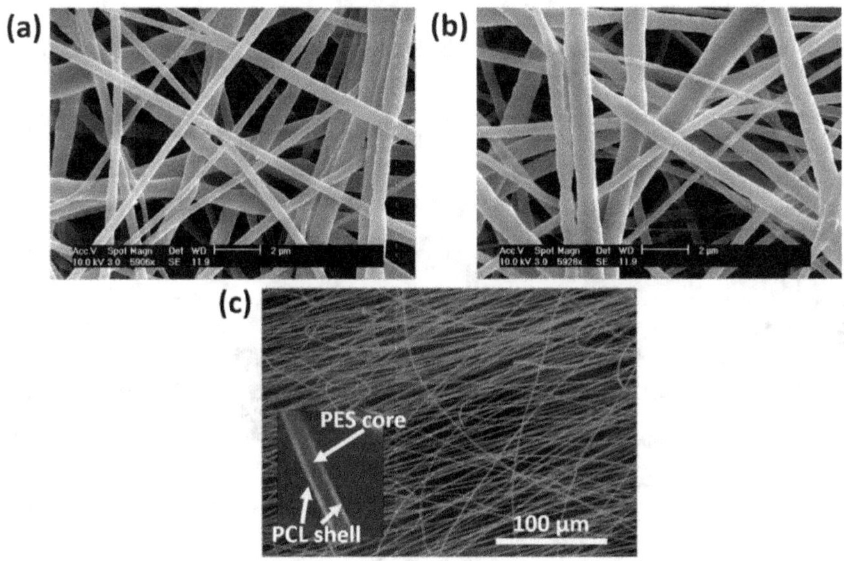

Figure 2.8 Scanning electron micrographs of (a) PES-PCL and (b) PSU-PCL fibers containing PtTFPP oxygen probe and (c) fluorescence image of PES(PtTFPP)-PCL fibers. Reproduced from Reference 59 with permission from American Chemical Society.

by PES-PCL fibers when exposed to 100% nitrogen and 100% oxygen, whereas, PSU-PCL fibers displayed a value of 8.9. In comparison, PdTFPP-comprising fibers had higher sensitivity because of the long lifetime of porphyrin. The resultant values of I_0/I_{100} were 80.6 and 106.7 for PES-PCL and PSU-PCL matrices, correspondingly. The response and recovery times were 0.12 and 0.11 s faster for PES-PCL

(0.24/0.39 s) and PSU-PCL (0.38/0.83 s) respectively as compared to platinum based porphyrin. Unexpectedly, these polymers with lower permeability of oxygen were effective for the measurement of high contents of oxygen by approximately 20% than PDMS. Each of the compositions exhibited Stern–Volmer plots with a linear behavior and the sensitivity could be customized through the selection of the sensing probe or matrix.

Dunst *et al.* [60] also observed that the resistance change of PEDOT at a given time depends on NO_2 gas concentration. PEDOT layer behaves like a total amount integrator. On plotting the resistance as a function of total amount of NO_2, one can obtain nearly linear dependence. In this work, it was noticed that in a measuring range, PEDOT exhibited total adsorption. Hence, PEDOT can be used as an accumulating sensing material, but more detailed studies need to be performed. In another study, Sekizkardes *et al.* [61] effectively

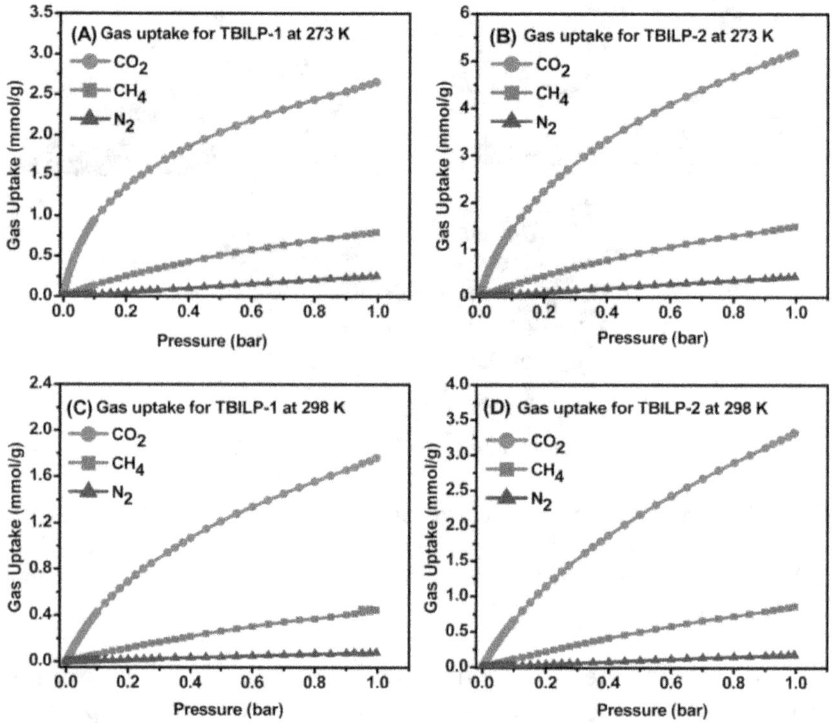

Figure 2.9 CO_2, CH_4, and N_2 gas adsorption isotherms for the polymers at 273 K (A, B) and 298 K (C, D). Reproduced from Reference 61 with permission from American Chemical Society.

incorporated both benzimidazole and triazine structural units inside organic polymers with porosity and investigated their effect on separation of CO_2 from landfill and flue gases. The triazine-based benzimidazole-linked polymers exhibited a high adsorption of CO_2 at 1 bar and 273 K (5.19 mmol g^{-1}/228 mg g^{-1}) (Figure 2.9). Ideal adsorbed solutions theory selectivity measurements exhibited high CO_2/N_2 selectivity of 62 at 298 K. The observed selectivity of CO_2/N_2 through triazine-based benzimidazole-linked polymers outclasses all the other porous organic polymers based on triazine described before. CO_2 selectivity and heats of adsorption exhibited that introduction of triazine units inside benzimidazole-linked polymers might increase selective adsorption of CO_2 over CH_4 and N_2. Triazine-based benzimidazole-linked polymers also revealed encouraging operational regenerability, capability and sorbent choice for CO_2 capture from flue and landfill gases. Figure 2.10 also shows the response of sensors made of molecularly imprinted polymer (MIP) coated on the quartz crystal microbalance (QCM) for the detection of l-menthol by having five repeated runs [62]. Table 2.1 also details other studies describing examples of gas sensors based on polymers.

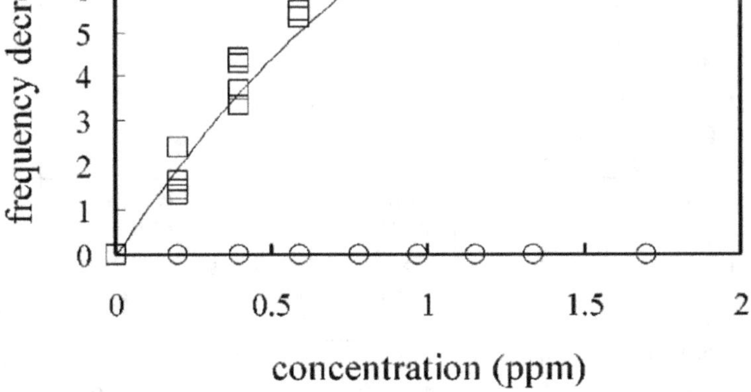

Figure 2.10 Response of the QCM sensors to menthol for the MIP coated (squares) and non-imprinted polymer coated (circles). Reproduced from Reference 62 with permission from American Chemical Society.

Table 2.1 Some examples of gas and odour sensors based on a variety of polymers

Sensor type	Polymers utilized	Areas of applications	Special features	Ref.
Gas sensor	Copolymers of poly (EDMA-co-MAA)	Sensing of terpene in environment	Piezoelectric sensor covered with polymer molecularly imprinted on it	[62]
Odour sensor	Poly [di(ethyleneglycol) adipate], polydimethylsilox-ane, poly[bis(cy-anoallyl) polysiloxane], poly [dimethylsiloxane-co-methyl (3-hydroxypropyl) siloxane]-g-poly(ethylene gly-col)3-aminopropyl ether, polydiphenoxyphos-pha-zene, hydroxyl ended polydimethylsilox-ane, polychloro-prene, beads of polystyrene, polyisobutylene	Detection of organic compounds with volatile nature	Sensor array	[63]
Gas sensor	Nafion	Identification of ethanol gas concentration	Polymer electrolyte membrane based fuel cell	[64]
Gas sensor	Poly(vinylchloride)	Sensing of gaseous NO_2 in air	In the sensor, a solid polymer electrode comprising of 10% PVC is used	[65]

Gas sensor	Poly(vinylalcohol)	Sensing of optical humidity	Methylene blue and crystal violet are integrated in PVA/H_3PO_4	[66]
Gas sensor	Propylene-butyl copolymer	Identification of xylene gas, toluene	Polymer film covered quartz resonator balance	[67]
Gas sensor	Poly[3-(butylthio)thiophene]	Gas sensing	Polymer films were fabricated through casting technique and LB deposition	[68]
Gas sensor	Polyaniline (PANI), acetic acid and polyaniline mixed film PANI-polystyrenesulfonic acid composite film	Detection of NO_2	Layers of polymer films prepared by self-assembly and Langmuir-Blodgett techniques	[69]
Odour sensor	Poly (4-vinyl phenol), poly (sulfone), poly (methyl methacrylate), poly (caprolactone), poly (N-vinyl pyrrolidone), poly (vinylidene fluoride), poly (ethylene-co-vinyl acetate), poly (ethylene glycol), poly (ethylene oxide) polyethylene	Odour identification	Array of polymer composites with conducting nature	[70]
Odour sensor	Polypyrrole, poly (3-methylthiophene),	Differentiate between various pure olive oils	Doping materials were utilized	[71]

In a gas flow, the amperometric sensors comprising of an electrochemical cell have the ability to respond electrochemically to vapors and gases [72,73]. The change in the electrodes and potential of electrodes is useful for the detection of gases. An improvement was observed by applying a catalytic micro-reactor [74]. This reactor contained a platinum strand which was heated, thus, leading the analytes to go through oxidation in order to detect the electrochemically unreactive materials. Use of polymers in such sensors can further help to tune the performance.

Polymer composites with conducting nature (with polymers such as PVC, PVA, PMMA, etc.) are also utilized for gas sensing. Hydrogen chloride (HCl) has been recognized as hazardous for work places having a 5 ppm of short-term contact limit. Nakagawa *et al.* [75] developed composite polymer films of alkoxy switched tetraphenylporphyrin–polymer in order to identify HCl in sub-ppm levels. Recovery behavior and sensor response could be improved by a glass transition temperature lower than sensing temperature in the matrix. Sensitivity towards HCl could be improved by imparting basicity to the material by providing alkoxy group. The changes imparted by varying HCl gas concentration were studied for Soret-and Q-bands. The use of a 5,10,15,20-tetra(40-butoxyphenyl)porphyrin - butylmethacrylate film improved selectivity towards HCl gas to sub-ppm levels. Kikuchi *et al.* [76] also prepared a copolymer based on tertiary amine moieties and phenylboronic acid. The copolymer was subsequently mixed with PVA to obtain a gel membrane complex. Glucose-responsive swelling changes of the complex were used for fabricating the electrodes to detect the change in the concentration of glucose. At physiological pH situations, the polymer complex-coated electrode detected the concentration of glucose with high sensitivity and selectivity. In Figure 2.11, the change in current by the complex for glucose concentration has been displayed at pH=7.4. In another study, Supriyatno *et al.* [77] fabricated mono-substituted tetraphenylporphin–polymer composite films and presented the opto-chemical recognition of HCl gas. The authors attained higher sensitivity up to sub-ppm levels of HCl by applying a poly(hexylmethacrylate) matrix in the composite.

Persaud and Pelosi described the sensor arrays based on conducting polymer for odour and gas detection based on polymers of thiophene, pyrrole, indole and aniline [78,79]. The effect of gases on the conductivity could also be observed in two ways [80]. Nucleophilic gases like ammonia, methanol and ethanol vapors were observed to

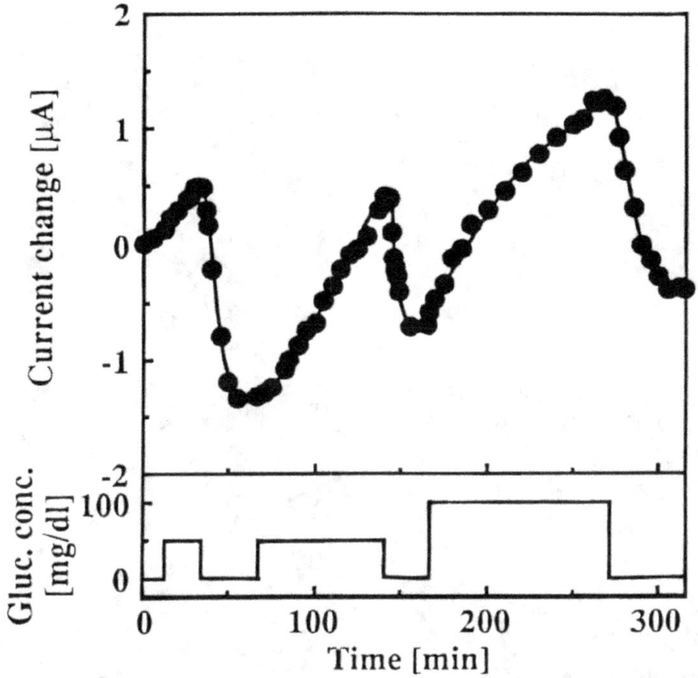

Figure 2.11 Current changes measured by complex membrane-coated platinum electrode in PBS at pH 7.4. Reproduced from Reference 76 with permission from American Chemical Society.

decrease the conductivity, while electrophilic gases like SO_2, PCl_3, NOx increased the conductivity.

Overall, polythiophene and derivatives [81,82], polyaniline and its composites [82,85-87] and polypyrrole [83,84] are the most studied conducting polymers for gas sensing applications. By electroploymerization of conducting polymers, functional combinations such as polypyrrole/polycarbonate, polythiophene(PTh)/polycarbonate (PC), polypyrrole (PPy)/polystyrene (PS), polythiophene/polystyrene, polyacrylonitrile (PAN)/polypyrrole (PPy), etc., can also be synthesized [88,89]. These polymers are characterized with properties like rapid adsorption/desorption kinetics, sensor response proportional to analyte concentration, low poisoning effects, optimum performance at low to ambient temperature, low power consumption, etc. Figure 2.12 also demonstrates the scanning electron images of polythiophene/polystyrene and polythiophene/polycarbonate films [89] from both electrode and solution sides.

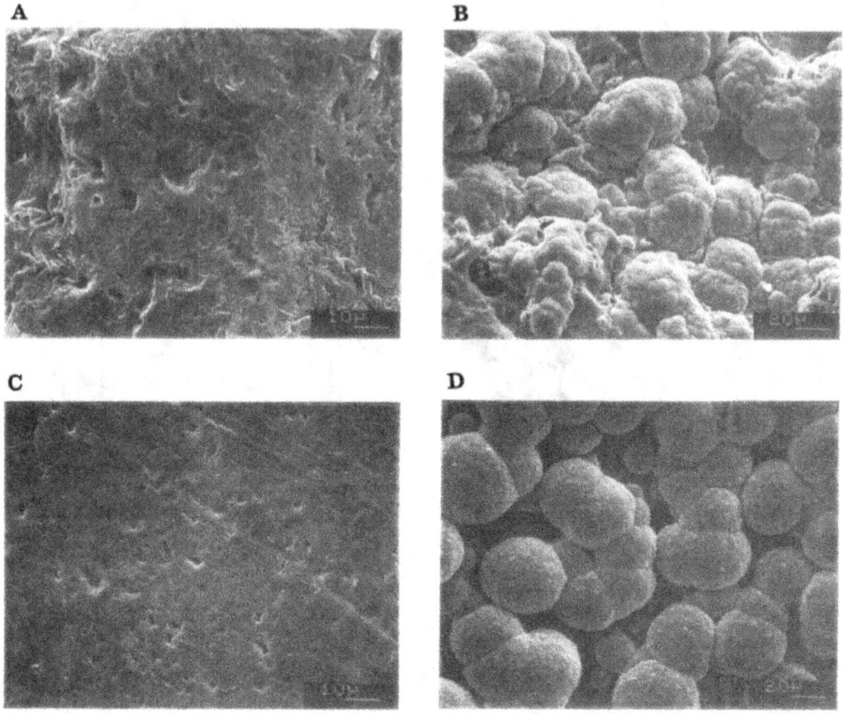

Figure 2.12 Scanning electron micrographs of (A) PTh/PS blend film (electrode side), (B) PTh/PS blend film (solution side), (C) PTh/PC blend film (electrode side), and (D) PTh/PC blend film (solution side). Reproduced from Reference 89 with permission from American Chemical Society.

Self-assembly techniques have been used for the detection of NO_2 and NH_3 gases using electroactive nanocomposite ultrathin films of polyaniline (PAN) and isopolymolybdic acid (PMA) [90]. The overall sensing process was based on doping-induced deposition of emeraldine base. Due to the high dependence of conductivity on the doping level, NH_3-sensing mechanism could be attributed to the de-doping of PAN by basic ammonia. By exposing the emeraldine base to NO_2, NO_2 acted as an oxidation doping agent which further resulted in the increase of conductivity. It was also observed that both electrical and optical properties of polyaniline with a p conjugation could be altered by the interaction of emeraldine salt (ES) and NH_3 gas [91]. The polaron density and band gap of the polymer decreased by the interaction of polymer and gas molecules. PANI composite was observed to be sensitive to even 10 ppm NH_3 concentration.

In another study, Chabukswar *et al.* [92] fabricated polyaniline with an acrylic acid doping which could be used as a NH_3 sensor from 1 to 600 ppm concentrations. Up to 58 ppm, a linear relation between the sensor response and concentration of the NH_3 gas was observed. The decrease in the electric resistance was explained by the removal of acrylic acid dopant from the site of proton removal using ammonia molecules. On the other hand, the composite with isopolymolybdic acid (PMA) exhibited a decreased relative response towards increase in ammonia concentration. Similarly, other studies have also reported diverse gas sensors [93-98].

Applying a novel interfacial polymerization technique, Virji *et al.* [99] reported the fabrication of nanofibers of polyaniline. The authors generated sensors based on nanofibers and compared the performance with the conventional polyaniline sensors. Figure 2.13 shows the scanning electron micrographs of polyaniline nanofibers cast from water and hexafluoro-2-propanol. Five different reaction mechanisms were analyzed: doping of base (NH_3), doping of acid (HCl), swelling (with $CHCl_3$), reduction (with N_2H_4) and conformational changes of polymer chain (due to CH_3OH). In each scenario, the polyaniline nanofibers accomplished improved behavior as compared to the traditional thin films. The small diameter, high surface area and porosity in the fibers increased the diffusion of dopants and molecules inside these nanofibers. In another study, Nylander *et al.* [15] investigated the performance of polypyrrole in gas sensing by generating polypyrrole permeated filter paper in presence of ammonia vapors. The authors observed that the sensor exhibited a linear behavior at room temperature with a response time of few minutes.

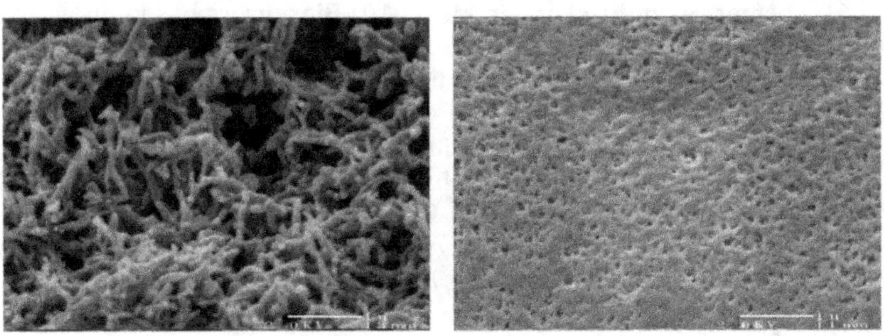

Figure 2.13 Scanning electron micrographs of films deposited on electrodes: (left) polyaniline nanofiber film cast from water and (right) polyaniline film cast from hexafluoro-2-propanol. Reproduced from Reference 99 with permission from American Chemical Society.

2.6 Conclusions

Polymers are immensely useful candidates for the development of advanced gas sensors. Beneficial features of polymers along with the ability to fine tune their performance profiles have led to significant research attention for fabricating sensors based on a large variety of polymers. With developments continuing at a fast pace in this direction, commercial applications of such sensors at a large scale can be envisaged in the coming years.

References

1. Lakard, B., Carquigny, S., Segut, O., Patois, T., and Lakard, S. (2015) Gas sensors based on electrodeposited polymers. *Metals*, **5**, 1371-1386.
2. Cosnier, S., and Holzinger, M. (2011) Electrosynthesized polymers for biosensing. *Chemical Society Reviews*, **40**, 2146-2156.
3. Albanese, D., Malvano, F., Sannini, A., Pilloton, R., and di Matteo, M. A. (2014) Doped polyaniline modified electrode amperometric biosensor for gluconic acid determination in grapes. *Sensors*, **14**, 11097-11109.
4. Nambiar, S., and Yeow J. T. W. (2011) Conductive polymer-based sensor for biomedical applications. *Biosensors and Bioelectronics*, **26**, 1825-1832.
5. Bobacka J., and Ivaska A. (2010) Chemical sensors based on conducting polymers. In: *Electropolymerization: Concepts, Materials and Applications*, Cosnier S., and Karyakin, A. (eds.), Wiley VCH, Germany, pp. 173-187.
6. Cosnier, S., and Holzinger, M. (2010) Biosensors based on electropolymerized films. In: *Electropolymerization: Concepts, Materials and Applications*, Cosnier, S., and Karyakin, A. (eds.), Wiley VCH, Germany, pp. 189-213.
7. Korostynska, O., Arshak, K., Gill, E., and Arshak, A. (2007) Review on state-of-the-art in polymer based pH sensors. *Sensors*, **7**, 3027-3042.
8. Sulka, G. D., Hnida, K., and Brzozka, A. (2012) pH sensors based on polypyrrole nanowire arrays. *Electrochimica Acta*, **104**, 536-541.
9. Braiek, M., Barhoumi, H., Maaref, A., and Renault, N. J. (2011) Elaboration and characterization of pH sensor based on polypyrrole nanowires. *Sensor Letters*, **9**, 2154-2157.
10. Lindfors, T., and Ivaska, A. (2002) pH sensitivity of polyaniline and its substituted derivatives. *Journal of Electroanalytical Chemistry*, **531**, 43-52.

11. Gu, F., Zhang, L., Yin, X., and Tong, L. (2008) Polymer single-nanowire optical sensors. *Nano Letters*, **8**, 2757-2761.
12. Emadi, T. A., Shafai, C., Freund, M. S., Thomson, D. J., Jayasz, D. S. and Whitex, N. D. G. (2009) Development of a Polymer-based Gas Sensor -humidity and CO_2 Sensitivity. In: *Proceedings of the 2nd Microsystems and Nanoelectronics Research Conference (MNRC)*, Canada, pp. 112-115.
13. Hagleitner, C., Lange, D., Hierlemann, A., Brand, O., and Baltes, H. (2002) CMOS single-chip gas detection system comprising capacitive, calorimetric and mass-sensitive microsensors. *IEEE Xplore: IEEE Journal of Solid-State Circuits*, **37**, 1867-1878.
14. Liu, X., Cheng, S., Liu, H., Hu, S., Zhang, D., and Ning, H. (2012) A survey on gas sensing technology. *Sensors*, **12**, 9635-9665.
15. Nylander, C., Armgarth, M., and Lundstrom, I. (1983) An ammonia detector based on a conducting polymer. *Analytical Chemistry Symposium Series*, **17**, 203-207.
16. Marini, M., Pilati, F., and Pourabbas, B. (2008) Smooth surface polypyrrole-silica core-shell nanoparticles: Preparation, characterization and properties. *Macromolecular Chemistry and Physics*, **209**, 1374-1380.
17. Hu, J., Wu, X., and Zeng, W. (2011) Formaldehyde sensor based on polypyrrole/β-cyclodextrin. *Journal of Controlled Release*, **152**, 211-213.
18. Oueiny, C., Berlioz, S., and Perrin, F. X. (2014) Carbon nanotube-polyaniline composites. *Progress in Polymer Science*, **39**, 707-748.
19. Gutierrez Pineda, E., Alcaide, F., Presa, M. J. R., Bolzan, A. E., Gervasi, C. A. (2015) Electrochemical preparation and characterization of polypyrrole/stainless steel electrodes decorated with gold nanoparticles. *ACS Applied Materials and Interfaces*, **7**, 2677-2687.
20. Shi, G. Q., Jin, S., Xue, G., and Li, C. (1995) A conducting polymer film stronger than aluminum. *Science*, **267**, 994-996.
21. Gilch, H. G., and Wheelwright, W. L. (1966) Polymerization of a-halogenated p-xylenes with base. *Journal of Polymer Science A*, **4**, 1337-1349.
22. Yang, Z., Hu, B., and Karasz, F. E. (1995) Polymer electroluminescence using ac or reverse dc biasing. *Macromolecules*, **28**, 6151-6154.
23. Jung, S. H., Kim, H. K., Kim, S. H., Kim, Y. H., Jeoung, S. C., and Kim, D. (2000) Palladium-catalyzed direct synthesis, photophysical properties, and tunable electroluminescence of novel silicon-based alternating copolymers. *Macromolecules*, **33**, 9277-9288.
24. McCullough, R. D. (1998) The chemistry of conducting polythiophenes. *Advanced Materials*, **10**, 93-116.
25. Nicolas-Debarnot, D., and Poncin-Epaillard, F. (2003) Polyaniline as a new sensitive layer for gas sensors. *Analytica Chimica Acta*, **475**,

1-15.

26. Lu, G. W., Qu, L. T., and Shi, G. Q. (2005) Electro chemical fabrication of neuron-type networks based on crystalline oligopyrene nanosheets. *Electrochimica Acta*, **51**, 340-346.
27. Reemts, J., Parisi, J., and Schlettwein, D. (2004) Electro chemical growth of gas-sensitive polyaniline thin films across an insulating gap. *Thin Solid Films*, **466**, 320-325.
28. Goto, H., Yoneyama, H., Togashi, F., Ohta, R., Tsujimoto, A., Kita, E., and Ohshima, K. (2008) Preparation of conducting polymers by electrochemical methods and demonstration of a polymer battery. *Journal of Chemical Education*, **85**, 1067-1070.
29. Cho, J. H., Yu, J. B., Kim, J. S., Sohn, S. O., Lee, D. D., and Huh, J. S. (2005) Sensing behaviors of polypyrrole sensor under humidity condition. *Sensors and Actuators B*, **108**, 389-392.
30. Brady, S., Lau, K. T., Megill, W., Wallace, G. G., and Diamond, D. (2005) The development and characterisation of conducting polymeric-based sensing devices. *Synthetic Metals*, **154**, 25-28.
31. Silverstein, M. S., Tai, H. W., Sergienko, A., Lumelsky, Y. L., and Pavlovsky, S. (2005) Poly HIPE: IPNs, hybrids, nanoscale porosity, silica monoliths and ICP-based sensors. *Polymer*, **46**, 6682-6694.
32. Prasad, G. K., Radhakrishnan, T. P., Kumar, D. S., and Krishna, M. G. (2005) Ammonia sensing characteristics of thin film based on polyelectrolyte templated polyaniline. *Sensors and Actuators B*, **106**, 626-631.
33. Tongpool, R., and Yoriya, S. (2005) Kinetics of nitrogen dioxide exposure in lead phthalocyanine sensors. *Thin Solid Films*, **477**, 148-152.
34. Ram, M. K., Yavuz, O., Lahsangah, V., and Aldissi, M. (2005) CO gas sensing from ultrathin nano-composite conducting polymer film. *Sensors and Actuators B*, **106**, 750-757.
35. Nohria, R., Khillan, R. K., Su, Y., Dikshit, R., Lvov, Y., and Varahramyan, K. (2006) Humidity sensor based on ultrathin polyaniline film deposited using layer-by-layer nano-assembly. *Sensors and Actuators B*, **114**, 218-222.
36. Agbor, N. E., Petty, M. C., and Monkman, A. P. (1995) Polyaniline thin-films for gas-sensing. *Sensors and Actuators B*, **28**, 173-179.
37. Stussi, E., Cella, S., Serra, G., and Venier, G. S. (1996) Fabrication of conducting polymer patterns for gas sensing by a dry technique. *Materials and Science Engineering C - Biomimetic Materials Sensors and Systems*, **4**, 27-33.
38. Ruangchuay, L., Sirivat, A., and Schwank, J. (2004) Selective conductivity response of polypyrrole-based sensor on flammable chemicals. *Reactive and Functional Polymers*, **61**, 11-22.
39. Sandberg, H. G. O., Backlund, T. G., Osterbacka, R., Jussila, S., Makela, T., and Stubb, H. (2005) Applications of an all-polymer solution-pro-

cessed high-performance, transistor. *Synthetic Metals*, **155**, 662-665.

40. Su, M., Fu, L., Wu, N. Q., Aslam, M., and Dravid, V. P. (2004) Individually addressed large-scale patterning of conducting polymers by localized electric fields. *Applied Physics Letters*, **84**, 828-830.
41. Li, G. F., Martinez, C., and Semancik, S. (2005) Controlled electrophoretic patterning of polyaniline from a colloidal suspension. *Journal of American Chemical Society*, **127**, 4903-4909.
42. Mabrook, M. F., Pearson, C., and Petty, M. C. (2006) Inkjet-printed polypyrrole thin films for vapour sensing. *Sensors and Actuators B*, **115**, 547-551.
43. Athawale, A. A., and Kulkarnil M. V. (2000) Polyaniline and its substituted derivatives as sensor for aliphatic alcohols. *Sensors and Actuators B*, **67**, 173-177.
44. Athawale, A. A., Bhagwat, S. V., and Katre, P. P. (2006) Nanocomposite of Pd-polyaniline as a selective methanol sensor. *Sensors and Actuators B*, **114**, 263-267.
45. Skotheim, T. A., Elsenbaumer, R. L., and Reynolds, J. R. (1998) *Handbook of Conducting Polymers*, 1st volume, 2nd edition, Marcel Dekker, USA.
46. Potje-Kamloth, K. (2010) Gas Sensing with conducting polymers. In: *Electropolymerization: Concepts, Materials and Applications*, Cosnier, S., Karyakin, A. (eds.), Wiley VCH, Germany, Chapter 8.
47. Hagleitner, C., Lange, D., Hierlemann, A., Brand, O., and Baltes, H. (2002) CMOS single-chip gas detection system comprising capacitive, calorimetric and mass-sensitive microsensors. *IEEE Journal of Solid-State Circuits*, **37**, 1867-1878.
48. Avramov, I. D., Rapp, M., Voigt, A., Stahl, U., and Dirschka, M. (2000) Comparative Studies on Polymer Coated SAW and STW Resonators for Chemical Gas Sensor Applications. In: *Proceedings of IEE/EIA International Frequency Control Symposium and Exhibition*, USA, pp. 58-65.
49. Wong, K. K. L., Tang, Z., Sin, J. K. O., Chan, P. C. H., Cheung, P. W., and Hiraoka, H. (1995) Study on Selectivity Enhancement of Tin Dioxide Gas Sensor using Non-conducting Polymer Membrane. In: *Proceedings of Hong Kong Electron Devices Meeting*, China, pp. 42-45.
50. Pelosi, P., and Persaud, K. C. (1988) *Nato ASI Series F: Computer and Systems Science*, Springer-Verlag, Germany.
51. Persaud, K. C., Jaggers, H. C., Shaw, J. S., Peacock, S. J., Nicholas, P., Bartlett, P. N., and Boden, G. F. (1991) Odour detection using sensor arrays. *Analytical Proceedings*, **28**, 339-344.
52. Persaud K. C. (1992) Electronic gas and odor detectors that mimic chemoreception in animals. *Trac-Trends in Analytical Chemistry*, **11**, 61.
53. Persaud K. C. (2005) Polymer for chemical sensing. *Materials Today*,

8, 38-54.

54. Korotcenkova, G., and Cho, B. K. (2011) Instability of metal oxide-based conductometric gas sensors and approaches to stability improvement. *Sensors and Actuators B*, **156**, 527-538.
55. Munoz, B. C., Steinthal, G., and Sunshine, S. (1999) Conductive polymer carbon black composites-based sensor arrays for use in an electronic nose. *Sensor Reviews*, **19**, 300-305.
56. Dan, Y., Cao, Y., Mallouk, T. E., Johnson, A. T., and Evoy, S. (2007) Dielectrophoretically assembled polymer nanowires for gas sensing. *Sensors and Actuators B*, **125**, 55-59.
57. Ji, X., Yao, Y., Li, J., Yan, X., and Huang, F. (2013) A supramolecular cross-linked conjugated polymer network for multiple fluorescent sensing. *Journal of American Chemical Society*, **135**, 74-77.
58. Ding, J., and Day, M. (2006) Novel highly fluorinated poly(arylene ether-1,3,4-oxadiazole)s, their preparation, and sensory properties to fluoride anion. *Macromolecules*, **39**, 6054-6062.
59. Xue, R., Ge, C., Richardson, K., Palmer, A., Viapiano, M., and Lannutti, J. J. (2015) Microscale sensing of oxygen via encapsulated porphyrin nanofibers: Effect of indicator and polymer "core" permeability. *ACS Applied Materials and Interfaces*, **7**, 8606-8614.
60. Dunst, K., Karczewski, J., and Jasinski P. (2017) Nitrogen dioxide sensing properties of PEDOT polymer films. *Sensors and Actuators B*, **247**, 108-113.
61. Sekizkardes, A. K., Altarawneh, S., Kahveci, Z., Islamoglu, T., and El-Kaderi, H. M. (2014) Highly selective CO_2 capture by triazine-based benzimidazole-linked polymers. *Macromolecules*, **47**, 8328-8334.
62. Percival, C. J., Stanley, S., Galle, M., Braithwaite, A., Newton, M. I., McHale, G., and Hayes, W. (2001) Molecular-imprinted, polymer-coated quartz crystal microbalances for the detection of terpenes. *Analytical Chemistry*, **73**, 4225-4228.
63. Polikar, R., Shinar, R., Udpa, L., and Portar, M. D. (2001) Artificial intelligence methods for selection of an optimized sensor array for identification of volatile organic compounds. *Sensors and Actuators B*, **80**, 243-254.
64. Kim, K. C., Cho, S. M., and Choi, H. G. (2000) Detection of ethanol gas concentration by fuel cell sensors fabricated using a solid polymer electrolyte. *Sensors and Actuators B*, **67**, 194-198.
65. Hrncirova, P., Opekar, F., and Stulik K. (2000) An amperometric solidstate NO_2 sensor with a solid polymer electrolyte and a reticulated vitreous carbon indicator electrode. *Sensors and Actuators B*, **69**, 199-204.
66. Somani, P. R., Viswanath, A. K., Aiyer, R. C., and Radhakrishnan, S. (2001) Charge-transfer complex forming dyes incorporated in solid polymer electrolyte for optical humidity sensing. *Sensors and Actuators B*, **80**, 141-148.

67. Nanto, H., Dougami, N., Mukai, T., Habara, M., Kusano, E., Kinbara, A., Ogawa, T., and Oyabu, T. (2000) A smart gas sensor using polymer-film-coated quartz resonator microbalance. *Sensors and Actuators B*, **66**, 16-18.
68. Rella, R., Siciliano, P., Quaranta, F., Primo, T., Valli, L., Schenetti, L., Mucci, A., and Iarossi. D. (2000) Gas sensing measurements and analysis of the optical properties of poly[3-(butylthio)thiophene] Langmuir–Blodgett films. *Sensors and Actuators B*, **68**, 203-209.
69. Xie, D., Jiang, Y., Pan, W., Li, D., Wu, Z., and Li, Y. (2002) Fabrication and characterization of polyaniline-based gas sensor by ultra-thin film technology. *Sensors and Actuators B*, **81**, 158-164.
70. Doleman, B. J., and Lewis, N. S. (2001) Comparison of odor detection thresholds and odor discriminabilities of a conducting polymer composite electronic nose vs mamallian olfaction. *Sensors and Actuators B*, **72**, 41-50.
71. Guadarrama, A., Rodrıguez-Mendez, M. L., de Saja, J. A., Rios, J. L., Olias, J. M. (2000) Array of sensors based on conducting polymers for the quality control of the aroma of the virgin olive oil. *Sensors and Actuators B*, **69**, 276-282.
72. Stetter, J. R., Jurs, P. C., and Rose, S. L. (1986) Detection of hazardous gases and vapors: pattern recognition analysis of data from an electrochemical sensor array. *Analytical Chemistry*, **58**, 860-866.
73. Ben Rashed, A. B., Bull, D. R., and Harris, G. J. (1995) A method for correcting short-term fatigue in electrochemical gas sensors. *Sensors and Actuators B*, **24**, 248-251.
74. Stetter, J. R., Findlay, M. W., Maclay, G. J., Zhang, J., Vaihinger, S., and Gopel, W. (1990) Sensor array and catalytic filament for chemical analysis of vapors and mixtures. *Sensors and Actuators B*, **1**, 43-47.
75. Nakagawa, K., Sadaoka, Y., Supriyatno, H., Kubo, A., Tsutsumi, C., and Tabuchi, K. (2001) Optochemical HCl gas detection using alkoxy substituted tetraphenyl porphyrin–polymer composite films. Effects of alkoxy chain length on sensing characteristics. *Sensors and Actuators B*, **76**, 42-46.
76. Kikuchi, A., Suzuki, K., Okabayashi, O., Hoshino, H., Kataoka, K., Sakurai, Y., and Okano, T. (1996) Glucose-sensing electrode coated with polymer complex gel containing phenylboronic acid. *Analytical Chemistry*, **68**, 823-828.
77. Supriyatno, H., Nakagawa, K., and Sadaoka, Y. (2001) Optochemical HCl gas detection using mono-substituted tetraphenyl porphin–polymer composite films. *Sensors and Actuators B*, **76**, 36-41.
78. Persaud, K. C., and Pelosi, P. (1985) An approach to an artificial nose. *Transactions - American Society for Artificial Internal Organs*, **31**, 297-300.
79. Amrani, E. H., Ibrahim, S., Persaud, K. C. (1993) Synthesis, chemical characterisation and multifrequency measurements of poly N-(2-

pyridyl) pyrrole for sensing volatile chemicals. *Material Science Engineering*, **C1**, 17-22.

80. Slater, J. M., and Watt, E. J. (1991) Examination of ammonia-poly(pyrrole) interactions by piezoelectric and conductivity measurements. *Analyst*, **116**, 1125-1130.
81. Bartlett, P. N., and Ling-Chung, S. K. (1989) Conducting polymer gas sensors Part III: Results for four different polymers and five different vapours. *Sensors and Actuators*, **20**, 287-292.
82. Marsella, M. J., Carroll, P. J., and Swager, T. M. (1995) Design of chemoresistive sensory materials: polythiophene-based pseudo polyrotaxanes. *Journal of American Chemical Society*, **117**, 9832-9841.
83. Bruschi, P., Cacialli, F., Nannini, A., and Neri, B. (1994) Gas and vapour effects on the resistance fluctuation spectra of conducting polymer thin-film resistors. *Sensors and Actuators B*, **18-19**, 421-425.
84. Torsi, L., Pezzuto, M., Siciliano, P., Rella, R., Sabbatini, L., Valli, L., Zambonin, P. G. (1998) Conducting polymers doped with metallic inclusions: new materials for gas sensors. *Sensors and Actuators B*, **48**, 362-367.
85. Hirata, M., and Sun, L. (1994) Characteristics of an organic semiconductor polyaniline film as a sensor for NH_3 gas. *Sensors and Actuators A*, **40**, 159-163.
86. Unde, S., Ganu, J., and Radhakrishnan, S. (1996) Conducting polymer based chemical sensor: characteristics and evaluation of polyaniline composite films. *Advanced Materials for Optics and Electronics*, **6**, 151-157.
87. Ogura, K., Saino, T., Nakayama, M., and Shiigi, H. (1997) The humidity dependence of the electrical conductivity of a soluble polyaniline-poly(vinyl alcohol) composite film. *Journal of Material Chemistry*, **7**, 2363-2366.
88. Park, Y. H., and Han, M. H. (1992) Preparation of conducting polyacrylonitrile/ polypyrrole composite films by electrochemical synthesis and their electrical properties. *Journal of Applied Polymer Science*, **45**, 1973-1982.
89. Wang, H. L., Toppare, L., and Fernandez, J. E. (1990) Conducting polymer blends: polythiophene and polypyrrole blends with polystyrene and poly(bisphenol A carbonate). *Macromolecules*, **23**, 1053-1059.
90. Li, D., Jiang, Y., Wu, Z., Chen, X., and Li, Y. (2000) Self-assembly of polyaniline ultrathin films based on doping-induced deposition effect and applications for chemical sensors. *Sensors and Actuators B*, **66**, 125-127.
91. Nicho, M. E., Trejo, M., Garcıa-Valenzuela, A., Saniger, J. M., Palacios, J., and Hu, H. (2001) Polyaniline composite coatings interrogated by nulling optical-transmittance bridge for sensing low concentrations

of ammonia gas. *Sensors and Actuators B*, **76**, 18-24.

92. Chabukswar, V. V., Pethkar, S., and Athawale, A. A. (2001) Acrylic acid doped polyaniline as an ammonia sensor. *Sensors and Actuators B*, **77**, 657-663.
93. Ho, K. C., Hung, W. T. (2001) An amperometric NO_2 gas sensor based on Pt/Nafionw electrode. *Sensors and Actuators B*, **79**, 11-16.
94. Patois, T., Sanchez, J. B., Berger, F., Rauch, J. Y., Fievet, P., and Lakard, B. (2012) Ammonia gassensors based on polypyrrole films: influence of electrodeposition parameters, *Sensors and Actuators B*, **171-172**, 431-439.
95. Zampetti, E., Pantalei, S., Muzyczuk, A., Bearzotti, A., De Cesare, F., Spinella, C., and Macagnano, A. (2013) A high sensitive NO_2 gas sensor based on PEDOT-PSS/TiO_2 nanofibres. *Sensors and Actuators B*, **176**, 390-398.
96. Chen, X. P., Liang, Q. H., Jiang, J. K., Wong, C. K. Y., Leung, S. Y. Y., Ye, H. Y., Yang, D. G., and Ren, T. L. (2016) Functionalization-induced changes in the structural and physical properties of amorphous polyaniline: a first-principles and molecular dynamics study. *Scientific Reports*, **6**, 20621.
97. Ando, B., Baglio, S., Pasquale, G. D., Pollicino, A., Agata, S. D., Gugliuzzo, C., Lombardo, C., and Re, G. (2015) An inkjet printed CO_2 gas sensor. *Proceedia Engineering*, **120**, 628-631.
98. Adhikari, B., and Majumdar, S. (2004) Polymers in sensor applications. *Progress in Polymer Science*, **29**, 699-766.
99. Virji, S., Huang, J., Kaner, R. B., and Weiller, B. H. (2004) Polyaniline nanofiber gas sensors: Examination of response mechanisms. *Nano Letters*, **4**(3), 491-496.

3

Thermally Responsive Polymers and Their Nanocomposites: Properties and Applications

3.1 Introduction

Thermally responsive (or thermoresponsive) polymers are a class of advanced materials where the polymers' characteristics change according to external thermal stimuli. These changes are usually significant and discontinuous and often describe the changes in solubility. In other words, the thermoresponsive polymers exhibit a sudden volume phase transition at a particular temperature causing a change in the solvation state. Thus, these polymers usually exhibit a miscibility gap at their lower critical solution temperature (LCST) and upper critical solution temperature (UCST). The polymers which become insoluble upon heating have a LCST, while the polymers which become soluble upon heating have an UCST. The characteristic of thermal responsiveness results in a wide range of applications of thermally responsive polymers in biomedical field [1]. Even if these materials are not restricted to the solvation state changes in aqueous environments, the use of these materials in the biomedical field concentrates on the aqueous environment solvation. There are materials which show LCST or UCST behavior in organic solvents other than water [2]. The use of thermally responsive polymers in the biomedical applications relies on their ability to change according to localized temperature changes, without causing any tissue damage. These dramatic changes can be utilized for various applications such as drug delivery, tissue engineering, cancer therapy, etc. [3-5]. The change from room temperature is used as the stimuli for changing the physical properties of the polymer in various biomedical applications. Thermoresponsive polymers are occasionally coupled with other monomers like pH responsive monomers, thus, resulting in double responsive copolymers [6] (Figure 3.1). Ability of nanoparticles to induce localized temperature changes in the body also makes them useful

Liyamol Jacob and Vikas Mittal, The Petroleum Institute (part of Khalifa University of Science and Technology), Abu Dhabi, UAE*
**Current address: Bletchington, Wellington County, Australia*

materials for generating polymer nanocomposites with thermoresponsive polymers for biological applications [7]. Such materials with tunable properties exhibit further applications in areas such as modulated drug delivery, immunoassays, photo-thermal cancer therapy, etc. [8].

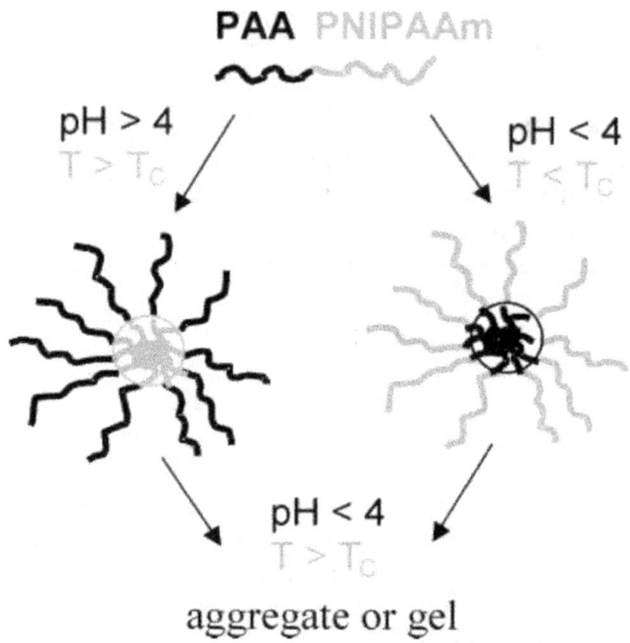

Figure 3.1 Aggregate formation for poly(N-isopropylacrylamide)-block-poly(acrylic acid) (PNIPAAm-*b*-PAA) in aqueous solution in dependence of pH and temperature. Reproduced from Reference 6 with permission from American Chemical Society.

3.2 Classification of Thermoresponsive Polymers based on Phase Change

As mentioned earlier, the thermoresponsive polymers can be characterized into LCST and UCST type polymers. Both categories exhibit different phase transition under temperature. The phase behavior of a particular polymer in a particular solvent depends upon various factors like the concentration of the polymer, presence of additives, molecular weight, temperature, etc. The phase diagram of a thermally responsive polymer shows one phase region and two phase region.

The position of one phase region and two phase region can be used to identify between LCST and UCST type thermoresponsive polymers [9].

3.3 Different Thermoresponsive Polymers and Nanocomposites

Broadly thermoresponsive polymers can be classified as natural and synthetic polymers. There are a large number of thermoresponsive polymers like poly(N-isopropylacrylamide), poly(N-vinylcaprolactam), polymethacrylamide, phosphazene derivatives [10-13], polypetides [14-21], polyamidomine [22,23], xycloglucan [24-26], poly(ethylene glycol), poly(propylene glycol) [27-35], chitosan and derivatives, cellulose and derivatives, gelatin and derivatives, etc. The thermoresponsive behavior of these polymers can be further enhanced/tuned by the formation of various nanocomposites and copolymers.

3.3.1 Poly(N-isopropylacrylamide) Copolymers and Nanocomposites

Poly(N-isopropylacrylamide) is the one of the most extensively studied thermoresponsive polymers with a LCST temperature around 32 °C [36]. This synthetic polymer exhibits sol-gel transition in aqueous environment at its LCST temperature. Its LCST can also be altered through changes in molecular weight, by the introduction of foreign molecules, etc. [37-51].

Mittal *et al.* [37] used surfactant free emulsion copolymerization to directly synthesize PS-PNIPAAm copolymer particles (Figure 3.2). In the second set of experiments, polystyrene particles with an atom transfer radical polymerization (ATRP) initiator shell were first synthesized. Monoliths with two different porosities were subsequently formed by grafting and crosslinking of PNIPAAm chains on these particles using the ATRP initiator (Figure 3.3). The particles were aggregated before grafted and the method was named as 'aggregate grafting' process. Swelling/deswelling behavior of both sets of materials was studied with respect to temperature and time. The same behavior of thermo-reversible hydrophobicity and hydrophilicity of the PNIPAAm chains was observed in the monoliths as the individual particles, though the extent and rate of swelling was hindered owing to the large degree of crosslinking among PNIPAAm brushes and space constraints to the chains. However, swelling/deswelling kinetics was still sufficiently fast in the monolith materials to use these supports for biological separations (e.g., proteins, viruses, etc.) driven solely by temperature changes.

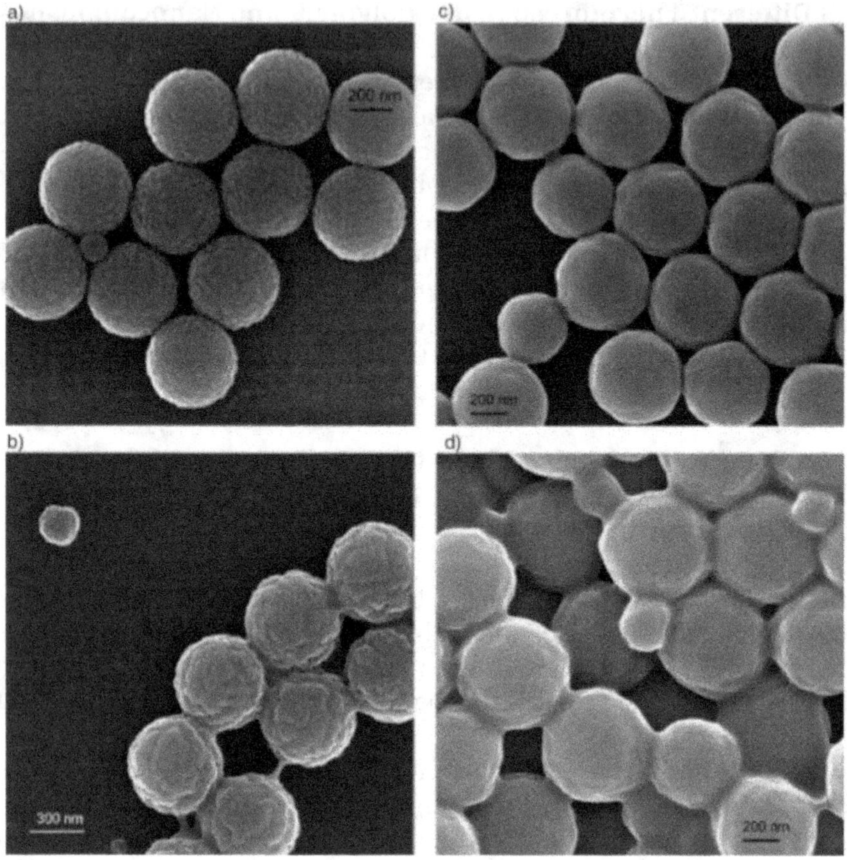

Figure 3.2 SEM images of PS-PNIPAAm copolymer particles synthesized using emulsion copolymerization. Reproduced from Reference 37 with permission from Wiley.

In another study, Chen *et al.* [47] generated nanoreactor with temperature-responsive poly(N-isopropylacrylamide) coated on external surface of mesoporous silica hollow spheres and Au nanoparticles at the internal surface. Figure 3.4 demonstrates the schematic of composite synthesis. The generated composite structure effectively separated Au and poly(N-isopropylacrylamide) from each other. As a result, the catalyst exhibited sharp on and off behavior during catalytic reduction reaction at temperatures below or above the LCST of the polymer.

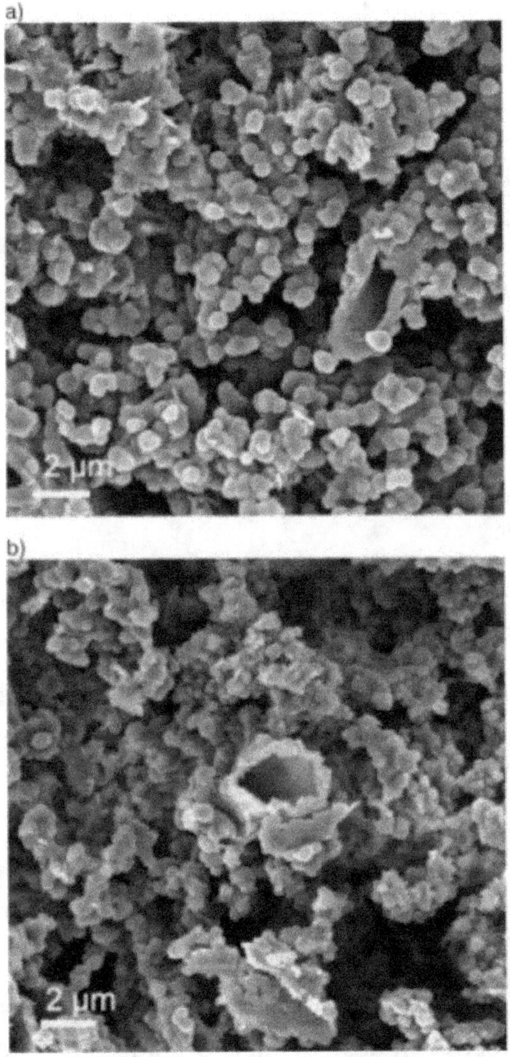

Figure 3.3 Scanning electron (SEM) micrographs of monoliths generated from crosslinked PNIPAAm brushes grafted from the surface of polystyrene particles. Reproduced from Reference 37 with permission from Wiley.

Shi *et al.* [50] synthesized thermo-, pH-, and light-responsive microgels based on poly(N-isopropylacrylamide-co- methacrylic acid)-

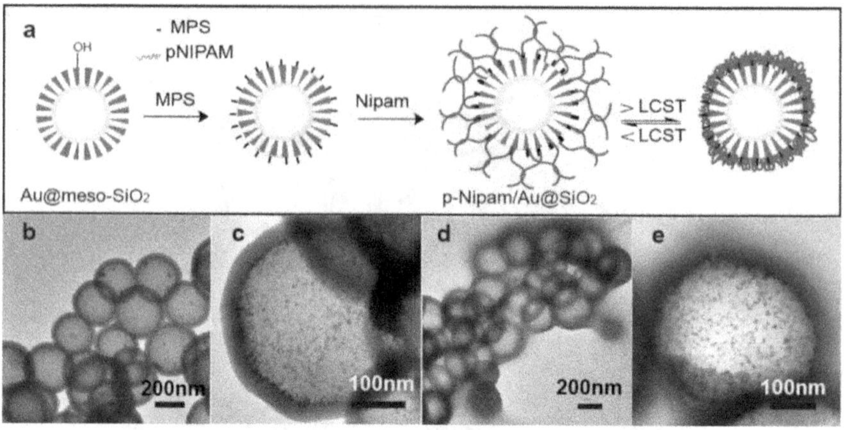

Figure 3.4 Schematic illustration for the synthesis process of poly(N-isopropylacrylamide) (PNIPAM)/Au@meso-SiO_2; (b,c) images of Au@meso-SiO_2 composite; (d,e) images of PNIPAM/Au@meso-SiO_2 composite. Reproduced from Reference 47 with permission from American Chemical Society.

Au (PNM-Au) hybrid using in-situ reduction method. Figure 3.5 shows the transmission electron images of these PNM-Au microgels. The microgels were observed to exhibit swelling/deswelling behavior in response to the stimuli such as surrounding temperature, pH, and light. In another study, Prabhakar *et al.* [51] utilized free radical polymerization for the preparation of poly(lactic acid) and poly(N-isopropylacrylamide)-co-acrylamide hydrogels from a blend solution. As synthesized hydrogel could be utilized both as a drug delivery agent and anti-bacterial material. Ravindra *et al.* [52] also utilized free radical polymerization for the synthesis of silver nanoparticle embedded poly(N-isopropylarclamide) hydrogels in aqueous medium with the crosslinking agent N,N'-methylenebis(acrylamide) (MBA). The hydrogels with additional curcumin loading were observed to be effective in destroying E-coli bacteria. The antibacterial properties of the silver nanoparticle embedded poly(N-isopropylarclamide) were also investigated by Bajpai *et al.* [53]. Cong *et al.* [54] synthesized hydrogel composite structure with thermoresponsive poly(N-isopropylacrylamide)/graphene/Au with excellent photothermal effect for water treatment. N-isopropylacrylamide and silylanized Mg/Al layered double hydroxides (SiLDHs) based thermoresponsive sorbents were prepared by free radical copolymerization method [55]. The material exhibited phase transfer at 40 °C and

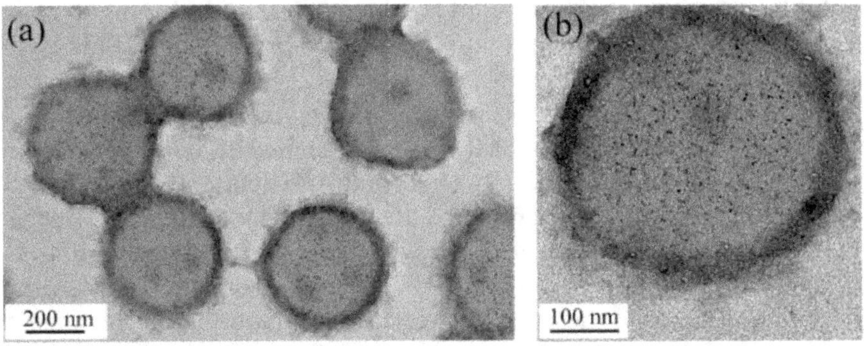

Figure 3.5 Transmission electron micrographs of PNM-Au microgels. Reproduced from Reference 50 with permission from American Chemical Society.

was recyclable and re-generable for the remediation of organic and inorganic pollutants from water.

In another study, temperature controlled manipulators based on the poly(N-isopropylacrylamide)-clay nanocomposite (NC) hydrogels were fabricated by the asymmetrical arrangement of clay over the polymer network by two step photo-polymerization [56]. The different hydrogel layers had different clay content which facilitated the thermoresponsive bending of the hydrogel composite and could be manipulated by altering the thickness of the polymer layers. The material was characterized by bending/unbending of the hydrogel structure as a response towards heating and cooling. Zhu *et al.* [57] fabricated poly(N-isopropylacrylamide)/graphene oxide (GO) hydrogel by in-situ γ-irradiation-assisted polymerization using water as a solvent. The high optical absorbance contributed to excellent photo-thermal properties for the designed composite. The material's behavior could be manipulated by near infrared (NIR) irradiation using laser. The NIR sensitivity of the composite increased with the amount of GO in the designed composite structure. Karimi *et al.* [58] synthesized thermoresponsive magnetic nanocomposites based on poly(N-isopropyl acrylamide-co-phthalocyanine) by in-situ dispersion polymerization. The percentage of Fe_3O_4 nanoparticles present in the nanocomposite significantly influenced the LCST of the system. Etika *et al.* [59] overcame the incompatibility of single walled carbon nanotubes and poly(N-isopropylacrylamide) in an aqueous medium using montmorillonite clay (MMT) as a compatibilizer. This method utilizing clay for generating affinity between nanotubes and water soluble polymers can be utilized for the fabrication of biomedical devices. In a study by Cavallrao *et al.* [60], curcumin delivery was facilitated by grafting poly(N-isopropylacrylamide) onto the external surface of halloysite. Temperature stimuli triggered the colloidal stability and wettability of the material. The presence of hollow cavity improved the biomaterial loading (Figure 3.6). The material was tested in gastro intestinal environment and exhibited excellent stability in acidic medium. Dong *et al.* [61] synthesized graphene-poly(N-isopropylacrylamide) hybrid materials by incorporating N-isopropylacrylamide by free radical polymerization technique on the surface of graphene sheets. 10-30% graphene loading helped in maintaining the LCST temperature of the polymer to 32 °C. Scherzinger *et al.* [62] also studied the effect of side chain substitution on the volume phase transition temperature of thermo-sensitive polymer microgels.

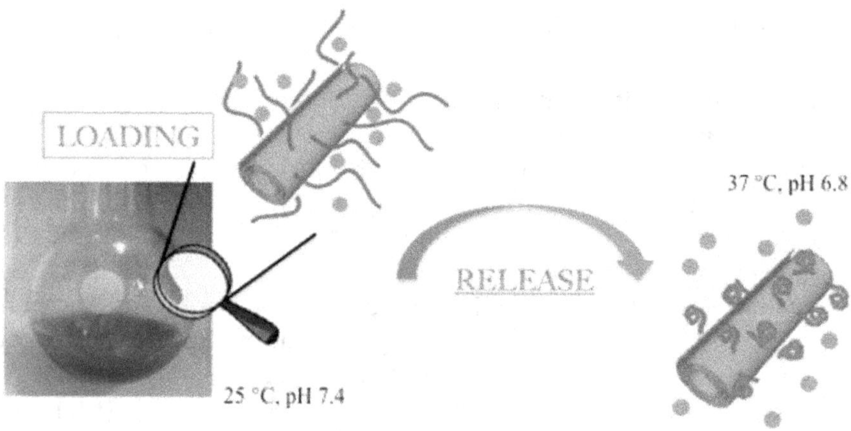

Figure 3.6 Schematic of loading and release of curcumin as a function of temperature. Reproduced from Reference 60 with permission from American Chemical Society.

3.3.2 Poly(N-vinylcaprolactam) Copolymers and Nanocomposites

Similar to poly(N-isopropylacrylamide), poly(N-vinylcaprolactam), a biocompatible thermoresponsive polymer, has been widely studied for use in a wide range of biomedical applications [63]. It can be fabricated with different types of architectures, viz. random, block, graft copolymers, hydrogels, etc. It has lower critical solution temperature in the range of 30-32 °C and has thermal properties comparable to PNIPAAm [64-68].

Pich *et al.* [69] reported the synthesis of temperature- and pH-responsive microgels based on poly(vinylcaprolactam-co-acetoacetoxy methacrylate) (VCL/AAEM) functionalized with vinylimidazole (VIm) (Figure 3.7). The sensitivity of microgels towards temperature and pH could be tailored by the variation in the VIm content. The authors observed that an increase in VIm content resulted in increased swelling of the microgels in acidic medium and enhancement of volume phase transition temperature. In another study, reversible addition fragmentation was also utilized to synthesize poly(ethylene glycol)-b-poly(N-vinylcaprolactam) (PEG-b-PNVCL) copolymers in DMF

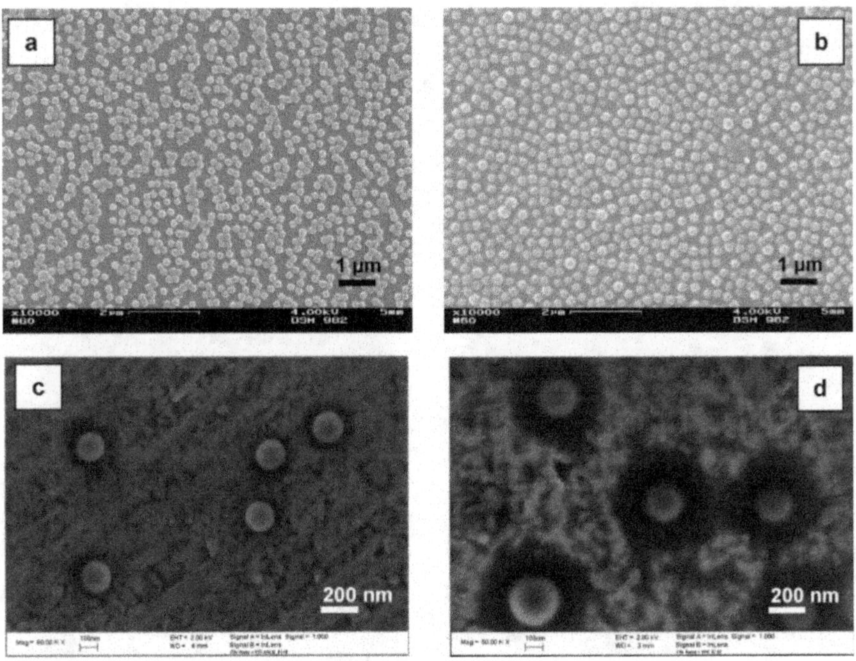

Figure 3.7 SEM (top) and FESEM (bottom) images of VCL/AAEM (a,c) and VCL/AAEM/Vim(2.9%) (b,d) microgels. Reproduced from Reference 69 with permission from American Chemical Society.

at 65 °C [70]. The polymer was grafted on the surface of gold nanoparticles forming a corona like architecture. The thermoresponsive material had excellent chemotherapeutic and phototherapeutic properties. Hurtgen *et al.* [71] synthesized symmetric triblock copolymers of PNVCL using a cobalt mediated radical polymerization process. It was the first study to use a cobalt mediated approach towards the synthesis of PNVCL thermoresponsive block copolymers. In another study, thermoresponsive hydrogel nanofibers based on PVCL were synthesized by copolymerizing it with hydroxymethyl acrylamide followed by electrospinning [72]. Owing to the quick and highly reversible temperature responses, these materials have the potential of use in microfluidic systems, biosensors and controlled drug delivery. Also, the cloud point temperature of PNVCL has been improvised by forming di-block and tri-block copolymers. The properties of such polymer largely depend on the percentage of constituent polymers and can be altered by the addition of salt additives [73]. Makhaeva *et al.* [74] also studied conformation changes of poly(vinylcaprolactam) and their complexes with ionic surfactants (cetylpyridinium chloride (CPC) and sodium dodecyl sulfate (SDS)). It was observed that the hydrodynamic size increased on adding ionic surfactant in case of higher polymer concentration, owing to intermolecular aggregation. In another study, precipitation polymerization with N,N'-bis(acryloyl) cystamine (BAC) as cross-linker, methacrylic acid (MAA) and polyethylene glycol (PEG) methyl ether methacrylate as co-monomers formed biodegradable microgels with PVCL [75] (Figure 3.8).

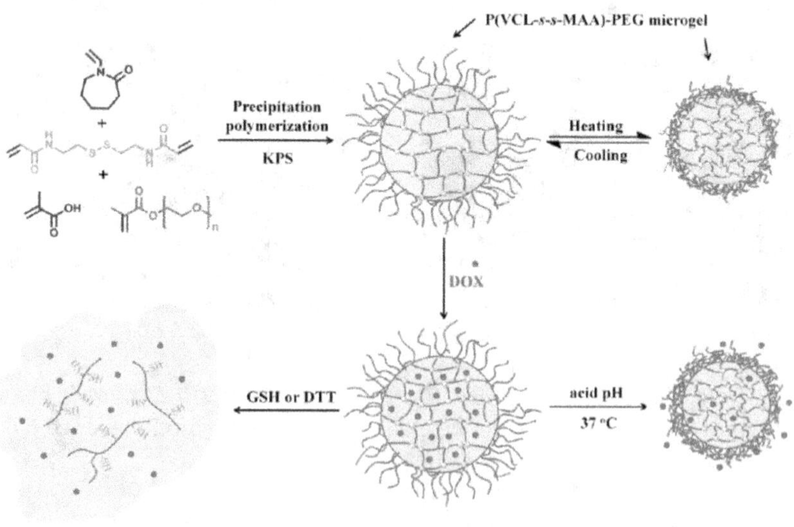

Figure 3.8 Schematic of the preparation, biodegradable behavior and stimuli-responsive drug release of PVCL-based microgels. Reproduced from Reference 75 with permission from American Chemical Society.

The material exhibited temperature sensitivity with varying MAA content and pH. Owing to the stable polymer chains which broke down under ambient conditions, the polymer was concluded to be useful as a non-toxic anti-tumor material.

3.3.3 Cellulose Derivatives and Nanocomposites

Cellulose is a water insoluble polysaccharide. The etherification of cellulose creates cellulose derivatives which are water soluble. Their chains exhibit thermally induced gelation under ambient conditions which vary greatly with the difference in the type of derivative and the mode of substitution [76]. Various composites have been generated by using different cellulose derivatives like methylcellulose and hydroxypropyl methylcellulose, which exhibit excellent thermoresponsive properties [77-82].

Dhar *et al.* [83] used microemulsion technique for the synthesis of cellulose derivatives based biodegradable and biocompatible polyampholyte microgels chitosan–carboxymethyl cellulose (CS-CMC) and chitosan–modified methyl cellulose (CS-ModMC). The materials could be used as an excellent drug carrier with good swelling properties based on the pH of the system. In another study, aqueous methylcellulose (MC) blended with phosphate-buffered saline hydrogels was used for cell transplantation by a direct intramuscular injection for ischemic diseases [84]. The material effectively overcame the issues regarding the retention of the transplanted cells in the targeted areas. Temperature changes above and below ~32 °C induced reversible sol-gel transformation in the hydrogel. Liu *et al.* [85] synthesized injectable hydrogel precursor solution by physical blend of xanthan gum (XG) and methylcellulose (MC) using water as a solvent. The material exhibited excellent shear thinning property at room temperature owing to the double helical architecture. The blend solution gelated with the increase of temperature above 37 °C forming methylcellulose networks. The storage modulus and the time taken for the material to gelate could be varied by varying the percentage of constituents in the composite. The material has the potential of use as an effective biodegradable and biocompatible drug delivery agent with long life time.

In another study, the addition of cellulose nanocrystals enhanced the properties of methylcellulose (MC) stabilized aqueous foams [86]. The foam was further modified using acrylamide (AM), 2-hydroxyethyl methacrylate (HEMA), or polyethylene glycol diacrylate

(PEGDA). The material could withstand high temperatures around 70 °C and exhibited fibrillary gel phase. Figure 3.9 also demonstrates the optical and scanning images of wet and dry foams. McKee *et al.* [87]

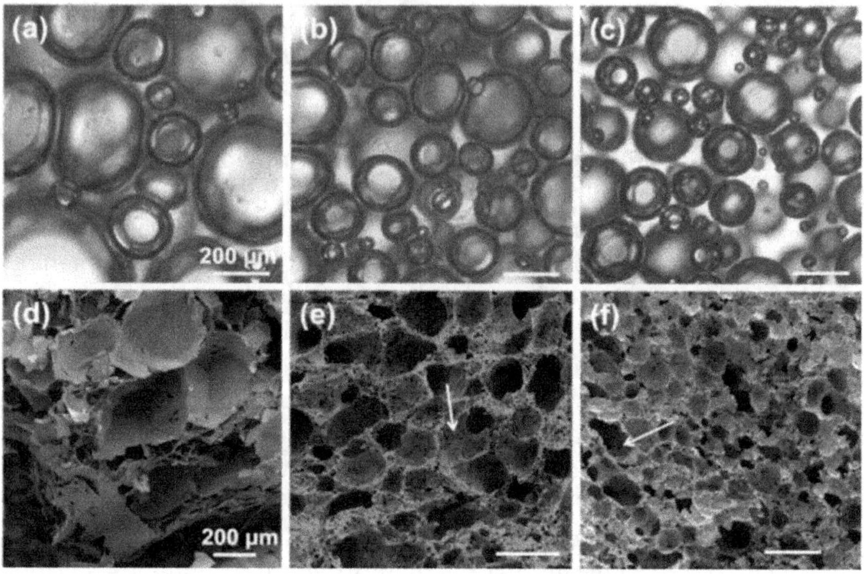

Figure 3.9 Optical micrographs of CNC-MC wet foams (a-c) and scanning electron images of the dry macroporous materials. Reproduced from Reference 86 with permission from American Chemical Society.

also synthesized thermoreversible hydrogels inspired by the plant cellulose binding in plant cell using CNC enforced MC. The materials exhibited different gelation (Figure 3.10) properties around 20 °C and 70 °C with a change in storage modulus.

3.3.4 Chitosan Derivatives and Nanocomposites

Similar to cellulose, chitosan is a natural linear polysaccharide. The material has been studied extensively for its drug delivery properties owing to its biocompatibility and biodegradability. It exhibits gelation abilities and is non-toxic. The gelation is observed at body temperature with increase in pH to 7.2 [88].

Chitosan-graft-PNIPAAm-graft-oligoproline (CSN-PRO) nanoporous foam with tunable porosity, pore structure and pore size was synthesized by Bernice *et al.* [89] using high internal phase emulsion templating method. The material exhibited dissolvability in aqueous

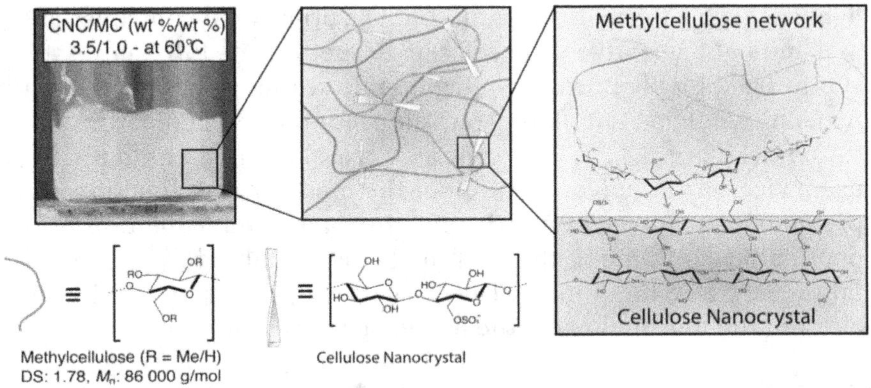

Figure 3.10 Representation of gelation for the sample CNC/MC 3.5 wt%/1.0 wt% at 60 °C. Reproduced from Reference 87 with permission from American Chemical Society.

environment above the LCST temperature of 34 °C which promises great applications in the field of tissue engineering and drug delivery. In another study, chitosan/β-glycerophosphate disodium salt (CS/GP) hydrogels were generated by introducing attapulgite (ATP) nanoparticles into the CS/GP hydrogels [90]. The material had enhanced mechanical properties compared to other composites and exhibited sol-gel transition at ambient conditions. Xu *et al.* [91] synthesized hybrid nanocomposites based on chitosan derivatives and layered double hydroxides. The material was studied for its drug delivering ability with the help of fluorescence and temperature sensitive environments. Yan *et al.* [92] synthesized glycerophosphate and hydroxyethyl cellulose modified chitosan based injectable hydrogels and studied their use for biologically active therapeutics.

3.3.5 Gelatin and Nanocomposites

Gelatin is a natural polymer which shows thermoresponsive behavior above 30 °C, where the chains assume a coil like structure. Gelatin is also functionalized with other polymers or used to generate nanocomposites for practical applications [93].

Daniel-da-Silva *et al.* [94] synthesized citrate encapsulated gold nanoparticles in gelatin-genipin matrix for the release of a model drug. Gold was effective in preventing the crosslinking between genipin and gelatin. The material was observed to exhibit thermoresponsive behavior in the temperature range feasible for living cells. The light triggering mechanism of Au could be utilized for controlled thermal and light sensitive drug release. Barbetta *et al.* [95] successfully synthesized highly porous biodegradable scaffolds, based on gelatin A3 using emulsion templating technique for tissue-engineering applications (Figure 3.11). Two different crosslinking techniques were used, viz. radical polymerization of the methacrylate functionalities previously introduced onto the gelatin chains and formation of isopeptide bridges among the gelatin chains promoted by the enzyme microbial transglutaminase. The use of these techniques significantly affected the morphology of the formed gelatin structure.

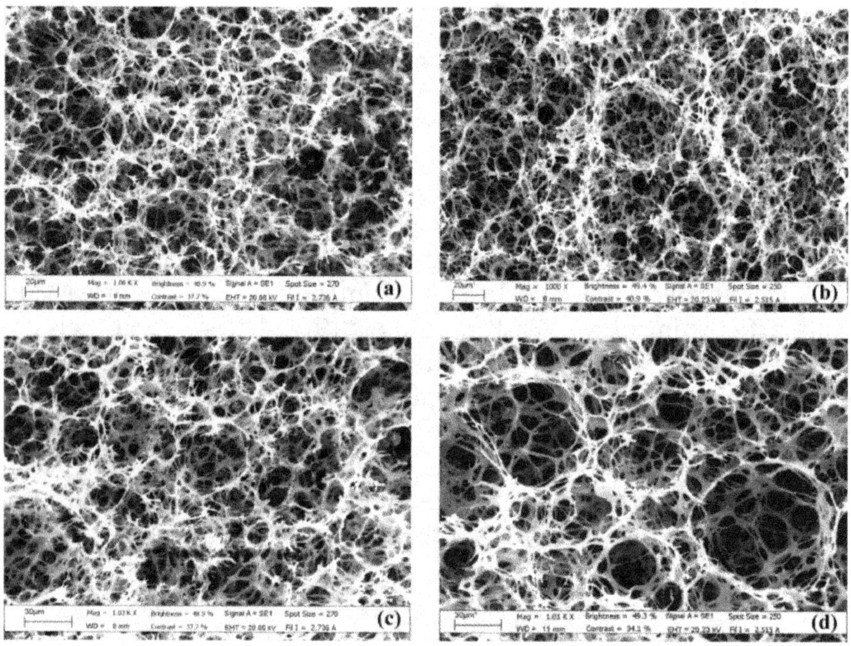

Figure 3.11 Scanning electron micrographs of gelatin A3 polyHIPE (high internal phase emulsion) solid foams. Reproduced from Reference 95 with permission from American Chemical Society.

In another study, Manjula *et al.* [96] utilized a green synthesis route using neem leaf extracts for the fabrication of gelatin-Ag nanoparticles- N-isopropylacrylamide hydrogels with the improved effect in the inactivation of bacteria with an additional thermosensitive property. Laponite-gelatin suspension was utilized by Li *et al.* [97] for the in-situ polymerization of acrylamide. Presence of gelatin in the hydrogel improved the pH sensitivity which resulted in the material with dual sensitivity. The material was highly thermally stable which has the potential to be utilized in various biological applications. Figure 3.12 demonstrates the micrographs of the hydrogels.

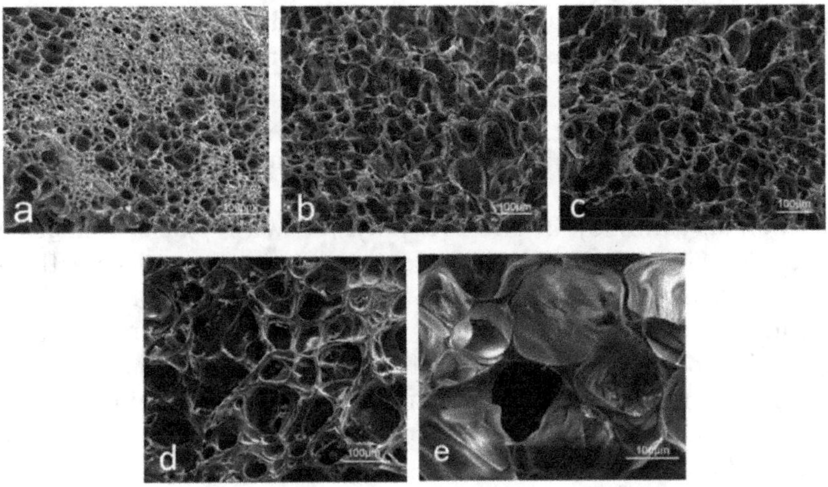

Figure 3.12 Scanning electron micrographs of gelatin/PAAm/Laponite hydrogels. Reproduced from Reference 97 with permission from American Chemical Society.

Gelatin factionalized thermoresponsive polycaprolactone (PCL) films were also used for cell recovery studies of gene-transfected cells [98]. Surface initiated ATRP of N-isopropylacrylamide (NIPAAm) and methacrylic acid sodium salt (MAAS) from the initiator-functionalized PCL surface was used for the primary fabrication of the nanocomposite brushes which was followed by incorporation of gelatin using carbodiimide chemistry. The incorporation of gelatin helped in cell adhesion above the LCST without affecting the cell detachment properties at lower temperatures. Lowering of temperature facilitated fast cell recovery. Thus, the generated materials can be utilized as intelligent adhesion modifiers for tissue engineering applications. Fatthollahipour *et al.* [99] utilized electrospinning method for the fabrication of poly(vinylalcohol) (PVA)/chitosan/lidocaine hydrochloride with dispersed gelatin nanoparticles. The system exhibited a nanofibrous core-cell architecture with dual drug delivery properties. Nanoprecipitation method was used for the preparation of gelatin nanoparticles which were then loaded with erythromycin antibiotic. Hyaluronic acid and gelatin functionalized laser-sintered porous polycaprolactone (PCL) was used to study soft/hard bi-phase scaffold for cartilage tissue engineering by Ming-Yih *et al.* [100]. In vitro chondrogenesis could be improved by using the generated material. Gandhi *et al.* [101] prepared tunable thermoresponsive gelatin nanogel with volume transition around 32 °C. Due to the random coil transition of the gelatin chains in the nanogel, thermally induced volume reduction was observed.

3.3.6 Miscellaneous Polymers and Nanocomposites

Poly(methacrylamide) is a UCST polymer with phase transition above 25 °C in aqueous environment. It has been used as a thermoresponsive polymer for various applications. Acid labile methacrylamide based copolymers were synthesized using free radical polymerization of N-(2-methoxy-1,3-dioxan-5-yl) methacrylamide (NMM) and N-(2-ethoxy-1,3-dioxan-5-yl)methacryl amide (NEM) [102]. The LCST of the copolymer could be shifted upwards by the addition of more hydrophilic NMM groups. The increase in pH had a positive effect on the hydrolysis of the polymer in acidic environments.

Huang *et al.* [103] synthesized enzyme immobilized thermoresponsive polyacrylamide polymers which were recyclable for enzymatic glycosylation. The product recovery from water could be achieved by thermal precipitation. Chu *et al.* [104] synthesized interpenetrating polymer networks comprising of polyacrylamide (PAAM) and poly(acrylic acid) (PAAC) which could be used as a thermoresponsive gating membrane with inverse gating characteristics. The opening of the membrane pores was driven by a decrease (rather than an increase) in the temperature. These membranes exhibited smart membrane actuators with a tunable thermoresponsive phase-transition behavior.

In another study, Li *et al.* [105] synthesized triblock copolymer gelators consisting of N-isopropylacrylamide and 2-(methacryloyloxy)ethyl phosphorylcholine by atom transfer radical polymerization with a bifunctional disulfide-based initiator. Under physiologically relevant environments, the material formed peptide degradable micellar gels. Kumashiro *et al.* [106] used spin coating to fabricate thread stripe-polyacrylamide on poly(N-isopropylacrylamide) surface for cell sheet recovery. The cells could be cultured on the patterned surface at 37 °C, and could be rapidly detached on lowering the temperature.

3.4 Applications of Thermoresponsive Polymers and Nanocomposites

The functional properties of thermoresponsive polymers pave the way for their advanced applications. The properties induced by forming hybrids/composite structures like dual responsivity (pH, light, fluorescence), biocompatibility, stability, reversibility, non-toxicity, etc., can be manipulated to enhance their spectrum of applications further. To date, the application of these polymers has been largely focused in the biomedical field. However, the formation of different structures like micelles, hydrogels, sheets, etc., can be used in other applications where particular structures are needed.

3.4.1 Biomedical Applications

Drug Delivery

The lack of physical and chemical instability as well as short life span limit the use of various materials in therapeutics. The disorder in the protein structure leads to unwanted agglomeration, precipitation, denaturation, etc., in the therapeutic carriers. Undesirable chemical changes including hydrolysis, deamination, oxidation and reduction are some of the limitations faced by therapeutic industry. The use of thermoresponsive polymers overcomes these issues by a great extent due to their properties such as stability, bio-activity and foldability [107]. The development of polymers which can respond to thermal stimuli which is already present in the biological systems is a great advancement in the field of pharmacology. This is made possible by a sharp nonlinear change in the properties of the carrier material by altering temperature. This response is characterized by the subsequent release of the drug molecules into the targeted region. It is important for the materials to remain unaffected in the body temperature region, but should be able to deliver the drug with a slight heating. Figure 3.13 shows one such potential system of the mono-layer protected metal clusters (MPC), where gold particles were protected with poly(N-isopropylacrylamide) (PNIPA) grafts [108]. For this purpose, MPC was first reacted with 1-mercapto-1-undecanol (MUD).

Ninomiya *et al.* [109] used ultrasound irradiation for the release of drugs into cancer cells. The drug delivery system was synthesized using liposomes modified with poly(NIPMAM-co-NIPAM) polymer. The modification provided the liposomes the capability to release the encapsulated calcein under 1 MHz ultrasound irradiation. In addition, the uptake of the drug by cells was enhanced by ultrasound irradiation.

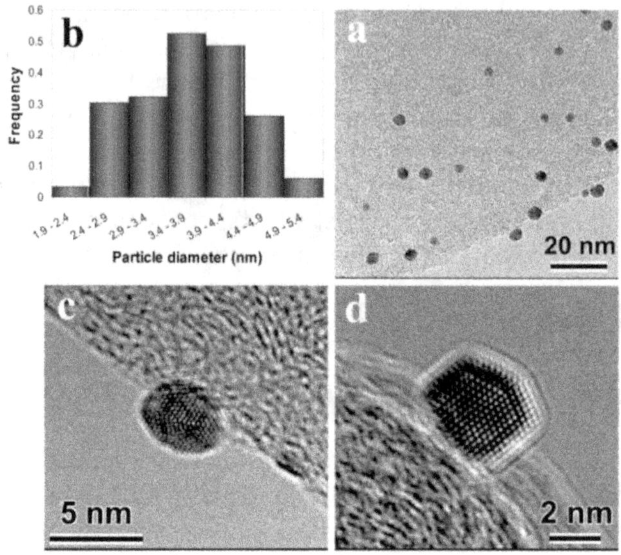

Figure 3.13 (a) Image of MPC-PNIPA (low magnification), (b) size distribution of MPC-PNIPA, (c) image of MPC- MUD (high magnification), (d) image of MPC-PNIPA (high magnification). Reproduced from Reference 108 with permission from American Chemical Society.

Polysaccharides maintain much interest in the drug delivery field because of their ability to form interpenetrating networks (IPN) and hydrogels with very efficient drug loading and release. These macromolecules are also of particular interest owing to their abundancy, opportunities to tailor the structure, availability of a large variety and origin from renewable sources [110-116]. Dragan *et al.* [117] synthesized multiresponsive IPN hydrogels of N,N-dimethylaminoethyl methacrylate (DMAEM) in presence of potato starch by crosslinking polymerization. The material exhibited sensitivity towards three different stimuli, viz. pH, temperature, ion nature and concentration. Diclofenac sodium (DS) was used as a model drug for the study of drug delivery. The drug was loaded in the system by swelling solvent sorption-evaporation methodology. The system was observed to work only in intestinal environment, not in gastric environment, thus, facilitating the easy movement of the drug through digestive system.

Like IPN, thermosensitive hydrogels also attained much interest in drug delivery systems due to low sol-gel transition temperature similar to the body temperature, which facilitates effective delivery of the drug at the targeted area. These materials can act as sustainable drug release agents by injectable hydrogel formation. Liu *et al.* [118] synthesized star shaped di-sulfide bonded polypeptide with dual sensitivity towards reduction temperature. The polypeptide formed the core in the star shaped architecture. The material could be deployed as sustainable drug carriers by in vivo injection. The thermoresponsive polymer was reported to have a number of advantages including long life span, targeted drug delivery, ease of application and simplicity in the formulation of drugs.

Tissue Engineering

Regeneration of functional tissue by replacing the old or dysfunctional form is known as tissue engineering. This is achieved by the use of systems containing living cells, a matrix for the living cells and biomolecules that help them regenerate in the particular environment. The system should be able to give support, provide proliferation and differentiation in the selected environment. The tissue engineering is a vast area of study which includes angiogenesis, wound healing, bone and cartilage regeneration, ocular diseases as well as mandible reconstruction. The regeneration and development of the specified tissue can be provided by fabricating tunable scaffolds for tissue engineering.

The study of thermoresponsive polymers for tissue engineering started as early as 1967 [119]. Adhesion and proliferation of the cells in a particular environment can be controlled by the use of thermoresponsive polymers as scaffolds. These materials undergo phase transition at a particular temperature, thus, releasing the cells at the site for regeneration and proliferation. Thermoresponsive polymers used for tissue engineering can be broadly classified into thermoresponsive insoluble surface grafts and soluble coatings.

Poly(N-isopropylacrylamide) is a widely studied thermoresponsive polymer for tissue engineering applications. As mentioned earlier, it converts into hydrophobic state from hydrophilic state characterized by quick precipitation upon heating. The polymer can be recovered by changing the temperature, while the released extracellular matrix remains unruffled [120]. Ohya *et al.* [121] synthesized thermoresponsive hyaluronans (HAs) of N-isopropylacrylamide on

HA by graft polymerization using dithiocarbamate iniferter (initiator, transfer agent and terminator) (Figure 3.14). The LCST temperature and phase change was unaffected by the factors like composition and M_n of both constituents and architecture. However, equilibrium transmittance of the aqueous solution above LCST exhibited an inverse proportional relationship with M_n and composition. Reduced attachment of endothelial cells was observed at physiological environments which makes it a good carrier in tissue engineering.

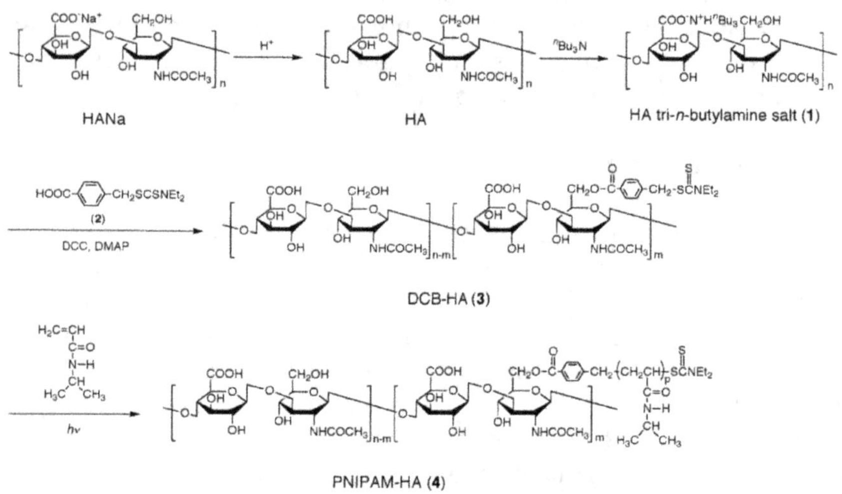

Figure 3.14 Synthetic route of polymer grafted hyaluronan. Reproduced from Reference 121 with permission from American Chemical Society.

Tan *et al.* [122] synthesized a series of thermoresponsive polymers based on thermoresponsive copolymer hydrogels. Materials were synthesized by the coupling of –COOH end capped poly(N-isopropylacrylamide) to aminated hyaluronic acid (AHA) through amide bonding by thermo-radical polymerization. The LCST of the material was around 30 °C regardless of the degree of grafting. The pore radius was dependent on the percentage of poly(N-isopropylacrylamide) in the interconnected porous hydrogels. Human adipose derived stem cells were encapsulated in the polymer and analysis was carried out for cytotoxicity. The copolymer was observed to be non-cytotoxic and preserved the entrapped cells. The copolymer was concluded to be excellent carrier for cells in tissue engineering applications. In another study, Huang *et al.* [123] synthesized hydrogel system based on poly(N-isopropylacrylamide), poly(D, L-lactic acid) and dextran to use as a scaffold for chondrocyte culture. The polymer exhibited properties like thermoresponsivity, porosity and partial biodegradation. The porous network in the system could be attributed to salt leaching. The polymer exhibited a low LCST of 32 °C. The material had a prolonged swelling life of 1 month in phosphate buffer solution. The chondrocytes isolated from chick sterna were incorporated in the hydrogels for a period of four weeks at 37 °C. Quantification performed using real-time reverse-transcriptase polymerase chain reaction for the messenger ribonucleic acid levels of the phenotypic markers proved that chondrocytes maintained their phenotype for four weeks and exhibited properties similar to in vivo development. Chondrocytes were collected from the hydrogel by lowering the temperature.

Gene Delivery

Gene therapy can be termed as the delivery of DNA to a target region, where it acts as a pro-drug for the synthesis of therapeutic protein [124]. Functional properties of thermoresponsive polymers can be utilized in macromolecule delivery, just like drug delivery to targeted areas. This function requires efficient control over biocompatibility, solubility and molecular weight of the polymers. Biodegradable polymers can further enhance the multi-functionality and safety of carriers, along with specificity. In case of gene delivery using a particular polymer, several other factors are also needed to be taken into account. The formation of polyplexes which can interact with the surrounding of the biological system as well as the problems arising during the internalization of the DNA into the cells are major issues faced while delivering DNA to targeted cells. Immunogenicity issues arising during the use of viral carrier also call for the use of non-viral vectors based on polymers for gene delivery [125-129]. Many thermoresponsive polymers have been studied for gene therapy as DNA release can be controlled by slight changes in the temperature surrounding the target cell region making these functional polymers competent as gene carriers.

Non-specific cellular uptake and cytocompatibility has attracted the use of polyplexes for multiple biological applications. Recently, core shell structure containing disulfide cross-linked poly(hydroxypropyl methacrylamide) (pHPMA) core and poly(ethylene glycol) (PEG) shell was fabricated by Novo *et al.* [130]. The core was encapsulated with plasmid DNA (pDNA) and the shell was decorated with folate molecules. The folate molecules helped in binding because of their affinity towards the receptors found in tumor cells. The system exhibited higher transfection abilities because of the enhanced cellular uptake. Introduction of targeting moiety on to the shells did not affect their cytocompatibility, thus, confirming the potential for large scale applications.

In a recent study, Calejo *et al.* [131] also synthesized a block copolymer poly(N-isopropylacrylamide)$_n$-block-poly((3acrylamidopropyl)trimethylammoniumchloride)$_m$ (PNIPAAM$_n$-*b*-PAMPTMA(+)$_m$) for use in in-vitro transfection studies of HeLa cells. The collapse of the polymer above its LCST resulted in the entrapment of the DNA molecules. The transfection efficiency of the polymer was affected by a number of factors like length of PNIPAAM chain and concentration of the polymer. The increase in the length of charged block had a negative impact on transfection efficiency, while the increase in concentration and length of PNIPAAM block led to its enhancement. The toxicity level was minimized by forming polyplexes at lower polymer to DNA charge ratio. The thermoresponsive phase changes resulted in the outward push of the charged block segments. Even at lower concentration, the polymer was able to effectively protect the DNA molecules.

In another study, Haladjova *et al.* [132] devised a novel approach for the synthesis of nano- and microcapsules using water as a solvent with the aid of thermoresponsive polymer templates (mesoglobules). The biomacromolecule attachment was simultaneously achieved during the synthesis of the polymer. The synthesis was completed in three different steps, viz. formation of mesoglobules; coating the seeds using radical polymerization; and subsequent core dissolution and removal by lowering the temperature of the reaction system. The covalent attachment of the biomacromolecules could be prevented by loading it during formation which exhibited significant advantage of the developed materials in gene delivery systems as reservoirs, carriers as well as transferring agents.

3.4.2 Other Applications

L-Proline moieties bound to a thermoresponsive polymer nanoreactor were used for the asymmetric reaction of aldol in water [133] (Figure 3.15). Thermoresponsive microgel also assisted in the process of Knoevenagel condensation reaction in water at ambient temperature conditions [134].

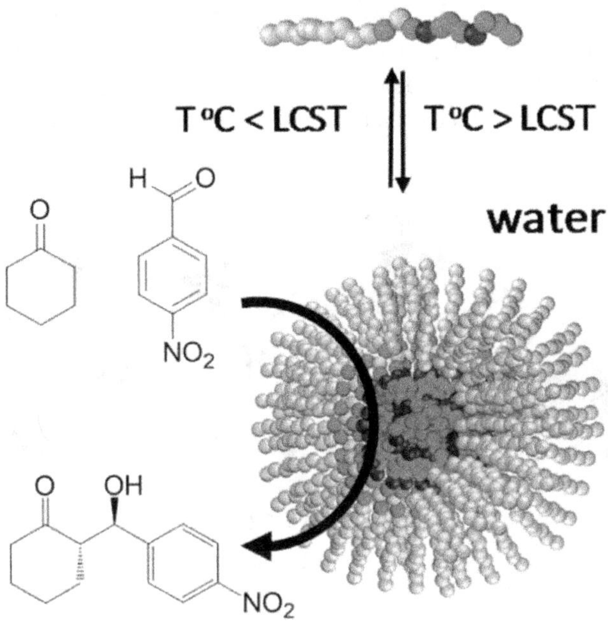

Figure 3.15 Representation of the asymmetric reaction of aldol in water. Reproduced from Reference 133 with permission from American Chemical Society.

The thermoresponsive polymers have been used in membrane separation industry as aqueous chromatography membranes for the separation of biological compounds [135-137]. Thermoresponsive polymer membranes facilitate selective permeability to molecules based on the temperature [138]. Mittal *et al.* [139] also reported thermally responsive poly(N-isopropylacrylamide) brushes grafted from polystyrene particles synthesized with surfactant free emulsion polymerization and functionalized with a thin shell of ATRP initiator on the surface. Cryo-scanning electron microscopy confirmed the growth of extremely long polymer layers from the surface, which otherwise seemed shrunken in the microscopy analysis owing to dehydration and possibly the effect of staining agent. The amount of grafted polymer also increased proportionally to the increase of the monomer concentration in the initial reaction system. The change in character from hydrophilic to hydrophobic with temperature and salt was found to be reversible and fast. The use of the developed materials as stationary phases for bio-separations was investigated. As shown in Figure 3.16, the adsorption process of tobacco mosaic virus was fully reversible with temperature, due to the characteristic of many biomolecules to adsorb on hydrophobic surfaces only. Thus, the observed findings successfully demonstrated a method for carrying out purification processes based on temperature switchable polymeric supports.

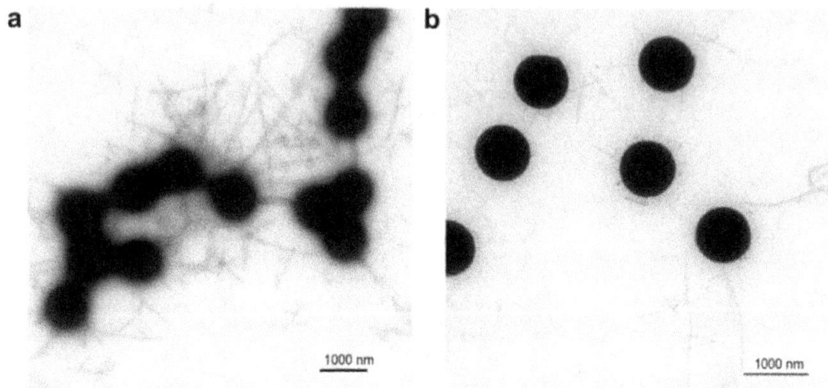

Figure 3.16 Transmission electron micrographs showing (a) adsorption and (b) desorption experiments of tobacco mosaic virus on the particles grafted with poly(N-isopropylamide). Reproduced from Reference 139 with permission from Elsevier.

3.5 Conclusions

Thermoresponsive polymers have attracted significant research attention over the last decade. Their ability to undergo phase change by the variation of temperature (and other external stimuli) along with other functional properties make them interesting candidates for various applications. In addition, these polymers offer other advantages such as large variety of derivatives, possibilities to tune the structure to generate copolymer as well as nanocomposites by incorporation of other functional materials, etc. Thus, their use in biomedical field for gene delivery, drug delivery, tissue engineering, along with membrane separation, chromatographic separations, etc., is growing rapidly. In addition, further detailed studies on the cytotoxicity and prolonged effect on environment would help to expand the spectrum of their applications. Overall, thermoresponsive polymers are a versatile class of materials which present an optimum set of unique properties suitable for many advanced applications.

References

1. Roy, D., Brooks, W. L. A., and Sumerlin, B. S. (2013) New directions in thermoresponsive polymers. *Chemical Society Reviews*, **42**, 7214-7243.
2. Wolff, M., Braun, D., and Nash, M. A. (2014) Detection of thermoresponsive polymer phase transition in dilute low-volume format by microscale thermophoretic depletion. *Analytical Chemistry*, **86**(14), 6797-6803.
3. Kohori, F., Sakai, K., Aoyagi, T., Yokoyama, M., Sakurai, Y., and Okano, T. (1998) Preparation and characterization of thermally responsive block copolymer micelles comprising poly(N-isopropylacrylamide-b-dl-lactide). *Journal of Controlled Release*, **55**(1), 87-98.
4. Chilkoti, A., Dreher, M. R., Meyer, D. E., and Raucher, D. (2002) Targeted drug delivery by thermally responsive polymers. *Advanced Drug Delivery Reviews*, **54**(5), 613-630.
5. Yamato, M., Konno, C., Utsumi, M., Kikuchi, A., and Okano, T. (2002) Thermally responsive polymer-grafted surfaces facilitate patterned cell seeding and co-culture. *Biomaterials*, **23**(2), 561-567.
6. Schilli, C. M., Zhang, M., Rizzardo, E., Thang, S. H., Chong Y. K., Edwards, K., Karlsson, G., and Muller, A. H. E. (2004) A new double-responsive block copolymer synthesized via RAFT polymerization: Poly(N-isopropylacrylamide)-block-poly(acrylic acid). *Macromolecules*, **37**(21), 7861-7866.
7. Hirsch, L. R., Gobin, A. M., Lowery, A. R., Tam. F., Drezek, R. A., Halas, N. J., and West J. L. (2006) Metal nanoshells. *Annals of Biomedical Engineering*, **34**(1), 15-22.
8. Gillet, R., Sakai, H., Nabae, Y., Hayakawa, T., and Kakimoto, M. (2016) Synthesis of hyperbranched-linear poly(N-isopropylacrylamide) polymers with a poly(siloxysilane) hyperbranched macroinitiator, and their application to cell culture on glass substrates. *Polymer Journal*, **48**, 1007-1012.
9. Gandhi, A., Paul, A., Sen, S. O., and Sen, K. K. (2015) Studies on thermoresponsive polymers: Phase behaviour, drug delivery and biomedical applications. *Asian Journal of Pharmaceutical Sciences*, **10**(2), 99-107.
10. Rothemund, S., and Teasdale, I. (2016) Preparation of polyphosphazenes: a tutorial review. *Chemical Society Reviews*, **45**, 5200-5215.
11. Sivagangi Reddy, N., Krishna Rao, K. S. V., Eswaramma, S., and Madhusudana Rao, K. (2016) Synthesis of dual responsive cyclotriphosphazene-based IPN hydrogels for controlled release of chemotherapeutic agent. *Polymers for Advanced Technologies*, **27**(3), 374-381.
12. Zhang, N., Salzinger, S., and Rieger, B. (2012) Poly(vinylphosphona-

te)s with widely tunable LCST: A promising alternative to conventional thermoresponsive polymers. *Macromolecules*, **45**(24), 9751-9758.

13. Tian, Z., Liu, X., Chen, C., and Allcock, H. R. (2012) Synthesis and micellar behavior of novel amphiphilic poly[bis(trifluoroethoxy)phosphazene]-co-poly[(dimethylamino)ethyl methacrylate] block copolymers. *Macromolecules*, **45**(5), 2502-2508.
14. Tao, X., Du, J., Wang, Y., and Ling, J. (2015) Polypeptoids with tunable cloud point temperatures synthesized from N-substituted glycine N-thiocarboxyanhydrides. *Polymer Chemistry*, **6**, 3164-3174.
15. Tao, X., Deng, Y., Shen, Z., and Ling, J. (2014) Controlled polymerization of N-substituted glycine N-thiocarboxyanhydrides initiated by rare earth borohydrides toward hydrophilic and hydrophobic polypeptoids. *Macromolecules*, **47**(18), 6173-6180.
16. Robinson, J. W., Secker, C., Weidner, S., and Schlaad, H. (2013) Thermoresponsive poly(N-C3 glycine)s. *Macromolecules*, **46**(3), 580-587.
17. Lahasky, S. H., Lu, L., Huberty, W. A., Cao, J., Guo, L., Garno, J. C., and Zhang, D. (2014) Synthesis and characterization of thermo-responsive polypeptoid bottlebrushes. *Polymer Chemistry*, **5**, 1418-1426.
18. Hu, B., Fu, W., and Zhao, B. (2016) Enhancing gelation of doubly thermosensitive hydrophilic ABC linear triblock copolymers in water by thermoresponsive hairy nanoparticles. *Macromolecules*, **49**, 15, 5502-5513.
19. Xuan, S., Lee, C.-U., Chen, C., Doyle, A. B., Zhang, Y., Guo, L., John, V. T., Hayes, D., and Zhang, D. (2016) Thermoreversible and injectable ABC polypeptoid hydrogels: Controlling the hydrogel properties through molecular design. *Chemistry of Materials*, **28**(3), 727-737.
20. Chan, B. A., Xuan, S., Horton, M., and Zhang, D. (2016) 1,1,3,3-Tetramethylguanidine-promoted ring-opening polymerization of N-butyl N-carboxyanhydride using alcohol initiators. *Macromolecules*, **49**(6), 2002-2012.
21. Birke, A., Huesmann, D., Kelsch, A., Weilbacher, M., Xie, J., Bros, M., Bopp, T., Becker, C., Landfester, K., and Barz, M. (2014) Polypeptoid-block-polypeptide copolymers: Synthesis, characterization, and application of amphiphilic block copolypept(o)ides in drug formulations and miniemulsion techniques. *Biomacromolecules*, **15**(2), 548-557.
22. Matai, I., and Gopinath, P. (2016) Chemically cross-linked hybrid nanogels of alginate and PAMAM dendrimers as efficient anticancer drug delivery vehicles. *ACS Biomaterials Science and Engineering*, **2**(2), 213-223.
23. Shi, K., Liu, Z., Wei, Y.-Y., Wang, W., Ju X.-J., Xie, R., and Chu, L.-Y. (2015) Near-infrared light-responsive poly(N-isopropylacrylamide)/graphene oxide nanocomposite hydrogels with ultrahigh

tensibility. *ACS Applied Materials and Interfaces*, 7(49), 27289-27298.

24. Chen, D., Guo, P., Chen, S., Cao, Y., Ji, W., Lei, X., Liu, L., Zhao, P., Wang, R., Qi, C., Liu, Y., and He, H. (2012) Properties of xyloglucan hydrogel as the biomedical sustained-release carriers. *Journal of Materials Science: Materials in Medicine*, **23**(4), 955-962.
25. Zhou, Q., Rutland, M. W., Teeri, T. T., and Brumer, H. (2007) Xyloglucan in cellulose modification. *Cellulose*, **14**(6), 625-641.
26. Hatton, F. L., Ruda, M., Lansalot, M., D'Agosto, F., Malmstrom, E., and Carlmark, A. (2016) Xyloglucan-functional latex particles via RAFT-mediated emulsion polymerization for the biomimetic modification of cellulose. *Biomacromolecules*, **17**(4), 1414-1424.
27. Vancoillie, G., Frank, D., and Hoogenboom, R. (2014) Thermoresponsive poly(oligo ethylene glycol acrylates). *Progress in Polymer Science*, **39**(6), 1074-1095.
28. Kafer, F., Liu, F., Stahlschmidt, U., Jerome, V., Freitag, R., Karg, M., and Agarwal, S. (2015) LCST and UCST in one: Double thermoresponsive behavior of block copolymers of poly(ethylene glycol) and poly(acrylamide-co-acrylonitrile). *Langmuir*, **31**(32), 8940-8946.
29. Cai, T., Marquez, M., and Hu, Z. (2007) Monodisperse thermoresponsive microgels of poly(ethylene glycol) analogue-based biopolymers. *Langmuir*, **23**(17), 8663-8666.
30. Shen, J., Chen, C., Fu, W., Shi, L., and Li, Z. (2013) Conformation-specific self-assembly of thermo-responsive poly(ethylene glycol)-b-polypeptide diblock copolymer. *Langmuir*, **29**(21), 6271-6278.
31. Warren, N. J., Mykhaylyk, O. O., Mahmood, D., Ryan, A. J., and Armes, S. P. (2014) RAFT aqueous dispersion polymerization yields poly(ethylene glycol)-based diblock copolymer nano-objects with predictable single phase morphologies. *Journal of American Chemical Society*, **136**(3), 1023-1033.
32. Song, X., Zhu, J.l., Wen, Y., Zhao, F., Zhang, Z.-X., and Li, J. (2017) Thermoresponsive supramolecular micellar drug delivery system based on star-linear pseudo-block polymer consisting of β-cyclodextrin-poly(N-isopropylacrylamide) and adamantyl-poly(ethylene glycol). *Journal of Colloid and Interface Science*, **490**, 372-379.
33. Mao, H., Pan, P., Shan, G., and Bao, Y. (2015) In situ formation and gelation mechanism of thermoresponsive stereocomplexed hydrogels upon mixing diblock and triblock poly(lactic acid)/poly(ethylene Glycol) copolymers. *Journal of Physical Chemistry B*, **119**(21), 6471-6480.
34. Anghelache, A., Teodorescu, M., Stanescu, P. O., Draghici, C., and Vuluga, D. M. (2014) Novel crosslinked thermoresponsive hydrogels with controlled poly(ethylene glycol)-poly(propylene glycol) multi block copolymer structure. *Colloid and Polymer Science*, **292**(4), 829-838.

35. Diao, B., Zhang, Z., Zhu, J., and Li, J. (2014) Biomass-based thermo-gelling copolymers consisting of lignin and grafted poly(N-isopropylacrylamide), poly(ethylene glycol), and poly(propylene glycol). *RSC Advances*, **4**, 42996-43003.
36. Zhu, C.-H., Hai, Z.-B., Cui, C.-H., Li, H.-H., Chen, J.-F., and Yu, S.-H. (2012) In situ controlled synthesis of thermosensitive poly(N-isopropylacrylamide)/Au nanocomposite hydrogels by gamma radiation for catalytic application. *Small*, **8**, 930-936.
37. Mittal, V., Matsko, N. B., Butte, A., and Morbidelli, M. (2008) Swelling deswelling behavior of PS-PNIPAAM copolymer particles and PNIPAAM brushes grafted from polystyrene particles & monoliths. *Macromolecular Materials and Engineering*, **293**, 491-502.
38. Mai, T. B., Tran, T. N., Islam, M. R., Park, J. M., and Lim, K. T. (2014) Covalent functionalization of silica nanoparticles with poly(N-isopropylacrylamide) employing thiol-ene chemistry and activator regenerated by electron transfer ATRP protocol. *Journal of Materials Science*, **49**, 1519-1526.
39. Karimi, A. R., Azadikhah, F., Rahimi, L., and Ghadimi, S. (2015) Fabrication of new Fe-phthalocyanine oligomer–magnetite hybrid magnetic nano particles and their effects on the LCST behavior of thermo-sensitive poly(N-isopropylacrylamide-co-acrylic acid) magnetic nanocomposites. *Colloids and Surfaces A: Physicochemical and Engineering Aspects*, **484**, 304-312.
40. Kuroiwa, K., Koga, Y., Ishimaru, Y., Nakashima, T., Hachisako, H., and Sakurai, S. (2016) Morphological control of hybrid amphiphilic poly(N-isopropylacrylamide)/metal cyanide complexes. *Polymer Journal*, **48**, 729-739.
41. Yang, H.-y., van Ee, R. J., Timmer, K., Craenmehr, E. G. M., Huang, J. H., Oner, F. C., Dhert, W. J. A., Kragten, A. H. M., Willems, N., Grinwis, G. C. M., Tryfonidou, M. A., Papen-Botterhuis, N. E., and Creemers, L. B. (2015) A novel injectable thermoresponsive and cytocompatible gel of poly(N-isopropylacrylamide) with layered double hydroxides facilitates siRNA delivery into chondrocytes in 3D culture. *Acta Biomaterialia*, **23**, 214-228.
42. Zhang, E., Wang, T., Hong, W., Sun, W., Liu, X., and Tong, Z. (2014) Infrared-driving actuation based on bilayer graphene oxide-poly(N-isopropylacrylamide) nanocomposite hydrogels. *Journal of Materials Chemistry A*, **2**, 15633-15639.
43. Islam, M. R., and Serpe, M. J. (2013) Penetration of polyelectrolytes into charged poly(N-isopropylacrylamide) microgel layers confined between two surfaces. *Macromolecules*, **46**(4), 1599-1606.
44. Abd El-Mohdy, H. L. (2013) Thermo-responsive behavior of radiation-induced poly(N-isopropylacrylamide)/polyethylene oxide nanocomposite. *Journal of Polymer Research*, **20**, 206.
45. Fernandez, V. V. A., Aguilar, J., Soltero, J. F. A., Moscoso-Sanchez F. J.,

Sanchez-Diaz J. C., Hernandez, E., Bautista, F., and Puig, J. E. (2016) Thermoresponsive poly(N-isopropylacrylamide) nanogels/poly(acrylamide) nanostructured hydrogels. *Journal of Macromolecular Science, Part A*, **53**(3), 152-159.

46. Chen, J. J., Ahmad, A. L., and Ooi, B. S. (2013) Poly(N-isopropylacrylamide-co-acrylic acid) hydrogels for copper ion adsorption: Equilibrium isotherms, kinetic and thermodynamic studies. *Journal of Environmental Chemical Engineering*, **1**(3), 339-348.
47. Chen, Z., Cui, Z.-M., Cao, C.-Y., He, W.-D., Jiang, L., and Song, W.-G. (2012) Temperature-eesponsive smart nanoreactors: Poly(N-isopropylacrylamide)-coated Au@mesoporous-SiO_2 hollow nanospheres. *Langmuir*, **28**(37), 13452-13458.
48. Kavitha, T., Kim, J.-O., Jang, S., Kim, D.-P., Kang, I.-K., and Park, S.-Y. (2016) Multifaceted thermoresponsive poly(N-vinylcaprolactam) coupled with carbon dots for biomedical applications. *Materials Science and Engineering: C*, **61**(1), 492-498.
49. Wang, R.-M., Liu, Q., Zhang, Y., Hong, Z., and Wang, H.-F. (2016) A thermo- and pH-responsive poly(N-isopropylacrylamide)–Mn–ZnS nanocomposite for controlled release and real-time photoluminescence tracking of doxorubicin. *RSC Advances*, **6**, 50985-50992.
50. Shi, S., Wang, Q., Wang, T., Ren, S., Gao, Y., and Wang, N. (2014) Thermo-, pH-, and light-responsive poly(N-isopropylacrylamide-co-methacrylic acid)–Au hybrid microgels prepared by the in situ reduction method based on Au-thiol chemistry. *Journal of Physical Chemistry B*, **118**(25), 7177-7186.
51. Prabhakar, M. N., Sudhakara, P., Subha, M. C. S., Chowdoji Rao, K., and Song, J. I. (2015) Novel thermoresponsive biodegradable nanocomposite hydrogels for dual function in biomedical applications. *Polymer-Plastics Technology and Engineering*, **54**(16), 1704-1714.
52. Ravindra. S., Varaprasad, K., Rajinikanth, V., Mulaba-Bafubiandi, A. F., and Ramam, K. V. S. (2013) Studies on curcumin loaded poly(N-isopropylacrylamide) silver nanocomposite hydrogels for antibacterial and drug releasing applications. *Journal of Macromolecular Science, Part A, Pure and Applied Chemistry*, **50**(12), 1230-1240.
53. Bajpai, S. K., Bajpai, M., and Sharma, L. (2011) In situ formation of silver nanoparticles in poly(N-isopropyl acrylamide) hydrogel for antibacterial applications. *Designed Monomers and Polymers*, **14**(4), 383-394.
54. Cong, H.-P., Qiu, J.-H., and Yu, S.-H. (2015), Thermoresponsive poly(N-isopropylacrylamide)/graphene/Au nanocomposite hydrogel for water treatment by a laser-assisted approach. *Small*, **11**, 1165-1170.
55. Chen, H., Qian, G., Ruan, X., and Frost, R. L. (2015) Abatement of aqueous anionic contaminants by thermo-responsive nanocomposites: (Poly(N-isopropylacrylamide))-co-silylanized magnesium /

aluminun layered double hydroxides. *Journal of Colloid and Interface Science*, **448**, 65-72.

56. Yao, C., Liu, Z., Yang, C., Wang, W., Ju, X.-J., Xie, R., and Chu, L.-Y. (2015) Poly(N-isopropylacrylamide)-clay nanocomposite hydrogels with responsive bending property as temperature-controlled manipulators. *Advanced Functional Materials*, **25**, 2980-2991.
57. Zhu, C.-H., Lu, Y., Peng, J., Chen, J.-F., and Yu, S.-H. (2012) Photothermally sensitive poly(N-isopropylacrylamide)/graphene oxide nanocomposite hydrogels as remote light-controlled liquid microvalves. *Advanced Functional Materials*, **22**, 4017-4022.
58. Karimi, A. R., Rahimi, L., Azadikhah, F., and Ghadimi, S. (2016) Preparation and lower critical solution temperature behavior investigation of new thermoresponsive poly(N-isopropylacrylamide-co-phthalocyanine) magnetic nanocomposites containing phthalocyanine-coated Fe_3O_4 hybrid. *Canadian Journal of Chemistry*, **94**, 723-731.
59. Etika, K. C., Liu, L., Cox, M. A., and Grunlan, J. C. (2016) Clay-mediated carbon nanotube dispersion in poly(N-Isopropylacrylamide). *Colloids and Surfaces A: Physicochemical and Engineering Aspects*, **489**, 19-26.
60. Cavallaro, G., Lazzara, G., Massaro, M., Milioto, S., Noto, R., Parisi, F., and Riela, S. (2015) Biocompatible Poly(N-isopropylacrylamide)-halloysite nanotubes for thermoresponsive curcumin release. *Journal of Physical Chemistry C*, **119**(16), 8944-8951.
61. Dong, J., Weng, J., and Dai, L. (2013) The effect of graphene on the lower critical solution temperature of poly(N-isopropylacrylamide). *Carbon*, **52**, 326-336.
62. Scherzinger, C., Balaceanu, A., Hofmann, C. H., Schwarz, A., Leonhard, K., Pich, A., and Richtering, W. (2015) Cononsolvency of mono- and di-alkyl N-substituted poly(acrylamide)s and poly(vinyl caprolactam). *Polymer*, **62**(7), 50-59.
63. Cortez-Lemus, N. A., and Licea-Claverie, A. (2016) Poly(N-vinylcaprolactam), a comprehensive review on a thermoresponsive polymer becoming popular. *Progress in Polymer Science*, **53**, 1-51.
64. Ou, R., Wei, J., Jiang, L., Simon, G. P., and Wang, H. (2016) Robust thermoresponsive polymer composite membrane with switchable superhydrophilicity and superhydrophobicity for efficient oil-water separation. *Environmental Science and Technology*, **50**(2), 906-914.
65. Nguyen, H. H., Brulet, A., Goudouneche, D., Saint-Aguet, P., Lauth-de Viguerie, N., and Marty, J.-D. (2015) The effect of polymer branching and average molar mass on the formation, stabilization and thermoresponsive properties of gold nanohybrids stabilized by poly(N-isopropylacrylamides). *Polymer Chemistry*, **6**, 5838-5850.
66. Indulekha, S., Arunkumar, P., Bahadur, D., and Srivastava, R. (2016)

Thermoresponsive polymeric gel as an on-demand transdermal drug delivery system for pain management. *Materials Science and Engineering: C*, **62**, 113-122.

67. Swamy, B. Y., Chang, J. H., Ahn, H., Lee, W.-K., and Chung, I. (2013) Thermoresponsive N-vinyl caprolactam grafted sodium alginate hydrogel beads for the controlled release of an anticancer drug. *Cellulose*, **20**, 1261-1273.
68. Lozinsky, V. I., Simenel, I. A., Kurskaya, E. A., Kulakova, V. K., Galeev, I. Y., Mattiasson, B., Grinberg, V. Y., and Khokhlov, A. R. (2000) Synthesis of N-vinylcarolactam polymers in water-containing media. *Polymer*, **41**(17), 6507-6518.
69. Pich, A., Tessier, A., Boyko, V., Lu, Y., Adler, H.-J. P. (2006) Synthesis and characterization of poly(vinylcaprolactam)-based microgels exhibiting temperature and pH-sensitive properties. *Macromolecules*, **39**, 7701-7707.
70. Liu, J., Detrembleur, C., De Pauw-Gillet, M.-C., Mornet, S., Duguet, E., and Jerome, C. (2014) Gold nanorods coated with a thermo-responsive poly(ethylene glycol)-b-poly(N-vinylcaprolactam) corona as drug delivery systems for remotely near infrared-triggered release. *Polymer Chemistry*, **5**, 799-813.
71. Hurtgen, M., Liu, J., Debuigne, A., Jerome, C., and Detrembleur (2012) Synthesis of thermo-responsive poly(N-vinylcaprolactam)-containing block copolymers by cobalt-mediated radical polymerization. *Journal of Polymer Science A, Polymer Chemistry*, **50**, 400-408.
72. Gonzalez, E., and Frey, M. W. (2017) Synthesis, characterization and electrospinning of poly(vinyl caprolactam-co-hydroxymethyl acrylamide) to create stimuli-responsive nanofibers. *Polymer*, **108**(13), 154-162.
73. Thomassin, J.-M., Mathieu, K., Kermagoret, A., Fustin, C.-A., Jerome, C., and Debuigne, A. (2015) Double thermo-responsive hydrogels from poly(vinylcaprolactam) containing diblock and triblock copolymers. *Polymer Chemistry*, **6**, 1856-1864.
74. Makhaeva, E. E., Tenhu, H., and Khokhlov, A. R. (1998) Conformational changes of poly(vinylcaprolactam) macromolecules and their complexes with ionic surfactants in aqueous solution. *Macromolecules*, **31**(18), 6112-6118.
75. Wang, Y., Nie, J., Chang, B., Sun, Y., and Yang, W. (2013) Poly(vinylcaprolactam)-based biodegradable multiresponsive microgels for drug delivery. *Biomacromolecules*, **14**(9), 3034-3046.
76. Jain, S., Sandhu, P. S., Malvi, R., and Gupta, B. (2013) Cellulose derivatives as thermoresponsive polymer: An overview. *Journal of Applied Pharmaceutical Science*, **3**(12), 139-144.
77. Wang, T., Chen, L., Shen, T., and Wu, D. (2016) Preparation and properties of a novel thermo-sensitive hydrogel based on chitosan/hydroxypropyl methylcellulose/glycerol. International Journal of Bio-

logical *Macromolecules*, **93**, Part A, 775-782.

78. Chen, C.-P., Hsieh, C.-M., Tsai, T., Yang, J.-C., and Chen, C.-T. (2012) Optimization and evaluation of a chitosan/hydroxypropyl methylcellulose hydrogel containing toluidine blue O for antimicrobial photodynamic inactivation. *International Journal of Molecular Sciences*, **16**(9), 20859-20872.
79. Calejo, M. T., Cardoso, A. M., Marques, E. F., Araujo, M. J., Kjoniksen, A.-L., Sande, S. A., de Lima, M. C., Jurado, A. S., and Nystrom, B. (2013) In vitro cytotoxicity of a thermoresponsive gel system combining ethyl(hydroxyethyl) cellulose and lysine-based surfactants. *Colloids and Surfaces B: Biointerfaces*, **102**, 682-686.
80. Ci, J., Kang, H., Liu, C., He, A., and Liu, R. (2017) Thermal sensitivity and protein anti-adsorption of hydroxypropyl cellulose-g-poly(2-(methacryloyloxy) ethyl phosphorylcholine). *Carbohydrate Polymers*, **157**, 757-765.
81. Tahara, Y., Sakiyama, M., Takeda, S., Nishimura, T., Mukai, S.-a., Sawada, S.-i., Sasaki, Y., and Akiyoshi, K. (2016) Self-assembled nanogels of cholesterol-bearing hydroxypropyl cellulose: A thermoresponsive building block for nanogel tectonic materials. *Langmuir*, **32**(47), 12283-12289.
82. Tian, Y., Ju, B., Zhang, S., and Hou, L. (2016) Thermoresponsive cellulose ether and its flocculation behavior for organic dye removal. *Carbohydrate Polymers*, **136**(20), 1209-1217.
83. Dhar, N., Akhlaghi, S. P., and Tam, K. C. (2012) Biodegradable and biocompatible polyampholyte microgels derived from chitosan, carboxymethyl cellulose and modified methyl cellulose. *Carbohydrate Polymers*, **87**(1), 101-109.
84. Huang, C.-C., Liao, Z.-X., Chen, D.-Y., Hsiao, C.-W., Chang, Y., and Sung, H.-W. (2014) Injectable cell constructs fabricated via culture on a thermoresponsive methylcellulose hydrogel system for the treatment of ischemic diseases. *Advanced Healthcare Materials*, **3**, 1133-1148.
85. Liu, Z., and Yao, P. (2015) Injectable thermo-responsive hydrogel composed of xanthan gum and methylcellulose double networks with shear-thinning property. *Carbohydrate Polymers*, **132**, 490-498.
86. Hu, Z., Xu, R., Cranston, E. D., and Pelton, R. H. (2016) Stable aqueous foams from cellulose nanocrystals and methyl cellulose. *Biomacromolecules*, **17**(12), 4095-4099.
87. McKee, J. R., Hietala, S., Seitsonen, J., Laine, J., Kontturi, E., and Ikkala, O. (2014) Thermoresponsive nanocellulose hydrogels with tunable mechanical properties. *ACS Macro Letters*, **3**(3), 266-270.
88. Bal, A., Ozkahraman, B., Acar, I., Ozyurek, M., and Guclu, G. (2014) Study on adsorption, regeneration, and reuse of crosslinked chitosan graft copolymers for Cu(II) ion removal from aqueous soluti-

ons. *Desalination and Water Treatment,* **52**(16), 3246-3255.

89. Oh, B. H. L., Bismarck, A., and Chan-Park, M. B. (2014) High internal phase emulsion templating with self-emulsifying and thermoresponsive chitosan-graft-PNIPAM-graft-oligoproline. *Biomacromolecules,* **15** (5), 1777-1787.
90. Wang, Q., and Chen, D. (2016) Synthesis and characterization of a chitosan based nanocomposite injectable hydrogel. *Carbohydrate Polymers,* **136**, 1228-1237.
91. Xu, T., Zhang, J., Chi, H., and Cao, F. (2016) Multifunctional properties of organic-inorganic hybrid nanocomposites based on chitosan derivatives and layered double hydroxides for ocular drug delivery. *Acta Biomaterialia,* **36**, 152-163.
92. Yan, J., Yang, L., Wang, G., Xiao, Y., Zhang, B., and Qi, N. (2010) Biocompatibility evaluation of chitosan-based injectable hydrogels for the culturing mice mesenchymal stem cells in vitro. *Journal of Biomaterials Applications,* **24**, 625-637.
93. Ross-Murphy, S. B. (1992) Structure and rheology of gelatin gels: recent progress. *Polymer,* **33**(12), 2622-2627.
94. Daniel-da-Silva, A. L., Salgueiro, A. M., and Trindade, T. (2013) Effects of Au nanoparticles on thermoresponsive genipin-crosslinked gelatin hydrogels. *Gold Bulletin,* **46**(1), 25-33.
95. Barbetta, A., Massimi, M., Devirgiliis, L. C., and Dentini, M. (2006) Enzymatic cross-linking versus radical polymerization in the preparation of gelatin polyHIPEs and their performance as scaffolds in the culture of hepatocytes. *Biomacromolecules,* **7**(11), 3059-3068.
96. Manjula, B., Varaprasad, K., Sadiku, R., Ramam, K., Reddy, G. V. S., and Raju, K. M. (2014) Development of microbial resistant thermosensitive Ag nanocomposite (gelatin) hydrogels via green process. *Journal of Biomedical Materials Research Part A,* **102**(4), 928-934.
97. Li, C., Mu, C., Lin, W., and Ngai, T. (2015) Gelatin effects on the physicochemical and hemocompatible properties of gelatin/PAAm/laponite nanocomposite hydrogels. *ACS Applied Materials and Interfaces,* **7**(33), 18732-18741.
98. Yuan, W., Zhao, N., Yu, B., Jiang, H., Ma, J., and Xu, F. (2014) Thermoresponsive gelatin-functionalized PCL film surfaces for improvement of cell adhesion and intelligent recovery of gene-transfected cells. *Science China Chemistry,* **57**(4), 586-595.
99. Fathollahipour, S., Abouei Mehrizi, A., Ghaee, A., and Koosha, M. (2015) Electrospinning of PVA/chitosan nanocomposite nanofibers containing gelatin nanoparticles as a dual drug delivery system. *Journal of Biomedical Materials Research Part A,* **103**(12), 3852-3862.
100. Lee, M.-Y, Tsai, W.-W., Chen, H.-J., Chen, J.-P., Chen, C.-H., Yeh, W.-L., and An, J. (2013) Laser sintered porous polycaprolacone scaffolds loaded with hyaluronic acid and gelatin-grafted thermoresponsive

hydrogel for cartilage tissue engineering. *Bio-Medical Materials and Engineering*, **23**(6), 533-543.

101. Gandhi, S. S., Yan, H., and Kim, C. (2014) Thermoresponsive gelatin nanogels. *ACS Macro Letters*, **3**(11),1210-1214.
102. Huang, X., Du, F., Ju, R., and Li, Z. (2007), Novel acid-labile, thermoresponsive poly(methacrylamide)s with pendent ortho ester moieties. *Macromolecular Rapid Communications*, **28**, 597-603.
103. Huang, X., Witte, K. L., Bergbreiter, D. E., and Wong, C.-H. (2001) Homogenous enzymatic synthesis using a thermo-responsive water-soluble polymer support. *Advanced Synthesis and Catalysis*, **343**(6-7), 675-681.
104. Chu, L.-Y., Li, Y., Zhu, J.-H., and Chen, W.-M. (2005) Negatively thermoresponsive membranes with functional gates driven by zipper-type hydrogen-bonding interactions. *Angewandte Chemie, International Edition*, **44**(14), 2124-2127.
105. Li, C., Madsen, J., Armes, S. P., and Lewis, A. L. (2006) A new class of biochemically degradable, stimulus-responsive triblock copolymer gelators. *Angewandte Chemie, International Edition*, **45**, 3510-3513.
106. Kumashiro, Y., Matsunaga, T., Muraoka, M., Tanaka, N., Itoga, K., Kobayashi, J., Tomiyama, Y., Kuroda, M., Shimizu, T., Hashimoto, I., Umemura, K., Yamato, M., and Okano, T. (2014) Rate control of cell sheet recovery by incorporating hydrophilic pattern in thermoresponsive cell culture dish. *Journal of Biomedical Materials Research Part A*, **102**(8), 2849-2856.
107. Sahoo, B., Devi, K. S., Banerjee, R., Maiti, T. K., Pramanik, P., and Dhara, D. (2013) Thermal and pH responsive polymer-tethered multifunctional magnetic nanoparticles for targeted delivery of anticancer drug. *ACS Applied Materials and Interfaces*, **5**(9), 3884-3893.
108. Raula, J., Shan, J., Nuopponen, M., Niskanen, A., Jiang, H., Kauppinen, E. I., and Tenhu, H. (2003) Synthesis of gold nanoparticles grafted with a thermoresponsive polymer by surface-induced reversible-addition-fragmentation chain-transfer polymerization. *Langmuir*, **19**(8), 3499-3504.
109. Ninomiya, K., Kawabata, S., Tashita, H., and Shimizu, N. (2014) Ultrasound-mediated drug delivery using liposomes modified with a thermosensitive polymer. *Ultrasonics Sonochemistry*, **21**(1), 310-316.
110. Matricardi, P., Di Meo, C., Coviello, T., Hennink, W. E., and Alhaique, F. (2013) Interpenetrating polymer networks polysaccharide hydrogels for drug delivery and tissue engineering. *Advanced Drug Delivery Reviews*, **65**(9), 1172-1187.
111. Guilherme, M. R., de Moura, M. R., Radovanovic, E., Geuskens, G., Rubira, A. F., and Muniz, E. C. (2005) Novel thermo-responsive membranes composed of interpenetrated polymer networks of alg-

inate-Ca^{2+} and poly(N-isopropylacrylamide). *Polymer*, **46**, 2668-2674.

112. Matricardi, P., Meo, C. D., Coviello, T., and Alhaique, F. (2008) Recent advances and perspectives on coated alginate microspheres for modified drug delivery. *Expert Opinion on Drug Delivery*, **5**, 417-425.
113. Oddo, L., Masci, G., Di Meo, C., Capitani, D., Mannina, L., Lamanna, R., De Santis, S., Alhaique, F., Coviello, T., and Matricardi, P. (2010) Novel thermosensitive calcium alginate microspheres: physico-chemical characterization and delivery properties. *Acta Biomaterialia*, **6**, 3657-3664.
114. Pescosolido, L., Vermonden, T., Malda, J., Censi, R., Dhert, W. J. A., Alhaique, F., Hennink, W. E., and Matricardi, P. (2011) In situ forming IPN hydrogels of calcium alginate and dextran-hema for biomedical applications. *Acta Biomaterialia*, **7**, 1627-1633.
115. Chen, S.-C., Wu, Y.-C., Mi, F.-L., Lin, Y.-H., Yu, L.-C., and Sung, H.-W. (2004) A novel pH-sensitive hydrogel composed of N, O-carboxymethyl chitosan and alginate cross-linked by genipin for protein drug delivery. *Journal of Controlled Release*, **96**(2), 285-300.
116. Zhang, G. Q., Zha, L. S., Zhou, M. H., Ma, J. H., and Liang, B. R. (2005) Preparation and characterization of pH- and temperature-responsive semi-interpenetrating polymer network hydrogels based on linear sodium alginate and crosslinked poly(N-isopropylacrylamide). *Journal of Applied Polymer Science*, **97**, 1931-1940.
117. Dragan, E. S., Loghin, D. F. A., Cocarta, A.-I., and Doroftei, M. (2016) Multi-stimuli-responsive semi-IPN cryogels with native and anionic potato starch entrapped in poly(N,N-dimethylaminoethyl methacrylate) matrix and their potential in drug delivery. *Reactive and Functional Polymers*, **105**, 66-77.
118. Liu, D.-L., Chang, X., and Dong, C.-M. (2013) Reduction- and thermosensitive star polypeptide micelles and hydrogels for on-demand drug delivery. *Chemical Communications*, **49**, 1229-1231.
119. Scarpa, J. S., Mueller, D. D., and Klotz, I. M. (1967) Slow hydrogen-deuterium exchange in a non-.alpha.-helical polyamide. *Journal of American Chemical Society*, **89**(24), 6024-6030.
120. da Silva, R. M., Mano, J. F., and Reis, R. L. (2007) Smart thermoresponsive coatings and surfaces for tissue engineering: switching cell-material boundaries. *Trends in Biotechnology*, **25**(12), 577-583.
121. Ohya, S., Nakayama, Y., and Matsuda, T. (2001) Thermoresponsive artificial extracellular matrix for tissue engineering: Hyaluronic acid bioconjugated with poly(N-isopropylacrylamide) grafts. *Biomacromolecules*, **2**(3), 856-863.
122. Tan, H., Ramirez, C. M., Miljkovic, N., Li, H., Rubin, J. P., and Marra, K. G. (2009) Thermosensitive injectable hyaluronic acid hydrogel for adipose tissue engineering. *Biomaterials*, **30**(36), 6844-6853.

123. Huang, X., Zhang, Y., Donahue, H. J., and Lowe, T. L. (2007) Porous thermoresponsive-co-biodegradable hydrogels as tissue-engineering scaffolds for 3-dimensional in vitro culture of chondrocytes. *Tissue Engineering*, **13**(11) 2645-2652.
124. Marshall, E. (1999) Gene therapy death prompts review of adenovirus vector. *Science*, **286**(5448), 2244-2245.
125. Boye, S. E., Boye, S. L., Lewin, A. S., and Hauswirth, W. W. (2013) Comprehensive review of retinal gene therapy. *Molecular Therapy*, **21**(3), 509-519.
126. Ginn, S. L., Alexander, I. E., Edelstein, M. L., Abedi, M. R., and Wixon, J. (2013) Gene therapy clinical trials worldwide to 2012 - an update. *The Journal of Gene Medicine*, **15**, 65-77.
127. Yla-herttuala, S. (2012) Endgame: glybera finally recommended for approval as the first gene therapy drug in the European Union. *Molecular Therapy*, **20**(10), 1831-1832.
128. Tachibana, M., Amato, P., Sparman, M., Woodward, J., Sanchis, D. M., Ma, H., Gutierrez, N. M., Tippner-Hedges, R., Kang, E., Lee, H. S., Ramsey, C., Masterson, K., Battaglia, D., Lee, D., Wu, D., Jensen, J., Patton, P., Gokhale, S., Stouffer, R., and Mitalipov, S. (2013) Towards germline gene therapy of inherited mitochondrial diseases. *Nature*, **493**, 627-631.
129. Aied, A., Greiser, U., Pandit, A., and Wang, W. (2013) Polymer gene delivery: overcoming the obstacles. *Drug Discovery Today*, **18**(21-22), 1090-1098.
130. Novo, L., Mastrobattista, E., van Nostrum, C. F., and Hennink, W. E. (2014) Targeted decationized polyplexes for cell specific gene delivery. *Bioconjugate Chemistry*, **25**(4), 802-812.
131. Calejo, M. T., Cardoso, A. M. S., Kjoniksen, A.-L., Zhu, K., Morais, C. M., Sande, S. A., Cardoso, A. L., de Lima, M. C. P., Jurado, A., and Nystrom, B. (2013) Temperature-responsive cationic block copolymers as nanocarriers for gene delivery. *International Journal of Pharmaceutics*, **448**(1), 105-114.
132. Haladjova, E., Toncheva-Moncheva, N., Apostolova, M. D., Trzebicka, B., Dworak, A., Petrov, P., Dimitrov, I., Rangelov, S., and Tsvetanov, C. B. (2014) Polymeric nanoparticle engineering: From temperature-responsive polymer mesoglobules to gene delivery systems. *Biomacromolecules*, **15**(12), 4377-4395.
133. Zayas, H. A., Lu, A., Valade, D., Amir, F., Jia, Z., O'Reilly, R. K., and Monteiro, M. J. (2013) Thermoresponsive polymer-supported l-proline micelle catalysts for the direct asymmetric aldol reaction in water. *ACS Macro Letters*, **2**(4), 327-331.
134. Seto, H., Imai, K., Hoshino, Y., and Miura, Y. (2016) Polymer microgel particles as basic catalysts for Knoevenagel condensation in water. *Polymer Journal*, **48**, 897-904.
135. Kikuchi, A., and Okano, T. (2002) Intelligent thermoresponsive pol-

ymeric stationary phases for aqueous chromatography of biological compounds. *Progress in Polymer Science*, **27**(6), 1165-1193.

136. Iwata, H., and Matsuda, T. (1988) Preparation and properties of novel environment-sensitive membranes prepared by graft polymerization onto a porous membrane. *Journal of Membrane Science*, **38**(2), 185-199.
137. Iwata, H., Oodate, M., Uyama, Y., Amemiya, H., and Ikada, Y. (1991) Preparation of temperature-sensitive membranes by graft polymerization onto a porous membrane. *Journal of Membrane Science*, **55**(1-2), 119-130.
138. Lequieu, W., and Du Prez, F. E. (2004) Segmented polymer networks based on poly(N-isopropyl acrylamide) and poly(tetrahydrofuran) as polymer membranes with thermo-responsive permeability. *Polymer*, **45**(3), 749-757.
139. Mittal, V., Matsko, N. B., Butte, A., and Morbidelli, M. (2007) Synthesis of temperature responsive polymer brushes from polystyrene latex particles functionalized with ATRP initiator. *European Polymer Journal*, **43**, 4868-4881.

4

Synthesis and Properties of Polymer Based (or Modified) Nanoparticles for Biomedical Applications

4.1 Introduction

Polymer nanoparticles based therapeutics have attained significant interest in various biomedical applications. Many polymers, being favorable biocompatible materials, play a key role in the area of smart drug delivery by delivering the therapeutic agents directly at the targeted site of action effectively. In order to promote safe and targeted drug delivery of the conventional medicines into the human body with reduced side-effects, biocompatible therapeutic nano-formulations are also being developed, such as polymeric capsules, spheroids colloids, core-shells, micelles, dendrimers, platelets, etc. Polymeric nanoparticles (PNPs) can be prepared by two main approaches, i.e., the polymerization of monomers and the dispersion of preformed polymers. Various synthetic techniques are involved in both approaches. In case of 'polymerization of monomers', PNPs are prepared by employing conventional emulsion, miniemulsion and micro-emulsion polymerization techniques. Other hetero-phase polymerization methods like interfacial and living/controlled radical polymerization are also involved. In case of 'dispersion of preformed polymers', PNPs can be achieved by using techniques such as solvent evaporation, salting out, dialysis, nanoprecipitation and supercritical fluid technology [1].

PNPs are used in smart drug delivery systems due to their key advantages, such as improved bioavailability, increased resistance time in the body and intended drug targeting to specific site of action. As a result, reduction in required drug dosage is achieved, along with safe delivery of drugs by reducing the risk of non-target cells and tissues from severe side effects of the therapeutic drugs [2]. Nano-particulate delivery system can be attained by maintaining the ideal requirements, i.e., surface chemistry, controlled particle size, enhanced per-

Naman Arora and Vikas Mittal, The Petroleum Institute (part of Khalifa University of Science and Technology), Abu Dhabi, UAE*
**Current address: Bletchington, Wellington County, Australia*

meation, solubility, flexibility, release of drug at specific site and particular activity at predetermined rate and time [3]. Many advanced multifunctional PNPs have been achieved with precise control over size, architecture, shape, surface charge and functionalization. However, there is a need to consider the careful design and control over the targeting properties of these nanocarriers because an improper selection of the chemistry of the building blocks of these nanocarriers and the loaded drug can deeply affect their pharmacokinetics, safety and intracellular fate [4].

This review highlights various techniques used to prepare PNPs (or polymer modified nanoparticles) and to further develop the key features for biomedical applications. It also covers the role of different polymeric nanoparticles employed in the biomedical field.

4.2 Polymer Nanoparticles and Preparation Techniques

Polymer nanoparticles are usually colloidal particles in the size range of 10-1000nm [5,6], having shapes as nanospheres and nanocapsules. Other types of nanocarriers have also been developed such as spheroids colloids, core-shells, micelles, dendrimers and platelets, etc. Nanospheres are defined as matrix particles where complete mass is solid and drug molecules get encapsulated within the particles or can be adsorbed at the surface. Nanocapsules are defined as a vesicular system which jointly acts as a reservoir. It consists of the solid material shell which surrounds the liquid core (water or oil), forming a cavity where substances can be entrapped [7].

As mentioned earlier, for the preparation of the PNPs, either 'polymerization of monomers' or 'dispersion of preformed polymers' methods can be employed. The selection of the preparation technique to generate PNPs is dependent on factors such as area of application, size requirement, type of polymeric system, etc. For example, in case of application in environmental and biomedical fields, polymeric system should be entirely free from any additives or reactants like organic solvents or surfactants. Therefore, considering the requirement of the system, 'rapid expansion of supercritical solution (RESS)' technique can be used because there is no requirement of surfactant or organic solvent to prepare PNPs in such technique.

4.2.1 Polymerization of Monomers

Polymerization method to prepare the PNPs leads to the achievement of desired properties for a specific application at the time of polymerization of the monomers. Various techniques like emulsion, miniemulsion, micro-emulsion and other hetero-phase polymerizations (interfacial and living/controlled radical) for the preparation of PNPs are discussed in this section. In addition, some representative examples of these techniques are also presented, however, these are by no means exhaustive in nature. More detailed reviews on these techniques have been reported elsewhere in literature.

Emulsion Polymerization

Emulsion polymerization is widely used for the production of polymers such as binders for nonwoven fabrics, adhesives, paints, additives for paper, etc. Emulsion polymers are often described as 'products by process' because the properties of the polymers can be determined during polymerization. On the basis of usage of surfactants, it can be categorized as conventional and surfactant free polymerization [8,9].

In conventional emulsion polymerization, surfactant, water, water-soluble initiator and monomer having low water solubility are the main components for the polymerization reaction. In this kind of polymerization, surfactant is dissolved in water until the critical micelle concentration (CMC) is achieved. For polymerization reaction, necessary sites are provided by the interior of the micelles. The polymerization propagates when a growing chain moves into a micelle and reacts with the monomer molecules. To sustain polymerization, monomer molecules move from the monomer droplets into the micelles (or growing particles) as monomer is consumed during polymerization occurring inside the micelles. PNPs are synthesized typically with ~100 nm in size, each containing many polymer chains after the polymerization reaction. For instance, Mittal *et al.* [10] reported polystyrene and surface functionalized polystyrene particles using emulsion polymerization (Figure 4.1). Yoon *et al.* [11] reported the preparation of poly(N-vinylcarbazole) nanoparticles by emulsion polymerization. Cheng *et al.* [12] also reported the SiO_2/poly(methylmethacrylate) (PMMA) composite particles by employing conventional emulsion polymerization.

Surfactant-free emulsion polymerization is an environmental friendly approach to prepare PNPs. The main feature of the techniques is the relatively clean surface of the particles. The main components of surfactant-free emulsion system are water, monomer and

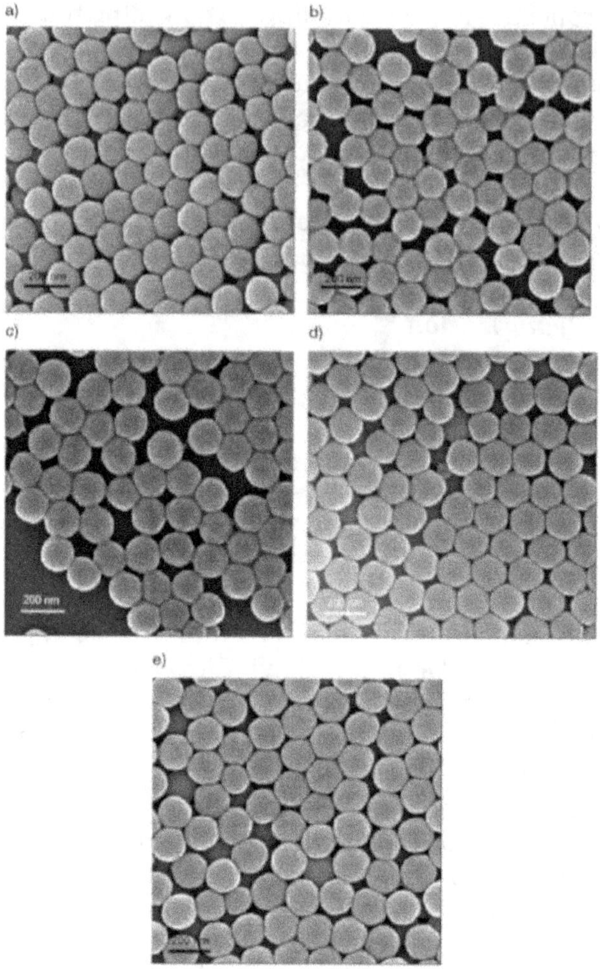

Figure 4.1 Scanning electron micrographs of polystyrene and surface functionalized particles generated using emulsion polymerization. Reproduced from Reference 10 with permission from Wiley.

a water-soluble initiator (peroxides or persulfates). The usage of ionizable initiators or ionic co-monomers stabilizes the PNPs in such polymerization systems. The homogeneous nucleation mechanism is dominant in case of surfactant-free emulsion polymerization [13-15]. Mittal *et al.* [16] utilized surfactant-free emulsion polymerization of styrene to synthesize narrowly dispersed polystyrene latex particles which served as seed for the subsequent surface functionalization processes (Figure 4.2). Chiu *et al.* [17] reported the preparation of

PMMA nanoparticles by applying surfactant-free emulsion polymerization. The polymerization of (methyl methacrylate) was directly initiated by a Cu^{2+}/HSO_3^- redox system. In another study, preparation of monodispersed PMMA nanoparticles with controlled sub-100 nm size have been reported by Camli *et al.* [18].

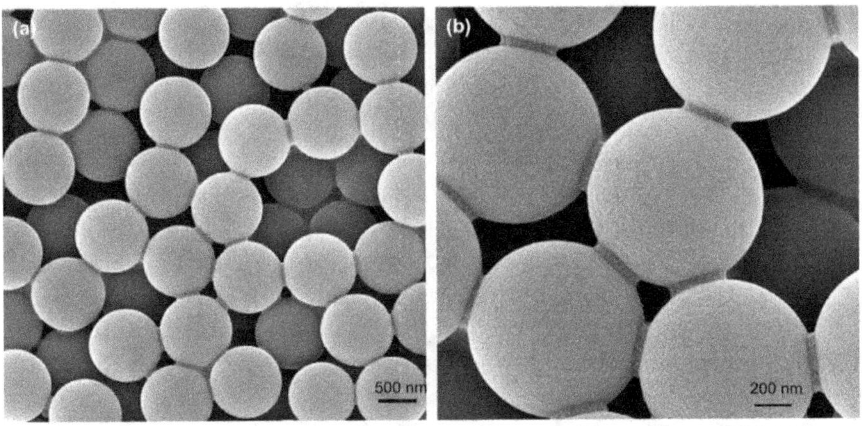

Figure 4.2 Scanning electron micrographs of polystyrene particles generated using surfactant free emulsion polymerization. Reproduced from Reference 16 with permission from Elsevier.

Miniemulsion Polymerization

Miniemulsions are typically stabilized by external agents (surfactants and co-surfactants) and need a high shear to achieve a steady state. Miniemulsions are thermodynamically unstable and have very low polydispersity. The stability period of a miniemulsion varies from hours to months. The typical particle size in these emulsions is 10-30 nm. The main components of the miniemulsion polymerization are monomer, surfactant, initiator, water and co-stabilizer. The utilization of the co-stabilizer (low molecular mass compound) and a high shear device (ultrasound) are the main differences between miniemulsion polymerization and emulsion polymerization. Wu *et al.* [19] reported the preparation of the polysaccharide-coated poly(*n*-butyl cyanoacrylate) nanoparticles by miniemulsion polymerization. In another study, Baruch-Sharon *et al.* [20] reported polychloromethylstyrene nanoparticles via miniemulsion polymerization, where hexadecane was used as co-stabilizer.

Micro-emulsion Polymerization

Colloidal polymer particles with high molar mass polymer chains can be achieved by both emulsion and micro-emulsion polymerization, which make them appear similar. However, these techniques are completely different with respect to kinetics. The commercial application of this method is limited due to its requirement of high ratio of surfactant to monomer. As a result, polymer formulations are dilute and particles are entirely covered with the surfactant molecules. The typical particle size of micro-emulsion is around 30-100 nm. Unlike emulsion and miniemulsion, micro-emulsion is thermodynamically stable. In micro-emulsion polymerization, both particle size and the average number of chains per particle are considered to be relatively lower as compared to emulsion polymerization. Three reaction rate intervals are exhibited by emulsion polymerization but only two are observed in the case of micro-emulsion polymerization. Reddy *et al.* [21] coated multi-walled carbon nanotubes (MWNTs) with the nanospheres of polyaniline by using micro-emulsion polymerization. In the presence of sodium dodecyl sulfate (SDS) (which acts as both dopant and surfactant), aniline polymerization was carried out with MWNTs. Shah *et al.* [22] reported the preparation of the copolymeric nanoparticles of methyl methacrylate and *N*-vinylcaprolactam via free radical polymerization. In this process, redox initiator like hydrogen peroxide and L-ascorbic acid were used in oil-in-water micro-emulsion containing SDS.

Interfacial Polymerization

Interfacial polymerization is one of the well-developed processes for the synthesis of polymer nanoparticles. This method involves step polymerization of two reactive monomers, which are dissolved respectively in two-phases, i.e., continuous phase and dispersed phase. The reaction occurs at the interface of the two liquids [23]. PNPs can be prepared by using interfacial cross-linking reactions such as radical polymerization [24], polyaddition and polycondensation [25,26]. For instance, Torini *et al.* [25] carried out the interfacial polycondensation encapsulation in miniemulsion to generate oily core polyurethane nanocapsules (Figure 4.3). Hexadecane was used as co-stabilizer. The microscopy analysis confirmed smooth external wall and spherical shape of the nanocapsules. Crespy *et al.* [26] reported the synthesis of polymeric hollow nanoreactors of polyuria (Figure 4.4),

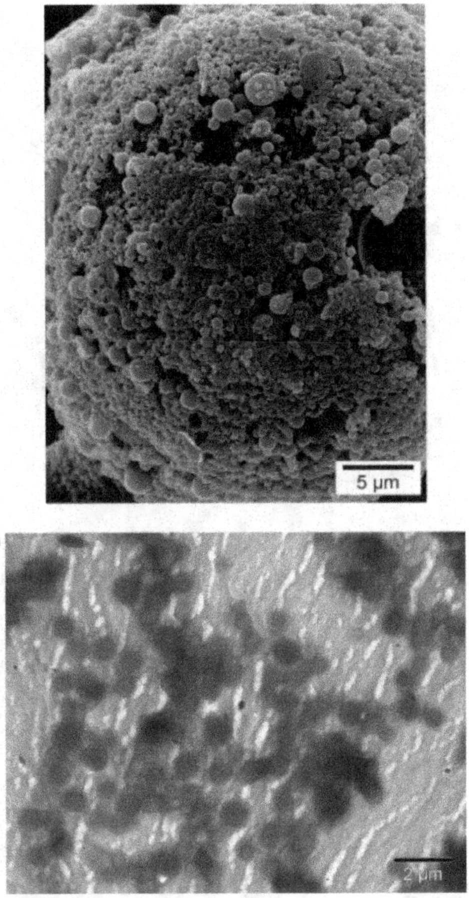

Figure 4.3 (a) Scanning and (b) transmission electron micrographs of polyurethane nanocapsules. Reproduced from Reference 25 with permission from American Chemical Society.

polythiourea and polyurethane by interfacial polycondensation or cross-linking in inverse miniemulsion.

Controlled/Living Radical Polymerization (C/LRP)

The lack of control on macromolecular architecture, molar mass and its distribution, and end functionalities are the primary limitations of radical polymerization. Uncontrollable and fast radical-radical termination reaction causes these limitations. Controlled/living radical polymerization have contributed to green chemistry and evidenced a

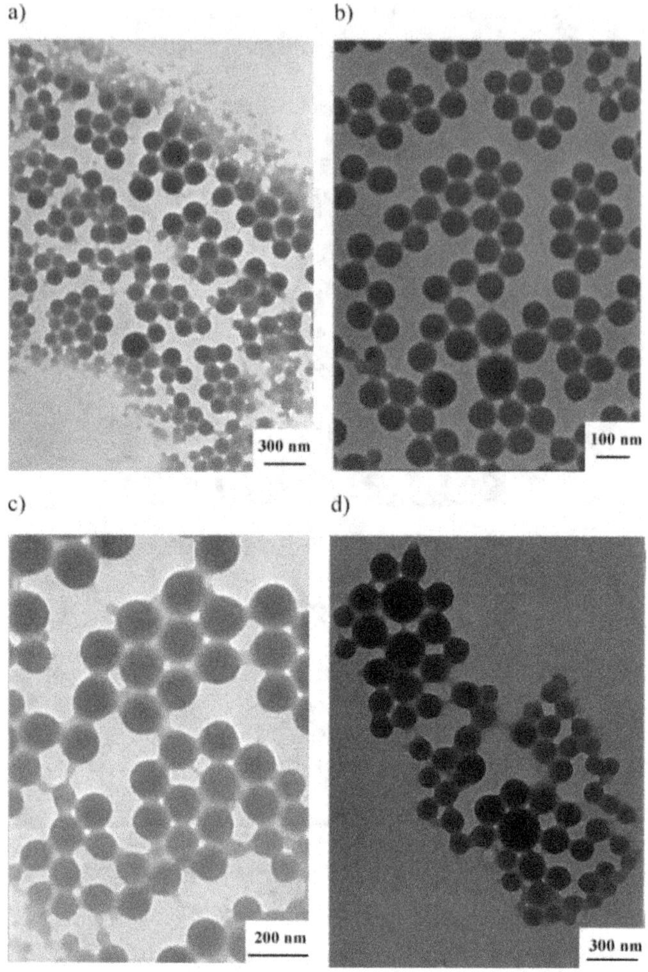

Figure 4.4 Transmission electron micrographs of polyurea nanoparticles. Reproduced from Reference 26 with permission from American Chemical Society.

significant growth in the field of pharmaceuticals and other medical applications. After the implementation of the C/LRP in industrial aqueous dispersed systems (like emulsion polymerization), PNPs with size distribution control and precise particle size have been achieved. Three methods are frequently followed among the available methods of the C/LRP, viz. reversible addition and fragmentation transfer chain polymerization (RAFT), atom transfer radical polymerization (ATRP) and nitroxide-mediated polymerization

(NMP). The size of the PNPs can be controlled by the nature and concentration of the monomer, initiator, emulsion type and control agent. Among these factors, the nature of control agent is a key factor to attain the particle size of the end product. Nicolas *et al.* [27] reported the preparation of nanostructured latex particles by using NMP. Min *et al.* [28] reported the one-pot synthesis of hairy nanoparticle by using ATRP, as shown in the schematic in Figure 4.5.

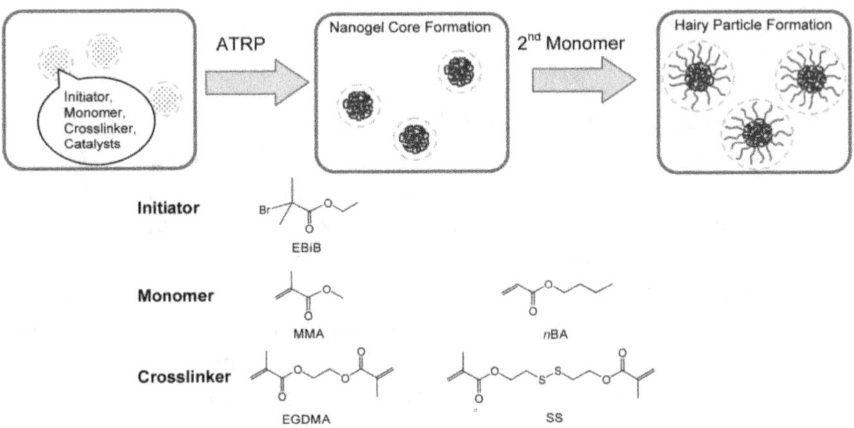

Figure 4.5 Schematic of the one-pot synthesis of hairy nanoparticle by ATRP. Reproduced from Reference 28 with permission from American Chemical Society.

4.2.2 Dispersion of Preformed Polymers

Many methods have been developed to prepare PNPs by performing dispersion of preformed polymers, such as solvent evaporation, nanoprecipitation, salting-out, dialysis and supercritical fluid technology. These methods are described in brief in below sections, with some representative examples.

Solvent Evaporation

PNPs from preformed polymers were firstly developed by solvent evaporation method. In this method, volatile solvents are used to generate polymer solutions and subsequently emulsions are formulated. Solvent for the polymer can be evaporated by diffusion through the continuous phase of the emulsion, which leads to the conversion of the emulsion into a nanoparticle suspension. Thus, high-speed homogenization or ultrasonication is used in this method, which is followed by evaporation of the solvent. Subsequently, ultracentrifugation is employed to collect the solidified nanoparticles which are later washed with distilled water to remove additives like surfactants. Finally, the product is subjected to lyophilization process [1,29]. Bilati *et al.* [30] reported poly(lactide-*co*-glycolide) nanocapsules, prepared by the water-in-oil-in-water solvent evaporation method.

Salting Out

Salting out process is another method to prepare the PNPs, which was first proposed by Bindschaedler *et al.* [31]. Salting out effect separates the water miscible solvent from aqueous solution. This process does not involve the use of chlorinated solvents and surfactants. Polymer solvents entirely miscible with water, e.g., acetone, are used in the formulation of the emulsion. Polymer and drug (to be encapsulated) are initially dissolved in a solvent (like acetone), which is later emulsified into an aqueous gel containing the salting-out agent (electrolytes, such as magnesium chloride or non-electrolytes such as sucrose) and a colloidal stabilizer such as polyvinylpyrrolidone or hydroxyethylcellulose. The formation of nanospheres is induced, when the oil/water emulsion is diluted with enough amount of water to increase the diffusion of acetone into the aqueous phase [29]. The role of the salting out agent is crucial as it decides the encapsulation efficiency for the drug. Later, cross-flow filtration eliminates the solvent and the salting out agent. Zweers *et al.* [32] reported the preparation of biodegradable polyester nanoparticles by employing the salting-out method. The authors observed that the size of spherical monodisperse particles was between 100 and 400 nm.

Nanoprecipitation

Nanoprecipitation is a basic, rapid and reproducible method, which was developed by Fessi *et al.* [33] for the preparation of the PNPs. This method is also known as solvent displacement method. The interfacial deposition of the polymer after the displacement of a semi-polar solvent from a lipophilic solution is the main principle of this method. The main components of the nanoprecipitation formulation are polymer, solvent, non-solvent and stabilizing agent. Polymer nanoparticles can be obtained by the gradual addition of organic phase into the aqueous phase under moderate stirring and vice-versa. Polymer nanoparticles with precise size and narrow distribution are formed instantly during the fast diffusion of the polymer solution into the non-solvent phase.

Dialysis

Dialysis is an effective method for the preparation of narrowly distributed PNPs [34]. In this method, polymer is dissolved in an organic solvent and kept inside a dialysis tube with particular molecular weight cut off. The displacement of the solvent inside the membrane is followed by the progressive aggregation of polymer due to a loss of solubility and formation of homogeneous suspensions of nanoparticles. The mechanism of formation of polymer nanoparticles by dialysis is usually considered to be similar to the nanoprecipitation method of PNP generation.

Supercritical Fluid Technology

The usage of supercritical fluids was encouraged to ensure the environmental friendly preparation of PNPs. This technology is capable to prepare PNPs without any trace of organic solvent, which avoids the disadvantages of the traditional methods. There are two principles to prepare PNPs by using supercritical fluids, rapid expansion of supercritical solution (RESS) and rapid expansion of supercritical solution into liquid solvent (RESOLV).

In RESS, a solution is formed by dissolving a solute into a supercritical fluid, followed by the rapid expansion of the solution across a capillary nozzle into ambient air. Well-dispersed particles are formed after the homogenous nucleation occurs. The homogenous nucleation is activated by the high degree of supersaturation which is accompanied by the rapid pressure reduction in the expansion. RESOLV involves the expansion of the supercritical solution into a liquid solvent, unlike RESS (i.e. ambient air).

In one of the studies, the preparation of poly (heptadecafluorodecyl acrylate) nanoparticles with an average size of less than 50 nm was reported by Meziani *et al.* [35]. The main drawback of RESS technique is that the prime products produced are not nanosized. In RESOLV, the liquid solvent apparently suppresses the particle growth in the expansion jet, thus, helping to achieve primarily nanosized particles.

4.3 Important Characteristics of PNPs Designed for Biomedical Applications

Many efforts have been made to design and develop the polymeric nanocarriers with modified physical, chemical and biological properties. The blood stability, bioavailability of nanoparticles, endocytic pathway, bio-distribution and intracellular distribution are dictated by the size, shape and surface characteristics of nanoparticles. In this section, important features of the polymer nanoparticles for biomedical applications have been discussed.

Stealth Nanocarriers

Interaction of nanoparticles with constituents of biological fluids can be avoided by using this strategy. In this strategy, a dense brush layer of neutral hydrophilic flexible polymers is coated on to the nanoparticles to prevent plasma proteins adsorption and avoid identification by the MPS (mononuclear phagocytic system) and subsequent clearance. It also prolongs the blood circulation time, thereby, allowing the accumulation of the nanoparticles at sites with impaired vasculature *via* the EPR (enhanced permeability and retention) effect. Polyethylene glycol (PEG) is mostly used for coating the nanoparticles. Chemically-active centers can be facilitated by the PEG chains for further functionalization of the nanoparticles. Additionally, aggregation of nanoparticles can be partially prevented by using PEG. By forming a neutral stabilizing interface, it obstructs the non-specific interactions with cells, so the neutral surface charge of the nanoparticles can be maintained. Poly(acrylic acid) (PAA), poly(*N*-vinylpyrrolidone) (PVP), poly(*N*-isopropylacrylamide) and poly(carboxybetaine) are the other hydrophilic polymers which have also been employed to coat the surface of the nanoparticles [4,36]. Figure 4.6 also demonstrates the synthesis methodology for the generation of PEGylated shell cross-linked knedel-like nanoparticles reported by Sun *et al.* [36].

Smart Ingredients

The surface of the core/shell structure of PNPs can be incorporated with specific functionalities which may facilitate their diffusion through many biological barriers and direct their bio-distribution to specific tissues, cells or organelles. Moreover, the responsiveness of these nanostructures to specific stimuli can be modulated and remotely controlled by smart ingredients. Stimuli-responsiveness (such as temperature, pH, oxidation, reduction, light, magnetic field, ultrasound, etc.) allows the controlled-drug release at specific sites or under applied conditions.

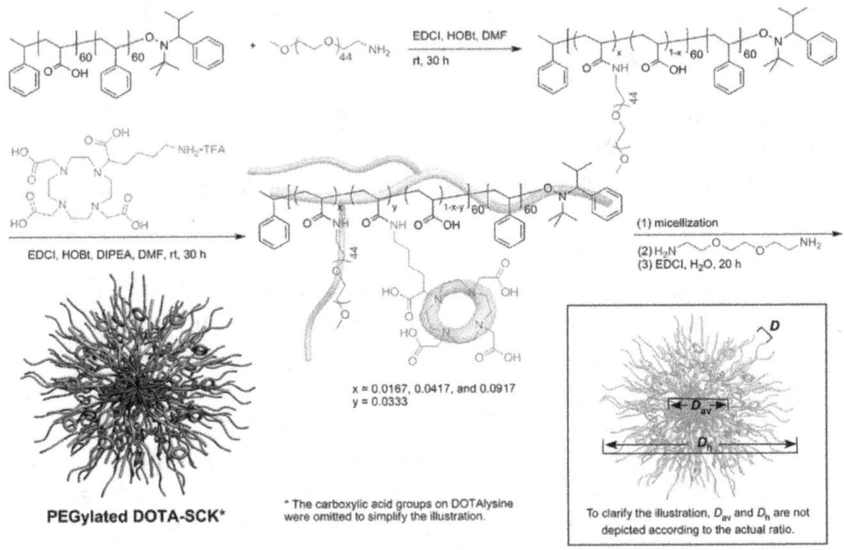

Figure 4.6 Schematic of the synthesis of PEGylated shell cross-linked knedel-like nanoparticles. Reproduced from Reference 36 with permission from American Chemical Society.

Skin and Mucus Permeation Enhancers

Skin and mucus are the external barriers for the delivery of PNPs to the specific target. These may manipulate the properties of the surface and stability of the nanoparticles before they reach the surface of the underlying tissues. Skin comprises of many layers with different thicknesses and structures, epidermis, stratum corneum (SC), dermis and subcutaneous tissues [37]. The main permeability barrier and the rate-limiting step for the delivery application through the skin is SC. The mucosal glands secrete the mucus (viscoelastic hydrogel) to protect the cells. The enzymatic activity, surface area, composition, thickness, pH, permeability and vascularity of mucus vary from one region to another and depend on the disease status. For instance, mucus turns thick and sticky in patients with cystic fibrosis [38]. Proteins, macromolecules, cells, bacteria, lipids, salts and cellular debris are also loaded into mucus gels. A nanoscopic network is formed by many components which contribute together to hinder the transportation of the nanoparticles. For the delivery of nanoparticles at required sites, mucociliary clearance is very important. Additionally, if binding takes place between the surface of nanoparticles and the mucus ingredients, it may lead to destabilization and aggregation of PNPs and displacement of their active load. Hydrolyzation of mucin or destabilization of SC can be achieved by many physical and chemical methods, which then enhance the permeation of delivery carriers through mucus and skin. Modification of nanocarriers with mucolytic, 'mucus dissolving', agents helps in decreasing the mucus viscosity by deteriorating the biopolymers (e.g. proteins) in the mucus mesh. It helps in reducing the interactions between the PNPs and mucus, thus, resulting in increased mobility and permeability of the PNPs. Based on the composition and delivery method of the PNPs, three main routes can be taken by the PNPs to penetrate the skin, such as transappendageal (hair follicles), intercellular (lipid bilayers) and transcorneocyte routes. The intercellular route is the most preferred route for the delivery of the nanocarriers. There is a requirement to incorporate chemical enhancers into the PNPs formulation to disrupt the lipid bilayers and subsequently create a path for the penetration of the nanocarriers [4,37]. Thus, several features can be added to the polymer nanoparticles to overcome different barriers.

Accumulation at a Particular Site

The surface of nanostructured carriers should be modified with a variety of targeting ligands in order to enhance their selective binding to the particular receptor. This approach reduces the exposure of normal cells to the drug and enhances an accumulation at the target sites [39]. Modification of the surface of the nanoparticles with targeting ligands can impact their immunogenicity, bio-distribution and clearance [40,41]. For instance, Figure 4.7 shows the synthesis of functionalized polymer nanoparticles with diagnostic/imaging units and targeting ligands for molecular targeting by controlled polymerization (RAFT) [41]. The results also confirmed the significant importance of controlling the ligand loading so as to achieve effective performance for therapeutic and imaging applications for the developed polymer nanoparticles.

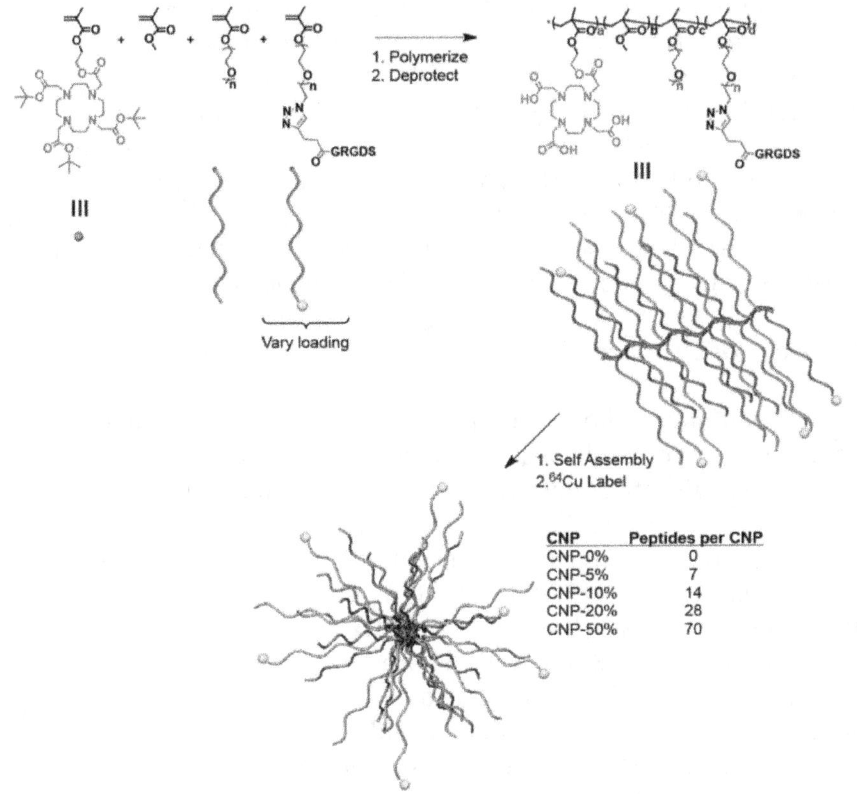

Figure 4.7 Synthesis of functionalized polymer particles. Reproduced from Reference 41 with permission from American Chemical Society.

Stimuli Responsiveness

Many strategies have been used to control the dissociation of nanoparticles and delivery of encapsulated drug at specific sites. There are many factors which can trigger the drug-release such as pH, oxidative or reductive conditions, temperature, light, etc. pH is the most typical stimuli for the purpose of drug delivery as there is a variation of pH between pathological sites and normal physiological sites. pH-sensitive ingredients can be incorporated into the nanoparticles to maintain the functionality of the nanostructured carrier at certain pH values, while it gets destabilized at other pH [42]. Also, endosomolytic polymers, lipids or fusogenic peptides can be incorporated into the formulation of the nanoparticles for endosomal destabilization and further cytoplasmic release. Incorporation of primary or tertiary amines at different ratios leads to the formation of bi-functional endosomolytic polymers [43-45]. Thioketal nanoparticles have also been synthesized by using poly(1,4-phenyleneacetone dimethylene thioketal) polymer for oral delivery of siRNA to inflammation sites present in the intestine [46]. The polymer consisted of thioketal linkages which were stable in acidic, basic and digestive environments, but got deteriorated at the inflammation sites to release its contents in tissues with high levels of reactive oxygen species.

Pathways of Delivery

In order to understand the appropriate pathways for the delivery of the nanostructured vehicles, it is critical to analyze various possibilities such as delivery of nanocarriers through skin, oral, ocular, nasal, pulmonary, buccal, etc. For designing the delivery route, one should be aware about the environmental conditions (like pH, temperature, oxidative or reductive conditions, etc.) at physiological sites. For example, the pH is basic (i.e. 7.8) in the case of eye mucus and acidic in other secretions (i.e. 3.5 to 4.5), whereas the pH of nasal mucus and lungs is neutral [38]. Usually, various PNPs are formulated in such manner that they get dissociated at acidic pH, thus, if these PNPs are delivered through the route which is acidic in nature, it can prove to be unfavorable due to the premature release of the carried drugs. For topical and mucosal delivery, semisolid matrices or scaffolds (like hydrogels or creams) are usually preferred which provide long-term contact with tissues and sustained delivery of encapsulated drugs. Drugs can be directly incorporated into the target organ by using auxiliary devices, such as catheter employed for delivery to the bladder, microneedles for topical delivery and aerosols for pulmonary delivery. Nanostructured carriers can be incorporated at the specific organ by employing these devices which increases the therapeutic benefits and reduces the toxicity to healthy tissues.

Size

Designing the PNPs with specific size, geometry, composition, charge, morphology and surface chemistry can deeply impact the circulation time, clearance, intracellular fate and selective tissue distribution. Aggregation and capillary occlusion is possible with large particles (>1 μm) that are commonly opsonized and accumulate in the liver and spleen. Small nanoparticles (<5 nm) are cleared quickly from the blood through extravasation or renal clearance. Rapid renal clearance is expected for nanoparticles smaller than 10 nm in diameter or for their dissociated polymer chains, if they are below the renal molecular weight cutoff size. Charges of the particle can also impact the renal filtration, large particles with negative charge can be excluded from excretion. Transportation of smaller nanocarriers is more convenient to lymph nodes through lymphatic drainage. PNPs with size (20-100 nm) have the maximum capacity for in vivo applications, due to their prolonged circulation in the blood. Renal and lymphatic clearances can be avoided by the nanoparticles due to their sufficient large size, whereas the size of these PNPs is still small enough to avoid opsonization. Additionally, nanoparticles within this size range are considered to be internalized by cells, as compared to smaller or larger particles [4,47].

Morphology

Different blood circulation time, solubilization capacities, biodistribution, intracellular fate, toxicity and cellular uptake can be exhibited by several morphologies and/or aspect ratios. The shape and dimensions of nanoparticles can impact their behavior both in vitro and in vivo applications. The PNPs mostly exhibit a spherical morphology which is the most usual shape. However, other morphologies are also achievable which depends on nanoparticle formulation and polymer configuration. It is usually observed that in comparison to elongated cylindrical shape, spherical nanoparticles can enter into cells to a greater extent. The PNPs of well-defined morphologies from block copolymers have been reported [48,49]. Both spherical and cylindrical shell crosslinked knedel-like nanoparticles (SCKs) of the similar formulation were conjugated with cell penetrating peptides (CPP) to study the impact of different shapes of nanoparticles on the cellular uptake [50]. Generally, with higher levels of CPP-functionalization, high cellular uptake was noticed and the effects of shape and size were crucial. For instance, the spherical SCKs with small size were quickly internalized by cells as compared to longer cylindrical nanoparticles. Conversely, cells took up the folate-functionalized spherical SCKs in lesser quantity, whereas greater quantity of folate-functionalized cylinders was taken up by the cells [51]. Here, cylinders had high potential to generate multivalent interactions (multiple receptors binding) owing to their longer dimensions and, hence, exhibited higher cellular uptake. In another study, the particle geometry was also confirmed to be an essential parameter which influenced the cellular uptake of the PNPs obtained through DeSimone's PRINT® (particle replication in non-wetting templates) technology [52]. Similar cylindrical shaped nanoparticles with different aspect ratios (i.e. 3 and 1) exhibited different cellular uptakes. Nanoparticles with higher aspect ratio were taken up more quickly. In addition, it is important to take into account other factors such as the transformation of morphologies of nanoparticles from one shape to another. The transformation between different morphologies can be controlled by adjusting polymer composition or by altering the ionic strength and pH of the solution or by addition of organic solvents [4,49].

Surface Chemistry

The surface of nanoparticles can be modified with different functional groups to alter their hydrophilicity, bio-distribution, surface charge, immunogenicity, in vivo circulation and intracellular bioavailability. At different stages, the surface charges affect the characteristics of nanoparticles such as blood stability, physical stability, cellular uptake, interactions with extracellular matrix, skin and mucus penetration and endocytosis. In blood, presence of excessive positive charge on the nanocarrier may result in opsonization, capillary occlusion, aggregation of particles and clearance with possible blood vessels. Interactions of positively- or negatively-charged nanoparticles with constituents of the extracellular matrix obstruct their penetration and may result in their clearance. At the cellular level, the surface charge is directly proportional to the cellular binding and association, due to the negatively-charged proteoglycans on the cell surface. With increasing zeta potential values, the uptake of nanoparticles generally increases. However, toxicity can be induced by excessive positive charge and immunological reactions can also be triggered. Several endocytic mechanisms can be followed by nanoparticles with different surface chemistries, exhibiting different intracellular trafficking pathways. The endocytosis process depends on the formulation, shape, size, charge and geometry of nanoparticles as well as cell density/type. It is generally complex to identify the key imparting factor because variation in any of the factors can impact the characteristics of other factors. For example, if positive/negative charge ratio of nanoparticles is increased, the morphology, stability and size of nanoparticles can be affected [4,53].

4.4 Various Polymers Based/Modified Nanoparticles for Biomedical Applications

Acharya *et al.* [54] reported the preparation of rapamycin loaded poly(lactide-co-glycolide) (PLGA) nanoparticles by using oil-in-water (O/W) single emulsion solvent evaporation technique. The rapamycin loaded PNPs were surface conjugated with antibodies to epidermal growth factor receptor (EGFR-Rapa-PNPs). In order to analyze the anticancer capability of the composition, in vitro cytotoxicity of EGFR-Rapa-PNPs was evaluated on malignant MCF 7 breast cancer cell lines. Overall, the results confirmed that spherical and monodispersed (~100 nm) PLGA nanoparticles were successfully loaded with rapamycin, having a loading capacity of 160 ± 5.8 μg drug per mg of nanoparticle. Liu *et al.* [55] employed the (O/W) single emulsion technique to prepare the doxorubicin loaded poly(ω-pentadecalactone-co-ρ-dioxanone) poly(PDL-co-DO) nanoparticles. For a time period of 20-60 days, the drug loaded PNPs exhibited controlled and continuous release of the drug. Liu *et al.* [56] also reported biodegradable copolyester carriers for camptothecin delivery, where camptothecin loaded poly(ω-pentadecalactone-co-butyleneco-succinate) (PPBS) nanoparticles were synthesized via (O/W) single emulsion method, exhibiting a significant encapsulation capacity of up to 96%. Camptothecin loaded PBBS nanoparticles were observed to be spherical in shape with a smooth surface morphology and had a zeta potential of around -10mV. Cirpanli *et al.* [57] reported the preparation of poly(lactide-co-glycolide) or poly(caprolactone) (PLGA or PCL) nanoparticles and amphiphilic cyclodextrin nanoparticles via nanoprecipitation method for the delivery of camptothecin for cancer therapy. The authors observed that amphiphilic cyclodextrin nanoparticles were more promising for an effective delivery of camptothecin as compared to the PNPs.

In another study, Zhang *et al.* [58] synthesized CPβCDs-insulin-loaded alginate/chitosan nanoparticles in a modified three-step procedure for an effective release of insulin and its protection against degradation in simulated gastric fluid (SGF). Insulin was complexed with cationic β-cyclodextrin polymer (CPβCDs) via both electrostatic attraction and inclusion, which were further encapsulated into alginate/chitosan nanoparticles. It was observed that the PNPs had mean size lower than 350 nm with loading capacity of insulin up to 87%. Potineni *et al.* [59] prepared poly(ethylene oxide) (PEO)-modified poly(β-amino ester) nanoparticles by using solvent displacement method in presence of pluronic F-108 for tumor-selective paclitaxel delivery. In another study, Bilensoy *et al.* [60] employed the nanoprecipitation technique to synthesize amphiphilic cyclodextrin nanocapsules (~500 nm) and nanospheres (~150 nm) which were loaded with paclitaxel. Cyclodextrin nanoparticles loaded with paclitaxel were observed to be promising alternative for cancer chemotherapy with reduced side effects, along with increased bioavailability of the drug. Bilensoy *et al.* [61] also reported tamoxifen citrate loaded β-cyclodextrin nanoparticles by using the nanoprecipitation technique.

Shenoy *et al.* [62] used solvent displacement method to prepare poly(ethylene oxide)-modified poly(caprolactone) (PEO-PCL) nanoparticles which were capable of targeted delivery of tamoxifen in breast cancer cells. In another study, Hamoudeh *et al.* [63] used (O/W) single emulsion solvent evaporation technique to prepare radioactive dirhenium decacarbonyl-loaded poly(L-lactic acid) (PLLA) nanoparticles for radionuclide intra-tumoral therapy. Hu *et al.* [64] also prepared Fe_3O_4/PLLA composite nanomaterials with an average size of 200 nm, by employing (O/W) emulsion solvent evaporation method. Tamoxifen was encapsulated into PLLA matrix to evaluate its in-vitro anticancer activity against MCF-7 breast cancer cells. The nanocomposites were also used as contrast agents for magnetic resonance imagining due to their superparamagnetic property.

Arias *et al.* [65] synthesized carbonyl iron/poly(butyl cyanoacrylate) core-shell nanoparticles for loading and delivery of 5-fluorouracil and ftorafur. Similarly, Arias *et al.* [66] also reported the preparation of tegafur (antitumor) loaded polymeric colloidal nanospheres of poly(alkyl cyanoacrylate) with a magnetic core, via emulsion polymerization method. Furthermore, the authors reported the delivery of 5-fluorouracil which was loaded onto core-shell colloidal nanospheres, consisting of a magnetic core and biodegradable polymeric shell [67]. Wang *et al.* [68] also synthesized magnetic chitosan-5-fluorouracil nanoparticles for targeting drug delivery system. The microscopy analysis revealed that the magnetic PNPs were spherical in shape and the size was observed within a range of 50- 60 nm. The drug loading rate and drug package rate of the magnetic PNPs were 13.4% and 25.8%, respectively. Arias *et al.* [69-74] also reported polymer nanoparticles based drug carriers such as i) ftorafur and diclofenac sodium loaded poly(ε-caprolactone) nanoparticles via interfacial polymer deposition method for the application of drug delivery, ii) 5-fluorouracil loaded iron/ethylcellulose (core/shell) nanoparticles using emulsion solvent evaporation method for cancer targeting application, iii) ftorafur loaded iron oxide/chitosan (core/shell) nanoparticles via coacervation method for drug delivery system, iv) diclofenac sodium loaded iron/ethylcellulose (core/shell) nanoparticles by using emulsion solvent evaporation method for the treatment of arthritis, v) spherical colloidal nanoparticles comprising of a carbonyl iron magnetic core and a biodegradable poly(butylcyanoacrylate) (PBCA) shell prepared via emulsion polymerization method for magnetic carrier technology, vi) magnetite/poly(ethyl-2-cyanoacrylate) nanoparticles via emulsion polymerization method for drug delivery application.

Balakrishnan *et al.* [75] reported the preparation of polymer coated magnetic nanoparticles for cancer treatment and targeted drug delivery. For the preparation of NPs, $FeCl_2$ solution was chemically reduced with sodium borohydride and subsequently coated with amine-terminated polyethylene glycol (a-PEG). In another study, Ma *et al.* [76] prepared superparamagnetic iron oxide/alginate nanoparticles (SPIO-alginate) via modified co-precipitation method. The nanoparticles were employed as magnetic resonance (MR) agents for the detection of liver cancer. Pharmacokinetics studies and distribution of SPIO-alginate nanoparticles in rats suggested that the MR agents accumulated in the liver and spleen due to presence of phagocytosing cells in these organs. Kumagai *et al.* [77] reported the synthesis of PEG-coated iron oxide nanoparticles (PEG- β-FeOOH) by hydrolysis of $FeCl_3{\cdot}6H_2O$ in water. Subsequently, the nanoparticles were treated with poly(ethylene glycol)-poly(aspartic acid) (PEG-PAsp) block copolymer. FT-IR analysis confirmed that PEG-PAsp molecules were multi-valently bound to the surface of the iron oxide nanoparticles through the coordination between the carboxylic acid residues present in the PAsp segment of the block copolymer and Fe on the surface of the β-FeOOH nanoparticles. Tumor-carrying mice were used in in-vivo magnetic resonance imaging (MRI) experiments. The results exhibited that the PEG coated nanoparticles have the potential to be used as tumor-selective MRI contrast agents.

Fan *et al.* [78] employed solvent-free atom transfer radical polymerization to prepare the poly(poly(ethyleneglycol) monomethacrylate)-grafted magnetic nanoparticles (P(PEGMA)-grafted MNPs) for the potential application in MRI of specific biotargets. Fourier transform infrared spectroscopy analysis confirmed the successful grafting of P(PEGMA) on the surface of MNPs. It was noticed that the FTIR spectrum of P(PEGMA)-grafted MNPs displayed the strong bands at 1110 cm^{-1}, 1726 cm^{-1} and 2900 cm^{-1}, which were attributed to the C-O-C ether stretching, C=O stretching and CH_2 stretching, respectively. These absorption bands confirmed the modification of the MNPs with PEGMA. Furthermore, the absence of C=C stretching vibration at 1630 cm^{-1}, which represents the characteristic band of PEGMA monomer, evidenced the successful polymerization of PEGMA monomer onto the MNPs. Ren *et al.* [79] reported the preparation of the Fe_3O_4/PLA-PEG (poly-D,L-lactide-co-polyethylene glycol) magnetic nanoparticles via phase separation method for the drug targeting and drug controlled release applications. The magnetic nanoparticles were observed to be spherical in structure with a size range of 100-200 nm. In another study, Venkatraman *et al.* [80] employed solvent evaporation and solvent dialysis method to synthesize micelle-like nanoparticles of PLA-PEG-PLA (poly (lactic acid)-poly (ethylene glycol)-poly (lactic acid)) triblock copolymer for the encapsulation of paclitaxel and 5-fluorouracil. Both anti-cancer drugs (paclitaxel and 5-fluorouracil) were successfully loaded into micellar nanoparticles for efficient drug delivery. Another system of triblock copolymer based nanoparticles was reported by Matsumoto *et al.* [81]. In this work, progesterone was loaded onto poly(L-lactide)-poly(ethylene-glycol)-poly(L-lactide) (PLA-PEG-PLA) copolymer nanoparticles for the efficient and controlled released delivery. The encapsulation efficiency for drug was around 70% and the mean diameter of the PNPs was less than 335 nm.

Lopera *et al.* [82] reported the preparation of magnetite/poly (lactic acid) (PLA) core/shell nanoparticles by using water-in-oil (W/O) double emulsion method for use in magnetically targeted biomedical applications. Figure 4.8 also demonstrates the transmission electron images of magnetite, PLA and composite nanoparticles. Nicoli *et al.* [83] reported triptorelin loaded poly(lactide-c--glycolide) (PLGA) nanospheres, prepared by using double emulsion solvent evaporation method and subsequently used for transdermal iontophoretic administration. In another study, Redhead *et al.* [84] synthesized the rose bengal loaded PLGA nanoparticles by applying interfacial deposition method for drug delivery application. After the incorporation of rose bengal drug into the PNPs, the surface of the nanoparticles was further modified with poloxamer 407 and poloxamine 908 for achieving the steric stabilizing layer of PEG on the surface. Yang *et al.* [85] also synthesized magnetic poly(ethyl-2-cyanoacrylate) (PECA) nanoparticles by using interfacial polymerization method. Anti-cancer drugs (like cisplatin and gemcitabine) were loaded into NPs for the application of targeted drug delivery.

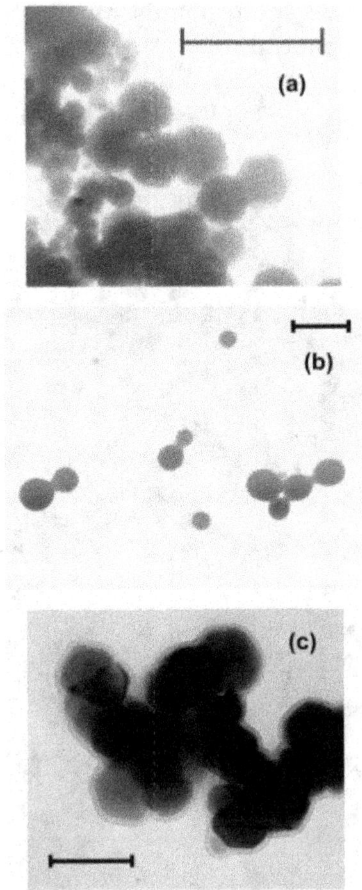

Figure 4.8 Transmission electron images of (a) magnetite, (b) PLA, and (c) composite particles (bar lengths: 500 nm (a), 4 μm (b), and 200 nm (c)). Reproduced from Reference 82 with permission from American Chemical Society.

Chatterjee *et al.* [86] reported the synthesis of polyethylene magnetic nanoparticles by using non-solvent-temperature-induced crystallization coupled with ultrasonication. These MNPs were effectively coated with ligands (i.e. avidin) and further used for biomedical application. In another study, Yuan *et al.* [87] used poly[2-(methacryloyloxy) ethyl phosphorylcholine]-block-(glycerol monomethacrylate)] ($PMPC_{30}$-$PGMA_{30}$) double-hydrophilic diblock copolymers to prepare sterically stabilized magnetic nanoparticles. The sterically stabilized sols were used as improved contrast agents for application in MRI. Figure 4.9 shows the transmission electron image of the copolymer stabilized magnetite sol.

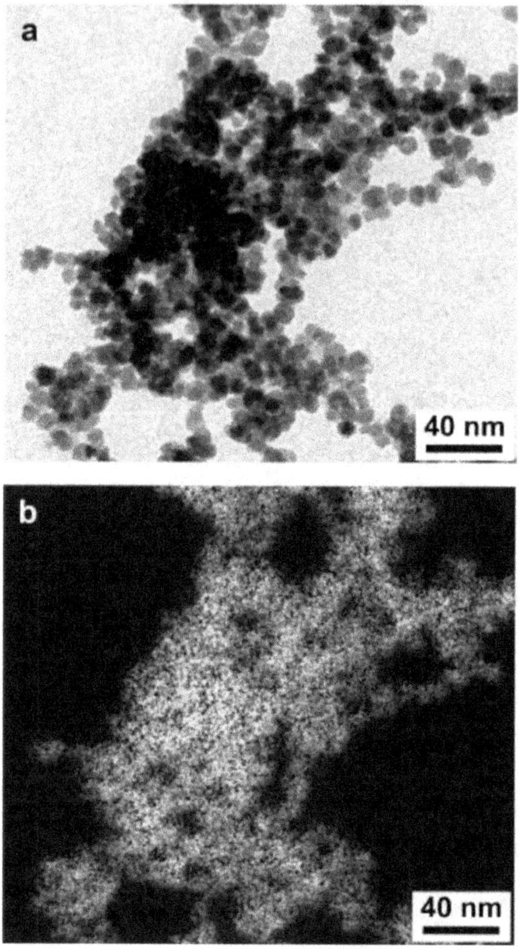

Figure 4.9 Transmission electron images for the $PMPC_{30}$-$PGMA_{30}$ stabilized magnetite sol: (a) conventional transmission electron image; (b) phosphorus map generated using electron spectroscopy imaging using transmission electron microscopy. Reproduced from Reference 87 with permission from American Chemical Society.

Aurich *et al.* [88] synthesized polyaspartate coated magnetite nanoparticles for the application of magnetic immunoassays and magnetic drug targeting. Chitosan modified magnetite nanoparticles were also synthesized by Kim *et al.* [89] for hyperthermia application.

The viability of L929 cells was observed to be 93.67% in the chitosan coated MNPs during cytotoxic test.

Poly-D,L-lactide-co-glycolide (PLGA) nanoparticles have been widely used for the preparation of nano-medicines because these particles hydrolyze in the body to yield biodegradable metabolite monomers, lactic acid and glycolic acid. Thus, various drugs have been encapsulated within PLGA nanoparticles, such as dexamethasone [90], thymopentin [91], 2-aminochromone [92], docetaxel [93], xanthones [94], 9-nitrocamptothecin [95], estradiol [96], paclitaxel [97], taxol [98] and etanidazole [99]. Poly-ε-caprolactone (PCL) deteriorates by hydrolysis of its ester linkages in physiological conditions, thus, has also been widely used for drug delivery applications. PCL based nanoparticles are mostly prepared by techniques such as nanoprecipitation, solvent evaporation and solvent displacement. Immobilization of many drugs has been achieved on the PCL nanoparticles such as clonazepam [100], saquinavir [101], docetaxel [102] and vinblastine [103]. Polylactic acid (PLA), being a biocompatible and biodegradable material, goes through scission in the body to monomeric units of lactic acid as a natural intermediate in carbohydrate metabolism. PLA based nanoparticles can be usually prepared by solvent evaporation, solvent displacement, salting out and solvent diffusion methods. These PNPs have also been loaded with various drugs for effective drug delivery such as hemoglobin [104], dexamethasone [105], ellagic acid [106], protein-C [107], zidovudine [108], oridonin [109], neurotoxin-I [110] and savoxepine [111].

The various studies reporting the development of polymeric nanoparticles (or polymer modified nanoparticles) for biomedical applications have also been summarized in Table 4.1.

4.5 Conclusions

The review focuses on the general description about the various methods to prepare polymer nanoparticles. Insights into the key features of the polymeric nanoparticles needed to be incorporated so as to fulfil the requirements for biomedical applications have also been provided. The review also presents recent studies reporting the development of polymer nanoparticles (or polymer modified nanoparticles) for these applications. Many studies demonstrating the use of such nanomaterials in real life conditions have been included. Overall, polymer (or polymer modified) nanoparticles are unique materials with large potential of useful applications in the medical field.

Table 4.1 Various polymer nanoparticles (or polymer modified nanoparticles) for biomedical applications

Material	Preparation method	Application	Encapsulated drugs or agents	Ref.
PLGA PNPs	Emulsion solvent evaporation	Breast cancer therapy	Rapamycin	54
(PDL-co-DO) copolyesters based PNPs	Oil-in-water single emulsion	Drug delivery	Doxorubicin	55
PPBS copolyesters based PNPs	Oil-in-water single emulsion	Camptothecin delivery for cancer therapy	Camptothecin	56
PLGA or PCL PNPs	Nanoprecipitation	Camptothecin delivery for cancer therapy	Camptothecin	57
CPβCDs-insulin-loaded alginate/chitosan nanoparticles	-	Delivery and protection of insulin in SGF	Insulin	58
PEO-modified poly (β-amino ester) PNPs	Solvent displacement	Tumor-selective paclitaxel delivery	Paclitaxel	59
Amphiphilic β-cyclodextrin PNPs	Nanoprecipitation	cancer chemotherapy	Paclitaxel	60
β-cyclodextrin PNPs	Nanoprecipitation	Show cytotoxicity against MCF-7 cells	Tamoxifen citrate	61
(PEO-PCL) PNPs	Solvent displacement	Breast cancer therapy	Tamoxifen	62
PLLA PNPs	(O/W) single emulsion solvent evaporation	Radionuclide intratumoral therapy	Dirhenium decacarbonyl	63

Fe_3O_4/PLLA Poly-nanocomposites	(O/W) emulsion solvent evaporation method	Magnetic drug targeting and controlled release	Tamoxifen	64
Carbonyl iron/poly(butyl cyanoacrylate) core-shell PNPs	Emulsion polymerization	Magnetic drug delivery for cancer treatment	5-Fluorouracil and Ftorafur	65
Magnetite/poly(alkyl cyanoacrylate) core-shell PNPs	Emulsion polymerization	Magnetic drug delivery for cancer treatment	Tegafur	66
Magnetite/poly(alkyl cyanoacrylate) core-shell PNPs	Emulsion polymerization	Drug delivery	5-Fluorouracil	67
Magnetic chitosan PNPs	-	Targeting drug delivery	5-Fluorouracil	68
Poly (ε-caprolactone) PNPs	Interfacial polymer disposition	Drug delivery	Ftorafur and Diclofenac sodium	69
Iron/ethylcellulose (core/shell) PNPs	Emulsion solvent evaporation	Cancer targeting	5-Fluorouracil	70
Iron oxide/chitosan (core/shell) PNPs	Coacervation method	Drug delivery system for cancer	Ftorafur	71
Iron/ethylcellulose (core/shell) PNPs	Emulsion solvent evaporation	Arthritis treatment	Diclofenac sodium	72
Carbonyl iron/PBCA core-shell PNPs	Emulsion polymerization	Magnetic carrier technology	-	73
Magnetite/poly (ethyl-2-cyanoacrylate) PNPs	Emulsion polymerization	Drug delivery system	-	74

Magnetic (a-PEG) PNPs	-	Cancer treatment	-	75
SPIO-alginate nanocomposites	Modified co-precipitation method	MR contrast agents for liver imaging	-	76
(PEG-coated- β-FeOOH) PNPs	Fuchs-Farthing method followed by stirring with $FeCl_3 \cdot 6H_2O$ aqueous solution	MR contrast agents for cancer imaging	-	77
(P(PEGMA)-grafted MNPs)	solvent-free atom transfer radical polymerization	MRI of specific biotargets	-	78
Fe_3O_4/PLA-PEG MNPs	Phase separation method	Drug targeting	-	79
PLA-PEG-PLA triblock copolymer based PNPs	Solvent evaporation and solvent dialysis	Drug delivery	5-Fluorouracil and Paclitaxel	80
PLA-PEG-PLA triblock copolymer based PNPs	Solvent evaporation	Controlled drug delivery	Progesterone	81
Magnetite/(PLA) core/shell PNPs	Double emulsion solvent evaporation	Magnetic targeting for biomedical applications	-	82
PLGA nanospheres	Double emulsion solvent evaporation	Transdermal delivery	Triptorelin	83
PLGA PNPs	Interfacial deposition method	Drug targeting	Rose bengal	84
PECA MNPs	Interfacial polymerization	Targeted delivery	Cisplatin and Gemcitabine	85

Polyethylene MNPs	Non-solvent-temperature-induced crystallization coupled with ultrasonication	-	Avidin	86
($PMPC_{30}$-$PGMA_{30}$) diblock copolymer based MNPs	Co-precipitation method	Magnetic resonance imaging	-	87
Polyaspartate coated MNPs	Co-precipitation method	Magnetic immunoassays or magnetic drug targeting	-	88
Chitosan coated MNPs	Co-precipitation method	Hyperthermia	-	89
PLGA PNPs	Solvent Evaporation; Double emulsion solvent evaporation; Solvent Evaporation; Emulsion solvent diffusion; Solvent displacement; Nanoprecipitation; Emulsion diffusion; Interfacial deposition; Solvent evaporation; Emulsification solvent evaporative	Drug delivery	Dexamethasone; Thymopentin; 2-Aminochromone; Docetaxel; Xanthones; 9-Nitrocamptothecin; Estradiol; Paclitaxel; Taxol; Etanidazole	90-99

Poly-ε-caprolactone (PCL) PNPs	Solvent evaporation; Solvent displacement; Nanoprecipitation; Emulsion method	Drug delivery	Clonazepam; Saquinavir; Docetaxel Vinblastine	100-103
Polylactic acid (PLA) PNPs	Double emulsion method; Spray drying; Emulsion diffusion evaporation; Double emulsion method; Solvent evaporation; Modified emulsion solvent diffusion; Double emulsion method; Salting out	Drug delivery	Hemoglobin; Dexamethasone; Ellagic acid; Protein-C; Zidovudine; Oridonin; Neurotoxin-I; Savoxepine	104-111

References

1. Rao, J. P., and Geckeler, K. E. (2011) Polymer nanoparticles: preparation techniques and size-control parameters. *Progress in Polymer Science*, **36**, 887-913.
2. Mudshinge, S. R., Deore, A. B., Patil, S., and Bhalgat, C. M. (2011) Nanoparticles: emerging carriers for drug delivery. *Saudi Pharmaceutical Journal*, **19**, 129-41.
3. Bennet, D., and Kim, S. (2014) Polymer nanoparticles for smart drug delivery. In: *Application of Nanotechnology in Drug Delivery*, Sezer, A. D. (ed.), IntechOpen, UK, DOI: 10.5772/58422.
4. Elsabahy, M., and Wooley, K. L. (2012) Design of polymeric nanoparticles for biomedical delivery applications. *Chemical Society Reviews*, **41**, 2545-61.
5. Kreuter, J. (1994) Nanoparticles. In: *Colloidal Drug Delivery Systems*,

Kreuter, J. (ed.), CRC Press, USA, Chapter 5.

6. Couvreur, P. (1987) Polyalkylcyanoacrylates as colloidal drug carriers. *Critical Reviews in Therapeutic Drug Carrier Systems*, **5**, 1-20.
7. Couvreur, P., Dubernet, C., and Puisieux, F. (1995) Controlled drug delivery with nanoparticles: current possibilities and future trends. *European Journal of Pharmaceutics and Biopharmaceutics*, **41**, 2-13.
8. Thickett, S. C., and Gilbert, R. G. (2007) Emulsion polymerization: state of the art in kinetics and mechanisms. *Polymer*, **48**, 6965-6991.
9. Asua, J. M. (2004) Emulsion polymerization: from fundamental mechanisms to process developments. *Journal of Polymer Science, Part A: Polymer Chemistry*, **42**, 1025-1041.
10. Mittal, V., Matsko, N. B., Butte, A., and Morbidelli, M. (2008) PNIPAAM grafted polymeric monoliths synthesized by the reactive gelation process and their swelling/deswelling characteristics. *Macromolecular Reaction Engineering*, **2**(3), 215-221.
11. Yoon, S.-J., Chun, H., Lee, M.-S., and Kim, N. (2009) Preparation of poly (N-vinylcarbazole) (PVK) nanoparticles by emulsion polymerization and PVK hollow particles. *Synthetic Metals*, **159**, 518-522.
12. Cheng, X., Chen, M., Zhou, S., and Wu, L. (2006) Preparation of SiO_2/PMMA composite particles via conventional emulsion polymerization. *Journal of Polymer Science, Part A: Polymer Chemistry*, **44**, 3807-3816.
13. Goodall, A., Wilkinson, M., and Hearn, J. (1977) Mechanism of emulsion polymerization of styrene in soap-free systems. *Journal of Polymer Science: Polymer Chemistry Edition*, **15**, 2193-2218.
14. Fitch, R. M., and Tsai, C. (1971) Particle formation in polymer colloids, III: Prediction of the number of particles by a homogeneous nucleation theory. *Polymer Colloids*, 73-102.
15. Hansen, F., and Ugelstad, J. (1979) Particle nucleation in emulsion polymerization. III. Nucleation in systems with anionic emulsifier investigated by seeded and unseeded polymerization. *Journal of Polymer Science: Polymer Chemistry Edition*, **17**, 3047-3067.
16. Mittal, V., Matsko, N. B., Butte, A., and Morbidelli, M. (2007) Functionalized polystyrene latex particles as substrates for ATRP: Surface and colloidal characterization. *Polymer*, **48**(10), 2806-2817.
17. Chiu, T. P., and Don, T. M. (2008) Synthesis and characterization of poly (methyl methacrylate) nanoparticles by emulsifier-free emulsion polymerization with a redox-initiated system. *Journal of Applied Polymer Science*, **109**, 3622-3630.
18. Camli, S. T., Buyukserin, F., Balci, O., and Budak, G. G. (2010) Size controlled synthesis of sub-100nm monodisperse poly (methylmethacrylate) nanoparticles using surfactant-free emulsion polymerization. *Journal of Colloid and Interface Science*, **344**, 528-532.
19. Wu, M., Dellacherie, E., Durand, A., and Marie, E. (2009) Poly (n-bu-

tyl cyanoacrylate) nanoparticles via miniemulsion polymerization (1): Dextran-based surfactants. *Colloids and Surfaces B: Biointerfaces*, **69**, 141-146.

20. Baruch-Sharon, S., and Margel, S. (2010) Synthesis and characterization of polychloromethylstyrene nanoparticles of narrow size distribution by emulsion and miniemulsion polymerization processes. *Colloid and Polymer Science*, **288**, 869-877.
21. Reddy, K. R., Sin, B. C., Yoo, C. H., Sohn, D., and Lee, Y. (2009) Coating of multiwalled carbon nanotubes with polymer nanospheres through microemulsion polymerization. *Journal of Colloid and Interface Science*, **340**, 160-165.
22. Shah, S., Pal, A., Gude, R., and Devi, S. (2010) Synthesis and characterization of thermo-responsive copolymeric nanoparticles of poly (methyl methacrylate-co-N-vinylcaprolactam). *European Polymer Journal*, **46**, 958-967.
23. Karode, S., Kulkarni, S., Suresh, A., and Mashelkar, R. (1998) New insights into kinetics and thermodynamics of interfacial polymerization. *Chemical Engineering Science*, **53**, 2649-2663.
24. Sarkar, D., El-Khoury, J., Lopina, S. T., and Hu, J. (2005) An effective method for preparing polymer nanocapsules with hydrophobic acrylic shell and hydrophilic interior by inverse emulsion radical polymerization. *Macromolecules*, **38**, 8603-8605.
25. Torini, L., Argillier, J., and Zydowicz, N. (2005) Interfacial polycondensation encapsulation in miniemulsion. *Macromolecules*, **38**, 3225-3236.
26. Crespy, D., Stark, M., Hoffmann-Richter, C., Ziener, U., and Landfester, K. (2007) Polymeric nanoreactors for hydrophilic reagents synthesized by interfacial polycondensation on miniemulsion droplets. *Macromolecules*, **40**, 3122-3135.
27. Nicolas, J., Ruzette, A.-V., Farcet, C., Gerard, P., Magnet, S., and Charleux, B. (2007) Nanostructured latex particles synthesized by nitroxide-mediated controlled/living free-radical polymerization in emulsion. *Polymer*, **48**, 7029-7040.
28. Min, K., Gao, H., Yoon, J. A., Wu, W., Kowalewski, T., and Matyjaszewski, K. (2009) One-pot synthesis of hairy nanoparticles by emulsion ATRP. *Macromolecules*, **42**, 1597-1603.
29. Reis, C. P., Neufeld, R. J., Ribeiro, A. J., and Veiga, F. (2006) Nanoencapsulation I. Methods for preparation of drug-loaded polymeric nanoparticles. *Nanomedicine: Nanotechnology, Biology and Medicine*, **2**, 8-21.
30. Bilati, U., Allemann, E., and Doelker, E. (2003) Sonication parameters for the preparation of biodegradable nanocapsules of controlled size by the double emulsion method. *Pharmaceutical Development and Technology*, **8**(1), 1-9.
31. Bindschaedler, C., Gurny, R., and Doelker, E. (1990) Process for Pre-

paring a Powder of Water-insoluble Polymer which can be Redispersed in a Liquid Phase, the Resulting Powder and Utilization Thereof. US patent US4968350A.

32. Zweers, M. L., Grijpma, D. W., Engbers, G. H., and Feijen, J. (2003) The preparation of monodisperse biodegradable polyester nanoparticles with a controlled size. *Journal of Biomedical Materials Research Part B: Applied Biomaterials*, **66,** 559-566.
33. Fessi, H., Puisieux, F., Devissaguet, J. P., Ammoury, N., and Benita, S. (1989) Nanocapsule formation by interfacial polymer deposition following solvent displacement. *International Journal of Pharmaceutics*, **55**, R1-R4.
34. Jeon, H.-J., Jeong, Y.-I., Jang, M.-K., Park, Y.-H., and Nah, J.-W. (2000) Effect of solvent on the preparation of surfactant-free poly (DL-lactide-co-glycolide) nanoparticles and norfloxacin release characteristics. *International Journal of Pharmaceutics*, **207**, 99-108.
35. Meziani, M. J., Pathak, P., Hurezeanu, R., Thies, M. C., Enick, R. M., and Sun, Y. P. (2004) Supercritical-Fluid Processing Technique for Nanoscale Polymer Particles. *Angewandte Chemie, International Edition*, **43**, 704-707.
36. Sun, G., Hagooly, A., Xu, J., Nystrom, A. M., Li, Z., Rossin, R., Moore, D. A., Wooey, K. L., and Welch, M. J. (2008) Facile, efficient approach to accomplish tunable chemistries and variable biodistributions for shell cross-linked nanoparticles. *Biomacromolecules*, **9**, 1997-2006.
37. Prausnitz, M. R., and Langer, R. (2008) Transdermal drug delivery. *Nature Biotechnology*, **26**, 1261-1268.
38. Lai, S. K., Wang, Y.-Y., and Hanes, J. (2009) Mucus-penetrating nanoparticles for drug and gene delivery to mucosal tissues. *Advanced Drug Delivery Reviews*, **61**, 158-171.
39. Shi, J., Xiao, Z., Kamaly, N., and Farokhzad, O. C. (2011) Self-assembled targeted nanoparticles: evolution of technologies and bench to bedside translation. *Accounts of Chemical Research*, **44**, 1123-1134.
40. Simard, P., and Leroux, J.-C. (2010) In vivo evaluation of pH-sensitive polymer-based immunoliposomes targeting the CD33 antigen. *Molecular Pharmaceutics*, **7**, 1098-1107.
41. Shokeen, M., Pressly, E. D., Hagooly, A., Zheleznyak, A., Ramos, N., Fiamengo, A. L., Welch, M. J., Hawker, C. J., and Anderson, C. J. (2011) Evaluation of multivalent, functional polymeric nanoparticles for imaging applications. *ACS Nano*, **5**, 738-747.
42. Elsabahy, M., Wazen, N., Bayo-Puxan, N., Deleavey, G., Servant, M., Damha, M. J., and Leroux, J.-C. (2009) Delivery of nucleic acids through the controlled disassembly of multifunctional nanocomplexes. *Advanced Functional Materials*, **19**, 3862-3867.
43. Zhang, K., Fang, H., Wang, Z., Li, Z., Taylor, J.-S. A., and Wooley, K. L. (2010) Structure-activity relationships of cationic shell-crosslinked knedel-like nanoparticles: shell composition and transfection effici-

ency/cytotoxicity. *Biomaterials*, **31**, 1805-1813.

44. Zhang, K., Fang, H., Wang, Z., Taylor, J.-S. A., and Wooley, K. L. (2009) Cationic shell-crosslinked knedel-like nanoparticles for highly efficient gene and oligonucleotide transfection of mammalian cells. *Biomaterials*, **30**, 968-977.
45. Zhang, K., Fang, H., Shen, G., Taylor, J.-S. A., and Wooley, K. L. (2009) Well-defined cationic shell crosslinked nanoparticles for efficient delivery of DNA or peptide nucleic acids. *Proceedings of the American Thoracic Society*, **6**, 450-457.
46. Wilson, D. S., Dalmasso, G., Wang, L., Sitaraman, S. V., Merlin, D., and Murthy, N. (2010) Orally delivered thioketal nanoparticles loaded with TNF-α–siRNA target inflammation and inhibit gene expression in the intestines. *Nature Materials*, **9**, 923-928.
47. Iversen, T.-G., Skotland, T., and Sandvig, K. (2011) Endocytosis and intracellular transport of nanoparticles: present knowledge and need for future studies. *Nano Today*, **6**, 176-185.
48. Lee, N. S., Lin, L. Y., Neumann, W. L., Freskos, J. N., Karwa, A., Shieh, J. J., Dorshow, R. B., and Wooley, K. L. (2011) Influence of nanostructure morphology on host capacity and kinetics of guest release. *Small*, **7**, 1998-2003.
49. Li, Z., Ma, J., Lee, N. S., and Wooley, K. L. (2011) Dynamic cylindrical assembly of triblock copolymers by a hierarchical process of covalent and supramolecular interactions. *Journal of the American Chemical Society*, **133**, 1228-1231.
50. Zhang, K., Fang, H., Chen, Z., Taylor, J.-S. A., and Wooley, K. L. (2008) Shape effects of nanoparticles conjugated with cell-penetrating peptides (HIV Tat PTD) on CHO cell uptake. *Bioconjugate Chemistry*, **19**, 1880-1887.
51. Zhang, K., Rossin, R., Hagooly, A., Chen, Z., Welch, M. J., and Wooley, K. L. (2008) Folate-mediated cell uptake of shell-crosslinked spheres and cylinders. *Journal of Polymer Science, Part A: Polymer Chemistry*, **46**, 7578-7583.
52. Gratton, S. E., Ropp, P. A., Pohlhaus, P. D., Luft, J. C., Madden, V. J., Napier, M. E., DeSimone, J. M. (2008) The effect of particle design on cellular internalization pathways. *Proceedings of the National Academy of Sciences*, **105**, 11613-11618.
53. Adler, A. F., and Leong, K. W. (2010) Emerging links between surface nanotechnology and endocytosis: impact on nonviral gene delivery. *Nano Today*, **5**, 553-569.
54. Acharya, S., Dilnawaz, F., and Sahoo, S. K. (2009) Targeted epidermal growth factor receptor nanoparticle bioconjugates for breast cancer therapy. *Biomaterials*, **30**, 5737-5750.
55. Liu, J., Jiang, Z., Zhang, S., Liu, C., Gross, R. A., Kyriakides, T. R., Saltzman, W. M. (2011) Biodegradation, biocompatibility, and drug delivery in poly (ω-pentadecalactone-co-p-dioxanone) copolyesters.

Biomaterials, **32**, 6646-6654.

56. Liu, J., Jiang, Z., Zhang, S., and Saltzman, W. M. (2009) Poly (ω-pentadecalactone-co-butylene-co-succinate) nanoparticles as biodegradable carriers for camptothecin delivery. *Biomaterials*, **30**, 5707-5719.
57. Cirpanli, Y., Allard, E., Passirani, C., Bilensoy, E., Lemaire, L., Calis, S., and Benoi, J. P. (2011) Antitumoral activity of camptothecin-loaded nanoparticles in 9L rat glioma model. *International Journal of Pharmaceutics*, **403**, 201-206.
58. Zhang, N., Li, J., Jiang, W., Ren, C., Li, J., Xin, J., and Li, K. (2010) Effective protection and controlled release of insulin by cationic β-cyclodextrin polymers from alginate/chitosan nanoparticles. *International Journal of Pharmaceutics*, **393**, 213-219.
59. Potineni, A., Lynn, D. M., Langer, R., and Amiji, M. M. (2003) Poly (ethylene oxide)-modified poly (β-amino ester) nanoparticles as a pH-sensitive biodegradable system for paclitaxel delivery. *Journal of Controlled Release*, **86**, 223-234.
60. Bilensoy, E., Gurkaynak, O., Dogan, A. L., and Hincal, A. A. (2008) Safety and efficacy of amphiphilic ß-cyclodextrin nanoparticles for paclitaxel delivery. *International Journal of Pharmaceutics*, **347**, 163-170.
61. Memisoglu-Bilensoy, E., Vural, I., Bochot, A., Renoir, J. M., Duchene, D., and Hincal, A. A. (2005) Tamoxifen citrate loaded amphiphilic β-cyclodextrin nanoparticles: In vitro characterization and cytotoxicity. *Journal of Controlled Release*, **104**, 489-496.
62. Shenoy, D. B., and Amiji, M. M. (2005) Poly (ethylene oxide)-modified poly (ε-caprolactone) nanoparticles for targeted delivery of tamoxifen in breast cancer. *International Journal of Pharmaceutics*, **293**, 261-270.
63. Hamoudeh, M., Salim, H., Barbos, D., Paunoiu, C., and Fessi, H. (2007) Preparation and characterization of radioactive dirhenium decacarbonyl-loaded PLLA nanoparticles for radionuclide intra-tumoral therapy. *European Journal of Pharmaceutics and Biopharmaceutics*, **67**, 597-611.
64. Hu, F., Neoh, K., and Kang, E. (2006) Synthesis and in vitro anti-cancer evaluation of tamoxifen-loaded magnetite/PLLA composite nanoparticles. *Biomaterials*, **27**, 5725-5733.
65. Arias, J. L., Linares-Molinero, F., Gallardo, V., and Delgado, A. V. (2008) Study of carbonyl iron/poly (butylcyanoacrylate)(core/shell) particles as anticancer drug delivery systems: Loading and release properties. *European Journal of Pharmaceutical Sciences*, **33**, 252-261.
66. Arias, J. L., Ruiz, M. A., Gallardo, V., and Delgado, A. V. (2008) Tegafur loading and release properties of magnetite/poly (alkylcyanoacrylate)(core/shell) nanoparticles. *Journal of Controlled Release*, **125**,

50-58.

67. Arias, J. L., Gallardo, V., Ruiz, M. A., and Delgado, A. V. (2008) Magnetite/poly (alkylcyanoacrylate)(core/shell) nanoparticles as 5-Fluorouracil delivery systems for active targeting. *European Journal of Pharmaceutics and Biopharmaceutics*, **69**, 54-63.
68. Wang, D.-S., Li, J.-G., Li, H.-P., and Tang, F.-Q. (2009) Preparation and drug releasing property of magnetic chitosan-5-fluorouracil nanoparticles. *Transactions of Nonferrous Metals Society of China*, **19**, 1232-1236.
69. Arias, J. L., Lopez-Viota, M., Saez-Fernandez, E., and Ruiz, M. A. (2010) Formulation and physicochemical characterization of poly (ε-caprolactone) nanoparticles loaded with ftorafur and diclofenac sodium. *Colloids and Surfaces B: Biointerfaces*, **75**, 204-208.
70. Arias, J. L., Lopez-Viota, M., Delgado, A. V., and Ruiz, M. A. (2010) Iron/ethylcellulose (core/shell) nanoplatform loaded with 5-fluorouracil for cancer targeting. *Colloids and Surfaces B: Biointerfaces*, **77**, 111-116.
71. Arias, J. L., Lopez-Viota, M., Saez-Fernandez, E., Ruiz, M. A., Delgado, A. V. (2011) Engineering of an antitumor (core/shell) magnetic nanoformulation based on the chemotherapy agent ftorafur. *Colloids and Surfaces A: Physicochemical and Engineering Aspects*, **384**, 157-163.
72. Arias, J. L., Lopez-Viota, M., Lopez-Viota, J., and Delgado, A. V. (2009) Development of iron/ethylcellulose (core/shell) nanoparticles loaded with diclofenac sodium for arthritis treatment. *International Journal of Pharmaceutics*, **382**, 270-276.
73. Arias, J., Gallardo, V., Linares-Molinero, F., and Delgado, A. (2006) Preparation and characterization of carbonyl iron/poly (butylcyanoacrylate) core/shell nanoparticles. *Journal of Colloid and Interface Science*, **299**, 599-607.
74. Arias, J., Gallardo, V., Gomez-Lopera, S., Plaza, R., and Delgado, A. (2001) Synthesis and characterization of poly (ethyl-2-cyanoacrylate) nanoparticles with a magnetic core. *Journal of Controlled Release*, **77**, 309-321.
75. Balakrishnan, S., Bonder, M. J., and Hadjipanayis, G. C. (2009) Particle size effect on phase and magnetic properties of polymer-coated magnetic nanoparticles. *Journal of Magnetism and Magnetic Materials*, **321**, 117-122.
76. Ma, H. L., Xu, Y. F., Qi, X. R., Maitani, Y., and Nagai, T. (2008) Superparamagnetic iron oxide nanoparticles stabilized by alginate: pharmacokinetics, tissue distribution, and applications in detecting liver cancers. *International Journal of Pharmaceutics*, **354**, 217-226.
77. Kumagai, M., Imai, Y., Nakamura, T., Yamasaki, Y., Sekino, M., Ueno, S., Hanaoka, K., Kikuchi, K., Nagano, T., Kaneko, E., Shimokado, K., and Kataoka, K. (2007) Iron hydroxide nanoparticles coated with

poly (ethylene glycol)-poly (aspartic acid) block copolymer as novel magnetic resonance contrast agents for in vivo cancer imaging. *Colloids and Surfaces B: Biointerfaces*, **56**, 174-181.

78. Fan, Q.-L., Neoh, K.-G., Kang, E.-T., Shuter, B., and Wang, S.-C. (2007) Solvent-free atom transfer radical polymerization for the preparation of poly (poly (ethyleneglycol) monomethacrylate)-grafted Fe_3O_4 nanoparticles: synthesis, characterization and cellular uptake. *Biomaterials*, **28**, 5426-5436.
79. Ren, J., Hong, H.-Y., Ren, T.-B., and Teng, X.-R. (2005) Preparation and characterization of magnetic PLA–PEG composite particles. *Materials Letters*, **59**, 2655-2658.
80. Venkatraman, S. S., Jie, P., Min, F., Freddy, B. Y. C., and Leong-Huat, G. (2005) Micelle-like nanoparticles of PLA–PEG–PLA triblock copolymer as chemotherapeutic carrier. *International Journal of Pharmaceutics*, **298**, 219-232.
81. Matsumoto, J., Nakada, Y., Sakurai, K., Nakamura, T., and Takahashi, Y. (1999) Preparation of nanoparticles consisted of poly (L-lactide)–poly (ethylene glycol)–poly (L-lactide) and their evaluation in vitro. *International Journal of Pharmaceutics*, **185**, 93-101.
82. Gomez-Lopera, S. A., Arias, J. L., Gallardo, V., and Delgado, A. V. (2006) Colloidal stability of magnetite/poly (lactic acid) core/shell nanoparticles. *Langmuir*, **22**, 2816-2821.
83. Nicoli, S., Santi, P., Couvreur, P., Couarraze, G., Colombo, P., and Fattal, E. (2001) Design of triptorelin loaded nanospheres for transdermal iontophoretic administration. *International Journal of Pharmaceutics*, **214**, 31-35.
84. Redhead, H., Davis, S., and Illum, L. Drug delivery in poly (lactide-co-glycolide) nanoparticles surface modified with poloxamer 407 and poloxamine 908: in vitro characterisation and in vivo evaluation. *Journal of Controlled Release*, **70**, 353-363.
85. Yang, J., Lee, H., Hyung, W., Park, S.-B., and Haam, S. (2006) Magnetic PECA nanoparticles as drug carriers for targeted delivery: synthesis and release characteristics. *Journal of Microencapsulation*, **23**,203-212.
86. Chatterjee, J., Haik, Y., and Chen, C.-J. (2002) Polyethylene magnetic nanoparticle: a new magnetic material for biomedical applications. *Journal of Magnetism and Magnetic Materials*, **246**, 382-391.
87. Yuan, J.-J., Armes, S., Takabayashi, Y., Prassides, K., Leite, C., Galembeck, F., and Lewis, A. (2006) Synthesis of biocompatible poly [2-(methacryloyloxy) ethyl phosphorylcholine]-coated magnetite nanoparticles. *Langmuir*, **22**, 10989-10993.
88. Aurich, K., Schwalbe, M., Clement, J. H., Weitschies, W., and Buske, N. (2007) Polyaspartate coated magnetite nanoparticles for biomedical applications. *Journal of Magnetism and Magnetic Materials*, **311**, 1-5.

89. Kim, D. H., Lee, S. H., Im, K. H., Kim, K. N., Kim, K. M., Shim, I. B., Lee, M. H., and Lee, Y.-K. (2006) Surface-modified magnetite nanoparticles for hyperthermia: Preparation, characterization, and cytotoxicity studies. *Current Applied Physics*, **6**(1), e242-e246.
90. Gomez-Gaete, C., Tsapis, N., Besnard, M., Bochot, A., and Fattal, E. (2007) Encapsulation of dexamethasone into biodegradable polymeric nanoparticles. *International Journal of Pharmaceutics*, **331**, 153-159.
91. Yin, Y., Chen, D., Qiao, M., Wei, X., and Hu, H. (2007) Lectin-conjugated PLGA nanoparticles loaded with thymopentin: ex vivo bioadhesion and in vivo biodistribution. *Journal of Controlled Release*, **123**, 27-38.
92. Labhasetwar, V., Song, C., Humphrey, W., Shebuski, R., and Levy, R. J. (1998) Arterial uptake of biodegradable nanoparticles: effect of surface modifications. *Journal of Pharmaceutical Sciences*, **87**, 1229-1234.
93. Esmaeili, F., Ghahremani, M. H., Ostad, S. N., Atyabi, F., Seyedabadi, M., Malekshahi, M. R., Amini, M., and Dinarvand, R. (2008) Folate-receptor-targeted delivery of docetaxel nanoparticles prepared by PLGA–PEG–folate conjugate. *Journal of Drug Targeting*, **16**, 415-423.
94. Teixeira, M., Alonso, M. J., Pinto, M. M., and Barbosa, C. M. (2005) Development and characterization of PLGA nanospheres and nanocapsules containing xanthone and 3-methoxyxanthone. *European Journal of Pharmaceutics and Biopharmaceutics*, **59**, 491-500.
95. Derakhshandeh, K., Erfan, M., and Dadashzadeh, S. (2007) Encapsulation of 9-nitrocamptothecin, a novel anticancer drug, in biodegradable nanoparticles: factorial design, characterization and release kinetics. *European Journal of Pharmaceutics and Biopharmaceutics*, **66**, 34-41.
96. Sahana, D., Mittal, G., Bhardwaj, V., and Kumar, M. (2008) PLGA nanoparticles for oral delivery of hydrophobic drugs: influence of organic solvent on nanoparticle formation and release behavior in vitro and in vivo using estradiol as a model drug. *Journal of Pharmaceutical Sciences*, **97**, 1530-1542.
97. Fonseca, C., Simoes, S., and Gaspar, R. (2002) Paclitaxel-loaded PLGA nanoparticles: preparation, physicochemical characterization and in vitro anti-tumoral activity. *Journal of Controlled Release*, **83**, 273-286.
98. Mu, L., and Feng, S. (2003) A novel controlled release formulation for the anticancer drug paclitaxel (Taxol®): PLGA nanoparticles containing vitamin E TPGS. *Journal of Controlled Release*, **86**, 33-48.
99. Jin, C., Bai, L., Wu, H., Teng, Z., Guo, G., and Chen, J. (2008) Cellular uptake and radiosensitization of SR-2508 loaded PLGA nanoparticles. *Journal of Nanoparticle Research*, **10**, 1045-1052.

100. Choi, C., Chae, S. Y., and Nah, J.-W. (2006) Thermosensitive poly (N-isopropylacrylamide)-b-poly (ε-caprolactone) nanoparticles for efficient drug delivery system. *Polymer*, **47**, 4571-4580.
101. Shah, L. K., and Amiji, M. M. (2006) Intracellular delivery of saquinavir in biodegradable polymeric nanoparticles for HIV/AIDS. Pharmaceutical Research, 23, 2638-2645.
102. Zheng, D., Li, X., Xu, H., Lu, X., Hu, Y., and Fan, W. (2009) Study on docetaxel-loaded nanoparticles with high antitumor efficacy against malignant melanoma. *Acta Biochimica et Biophysica Sinica*, **41**, 578-587.
103. Prabu, P., Chaudhari, A. A., Dharmaraj, N., Khil, M., Park, S., and Kim, H. (2009) Preparation, characterization, in-vitro drug release and cellular uptake of poly (caprolactone) grafted dextran copolymeric nanoparticles loaded with anticancer drug. *Journal of Biomedical Materials Research, Part A*, **90**, 1128-1136.
104. Sheng, Y., Yuan, Y., Liu, C., Tao, X., Shan, X., and Xu, F. (2009) In vitro macrophage uptake and in vivo biodistribution of PLA–PEG nanoparticles loaded with hemoglobin as blood substitutes: effect of PEG content. *Journal of Materials Science: Materials in Medicine*, **20**, 1881-1891.
105. Gomez-Gaete, C., Fattal, E., Silva, L., Besnard, M., and Tsapis, N. (2008) Dexamethasone acetate encapsulation into Trojan particles. *Journal of Controlled Release*, **128**, 41-49.
106. Sonaje, K., Italia, J., Sharma, G., Bhardwaj, V., Tikoo, K., Kumar, M. R. (2007) Development of biodegradable nanoparticles for oral delivery of ellagic acid and evaluation of their antioxidant efficacy against cyclosporine A-induced nephrotoxicity in rats. *Pharmaceutical Research*, **24**, 899-908.
107. Zambaux, M., Bonneaux, F., Gref, R., Dellacherie, E., and Vigneron, C. (1999) Preparation and characterization of protein C-loaded PLA nanoparticles. *Journal of Controlled Release*, **60**, 179-188.
108. Mainardes, R. M., Gremiao, M. P. D., Brunetti, I. L., Da Fonseca, L. M., and Khalil, N. M. (2009) Zidovudine-loaded PLA and PLA–PEG blend nanoparticles: Influence of polymer type on phagocytic uptake by polymorphonuclear cells. *Journal of Pharmaceutical Sciences*, **98**, 257-267.
109. Xing, J., Zhang, D., and Tan, T. (2007) Studies on the oridonin-loaded poly (D, L-lactic acid) nanoparticles in vitro and in vivo. *International Journal of Biological Macromolecules*, **40**, 153-158.
110. Cheng, Q., Feng, J., Chen, J., Zhu, X., and Li, F. (2008) Brain transport of neurotoxin-I with PLA nanoparticles through intranasal administration in rats: a microdialysis study. *Biopharmaceutics & Drug Disposition*, **29**, 431-439.
111. Leroux, J.-C., Allemann, E., De Jaeghere, F., Doelker, E., and Gurny, R. (1996) Biodegradable nanoparticles from sustained release formu-

lations to improved site specific drug delivery. *Journal of Controlled Release*, **39**, 339-350.

5

Polymers and Copolymers Utilized in Photovoltaic Applications

5.1 Introduction

In early 1990s, organic photovoltaics used to exhibit short life duration and relatively low efficiency as compared to the silicon-based photovoltaics [1]. However, in past few years, organic photovoltaics have exhibited significant enhancement in efficiency and operational life duration (to 20,000 h) [2-4]. Additionally, organic photovoltaics exhibit benefits associated with operations at low-light levels like indoor applications as compared to the inorganic silicon-based photovoltaics. Efficiency of organic photovoltaics is affected by many components such as polymer structure, film morphology, electron acceptor choice and the ratio of the accepter and polymer donor [5].

5.2 Factors Influencing Power Conversion Efficiency of Polymer Solar Cell Devices

Factors like uncomplicated synthesis and fabrication, cost effectiveness, low weight and potential to be assembled into sizeable area flexible devices have helped polymer solar cells to exhibit promising performance in recent years [6-30]. Combination of an acceptor and photoactive blend layer of a conjugated polymer donor constitutes the polymer solar cell, sandwiched between a cathode and an anode. Either cathode or anode should have transparency for constructing the polymer solar cell. The research effort is mainly focused on gaining enhanced power conversion efficiency (PCE) of polymer solar cells. Power conversion efficiency of polymer solar cells depends on the magnitude of short-circuit current density (J_{sc}), fill factor (FF) and open-circuit voltage (V_{oc}). The charge carrier mobility of the acceptor and donor materials and absorption band wavelength control J_{sc}, whereas V_{oc} is proportionally related to the difference between the

Naman Arora and Vikas Mittal, The Petroleum Institute (part of Khalifa University of Science and Technology), Abu Dhabi, UAE*
**Current address: Bletchington, Wellington County, Australia*

lowest unoccupied molecular orbital (LUMO) of the acceptor and the highest occupied molecular orbital (HOMO) of the donor [31-34]. Device fabrication conditions and charge carrier mobility decide the fill factor of the device. Therefore, for increasing PCE of the polymer solar cells, it is significant to gain to high value of J_{sc}, higher hole mobility for high J_{sc} and larger FF, and a low-lying HOMO energy level for obtaining higher open-circuit voltage by designing and synthesizing advanced conjugated polymer donor materials containing narrower band gap (broader absorption).

5.3 Importance of Generating Low Band Gap Polymers

The requirement behind preparing low band gap polymers is driven be the facts: i) band gap of around 1.4 eV is indicated by the Shockley-Quiesser equation which is considered to be an ideal for a single junction solar cell device [35], ii) tandem based polymer solar cells need an active material containing a band gap <1.5 eV in combination with the broad band gap material containing a band gap of 1.9 eV [26,36-39], iii) visible-transparent polymer solar cells is one of the applications which requires an active material that absorbs more near-infrared and ultra-violet light and absorbs minute visible light [40-43]. For fulfilling the requirements to achieve these objectives, various synthetic approaches have shown promising results with respect to narrowing the band gap of organic polymeric materials [44-46]. It should be noted that high PCE of devices cannot be guaranteed solely by a small band gap polymer. Appropriate arrangement of the HOMO and LUMO energy levels is significant for effective charge transfer to the electron acceptor material and for obtaining device's high open circuit voltage. There is also a requirement for enhancing J_{sc} and FF of the device through blending of the high charge carrier mobility and morphology with the acceptor material [15,31,47]. As of now, single junction devices exhibiting power conversion efficiency of around 7% have been developed by utilizing active material containing band gaps in the range 1.8-1.6 eV (mid-band gap polymer) [48-53]. Several attempts have been made for obtaining high performance polymers (band gap <1.5 eV), however, mid band gap polymers like thionothiophene are largely the best available candidates to date.

5.4 Various Polymers and Copolymers for Photovoltaic Devices

As mentioned earlier, polymer solar cell devices have exhibited fast development in recent years with enhanced PCE [27]. The V_{oc} of the device and the harvesting of solar light are mainly determined by the active layer of polymer solar cell. Various donor polymers have been synthesized due to their potential properties of optical absorption and low band gaps [54]. For instance, Figure 5.1 shows the synthesis of such functional polymers by the combination of indacenodithiophene (IDT) and two quinoxaline derivatives to form new PIDT-diphQ and PIDT-phanQ [54]. Usually, donor-acceptor (D-A) copolymers are also introduced with the units of acceptor which affects

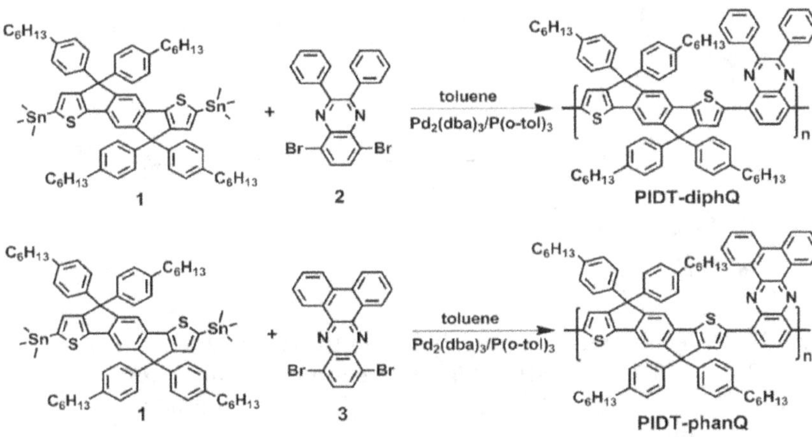

Figure 5.1 Synthesis of PIDT-diphQ and PIDT-phanQ polymers. Reproduced from Reference 54 with permission from American Chemical Society.

the HOMO level and subsequently influences the V_{oc} of polymer solar cell devices. Hence, new materials are being developed with low-lying HOMO levels. In one such study, Ye *et al.* [55] designed two-dimensional conjugated polymer acceptor (PBDTNDI-T) with naphthalene diimide (NDI) and benzo(1,2-b:4,5-b)dithiophene (BDT). In recent years, units of two dimensional conjugated BDT have been utilized as basic blocks in efficient donor polymers. Herein, the authors successfully developed two dimensional-conjugated BDT-based polymer acceptor (PBDTNDI-T), as shown in Figure 5.2. It was observed that nearly 3% of power conversion efficiency was exhibited by the polymer solar cell devices based on PBDTNDI-T. Hence, it could be concluded that affixation of side chains (alkylthiothiophene) to the π-conjugated polymer resulted in the successful conversion of polymer donor to polymer acceptor. Overall, this type of two-dimensional conjugated polymer, being a potential semiconductor, exhibits potential of significant performance for lithium batteries and transistors, along with solar cells.

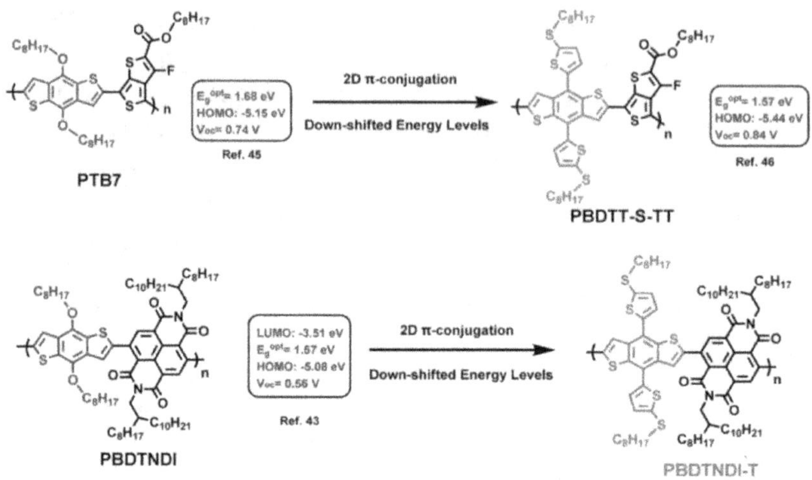

Figure 5.2 Molecular structures *of* the π-conjugated polymer donors (PTB7, PBDTT-S-TT, PBDTNDI) and the target polymer acceptor (PBDTNDI-T). Reproduced from Reference 55 with permission from American Chemical Society.

Zhang *et al.* [56] reported the preparation of a conjugated polymer with large band gap exhibiting significant performance for photovoltaic applications. The authors prepared PM6 copolymer based on 4,8-bis(5-(2-ethylhexyl)-4-fluorothiophen-2-yl)benzo[1,2-b:4,5-b′]-dithiophene (BDT-F) and 1,3-bis(thiophen-2-yl)-5,7-bis(2-ethylhexyl) benzo-[1,2-c:4,5-c′]dithiophene-4,8-dione (BDD). In this copolymer, BDT-F and BDD contributed as donor and acceptor materials, respectively. The polymer exhibited a large band gap of 1.80 eV and contained a deep level of HOMO energy, i.e. -5.45 eV. High power conversion efficiency (9.2%) and open-circuit voltage (0.98 V) with a film thickness around 75 nm were exhibited by PM6 based inverted PSC. Additionally, high average transmission (67%) in the visible range and PCE of 5.6% were exhibited by PM6 constituted semitransparent device. For the development of semitransparent single-junction polymer solar cell devices, PM6 was confirmed to be a suitable material. In another study, Liu *et al.* [57] reported the synthesis of conjugated crossed-benzodithiophene (electron-rich moiety) to generate donor-acceptor (D-A) copolymers. PCE of 3.74% and V_{oc} of 0.9 V were exhibited by the crossed-benzodithiophene and 4,7-di(thiophen-2-yl)-2,1,3-benzothiadiazole (DTBT) based polymer. The authors modified the BDT units (acceptor) with fluorinated and non-fluorinated moieties. The crossed-BDT and 4,7-di(thiophen-2-yl)-5,6-difluoro-2,1,3-benzothiadiazole (DTffBT) based polymers exhibited V_{oc} of around 1 V, which was gained due to low-lying HOMO energy level induced by fluorination. The authors also observed the blue-shift absorption and decrement in HOMO level in both fluorinated and non-fluorinated polymers.

Tandem structure is a potential route for enhancing the efficiency of polymer solar cells. Till few years back, it was not possible to develop tandem solar cells with high performance due to limited low-band gap polymers with high performance. You *et al.* [27] demonstrated the preparation of poly[2,7-(5,5-bis-(3,7-dimethyloctyl)-5H-dithieno[3,2-b:20,30-d]pyran)-alt-4,7-(5,6-difluoro-2,1,3-benzothia, which is a low band gap polymer with high performance exhibiting band gap of 1.38 eV. The polymer also exhibited deep HOMO level and high mobility due to which power conversion efficiency of 7.9% was exhibited by single-junction devices. Tandem solar cells with PCE of 10.6% were achieved by utilizing this polymer. Dou *et al.* [58] studied various low band gap polymers with alternating units of diketopyrrolopyrrole (DPP) and thienylbenzodithiophene (BDTT). The authors copolymerized BDTT unit with DPP unit containing furan (FDPP) (Figure 5.3), and PCE of 5% was achieved in a single junction solar cell by employing resultant polymer (PBDTT-FDPP) [59]. By converting furan into thiopene moiety, enhanced J_{sc} and FF were exhibited by PBDTT-DPP which further led to the enhanced PCE of 6.5% [26,59]. PCE of 8.6% was demonstrated by tandem polymer solar cells by applying PBDTT-DPP. The polymer was also utilized in visible-transparent polymer solar cells which led to a gain of 4% in power conversion efficiency with transparency of 60% in the visible region [26,43]. However, the power conversion efficiencies were limited due to the diminished range of absorption (≤ 850 nm) and poor external quantum efficiency (less than 50 %) in the near infrared region (NIR). Enhanced power conversion efficiencies could be achieved by both type of devices by reducing the band gap for harvesting large number of photons in the region of NIR and by enhancing the mobility of charge carrier of low band gap polymer. Subsequently, it was observed that replacement of atom of sulphur in the thiophene part of the DPP unit with an atom of selenium for making the selenophene-based DPP (SeDPP) unit could lead to decrement in band gap and increased the characteristics of charge transport in photovoltaic devices [60-63]. The authors reported the preparation of selenium-replaced polymer PBDTT-SeDPP, which exhibited efficiency of 7.2% in a polymer solar cell.

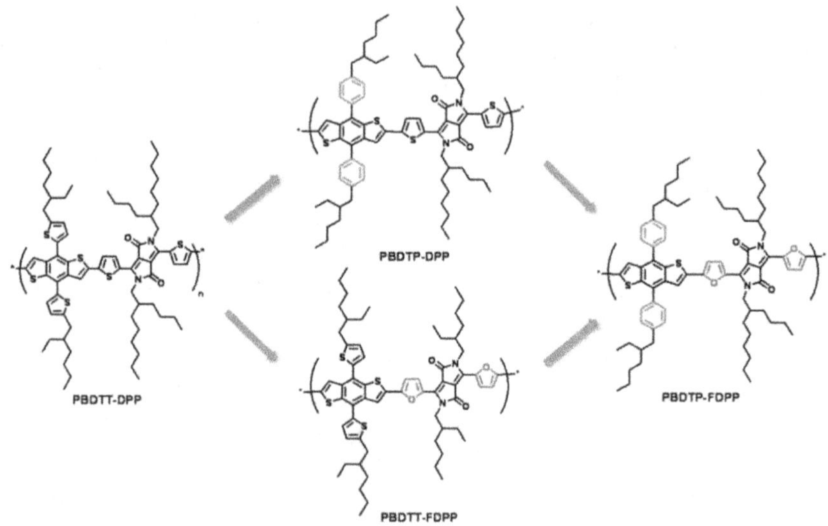

Figure 5.3 Chemical structures of PBDTT-DPP, PBDTP-DPP, PBDTT-FDPP, and PBDTP-FDPP. Reproduced from Reference 59 with permission from American Chemical Society.

Zhang *et al.* [64] designed integrated silole having silicon-bridged heptacyclic arene (ARSi) and naphthobisoxadiazole moieties for enhanced efficiency of polymer solar cells, thereby exhibiting PCE of 8.37%. In another study, Liu *et al.* [65] reported the preparation of conjugated polymers constituted of BDT and DTQx (2,3-diphenyl-5,8-di(thiophen-2-yl)quinoxaline) based units (Figure 5.4). The photovoltaic properties of polymer solar cells were characterized by using the developed polymers. It was observed that circuit current density and fill factor got affected to a small extent by fluorination process, but it was significanlty beneficial for enhancing device V_{OC}. Due to the synergistic effect of fluorination, one of the developed polymers achieved a high open-circuit voltage of 0.90 V. The results revealed that power conversion efficiencies of the polymer solar cells using the developed polymers were in the range of 5.63-8.55%, whereas the V_{OC} values ranged between 0.64-0.90 V.

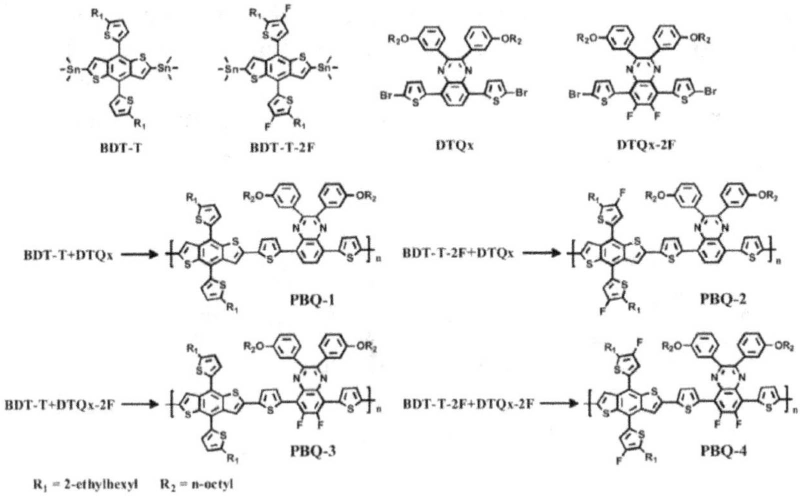

Figure 5.4 Molecular structures and synthesis strategies of the developed polymers. Reproduced from Reference 65 with permission from American Chemical Society.

Next-generation photovoltaic technology, where the polymer solar cells contain an active layer consisting of semiconducting polymer and derivative of fullerene acting as electron donor and electron acceptor, respectively, has also gained significant attention [6,66]. Duan *et al.* [67] utilized BDT and two 5,6-difluorobenzo(2,1,3)thiadiazole-based monomers for synthesizing state of the art semiconductor copolymers to develop solar cells (Figure 5.5). It was concluded that around 8% of power conversion efficiency was exhibited by random copolymers, which was higher than their binary parent polymers. Also, even after deposition of active layer from non-halogenated solvents, power conversion efficiency of 8% was achieved. In this study, the developed copolymers, thus, represented functional photovoltaic polymers for the generation of polymer solar cells. In another study, Nakano *et al.* [68] reported the synthesis of π-conjugated naphtho-[2,3-b:6,7-b]dithiophene-4,5,9,10-diimide (NDTI) based copolymers with units of different thiophene molecules. The copolymers were applied in field effect transistor and organic solar cells. The authors observed that NDTI could be promising for ambipolar organic semiconductors. Intramolecular donor-acceptor interaction could be induced by the integration of strong electron affinity of naphtho-[2,3-b:6,7-b]dithiophene-4,5,9,10-diimide with the electron rich thiophenes, which resulted in narrow energy gaps. Due to this, photo-response could be generated in the near infrared region in organic solar cells. Combination of low-lying LUMO and high-lying HOMO levels could lead to effective carrier injection of electrons and holes, enabling ambipolar operation possible in field effect transistor devices (Figure 5.6).

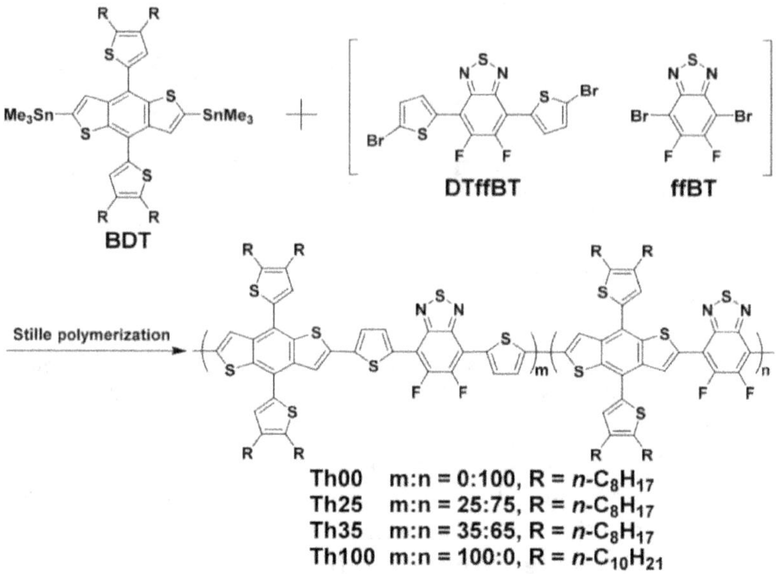

Figure 5.5 Synthesis schematic for the generation of polymers. Reproduced from Reference 67 with permission from American Chemical Society.

Liu *et al.* [69] reported the preparation and characterization of different wide band gap copolymers (PBT1-MP, PBT1-EH, PBT1-BO and PBT1-EH) with different alkyl side chains on benzodithiophene and benzodithiophene-4,8-dione as electron rich and electron deficient units, respectively. It was observed that photovoltaic, electrical and optical properties were affected due to the alkyl side chains. Additionally, these alkyl side chains had influence on molecular packing as well as crystallinity of polymers. The authors observed that long alkyl substituents on benzodithiophene-4,8-dione unit could lead to steric hindrance and twisted molecular geometry. As a result, high PCE of 10% was exhibited by PBT1-EH with bulky side chain on benzodithiophene-4,8-dione unit in fullerene as well as non-fullerene acceptor based solar cells. In another study, Zhou *et al.* [70] also reported the polymer solar cells based on naphthodithiophene diimide

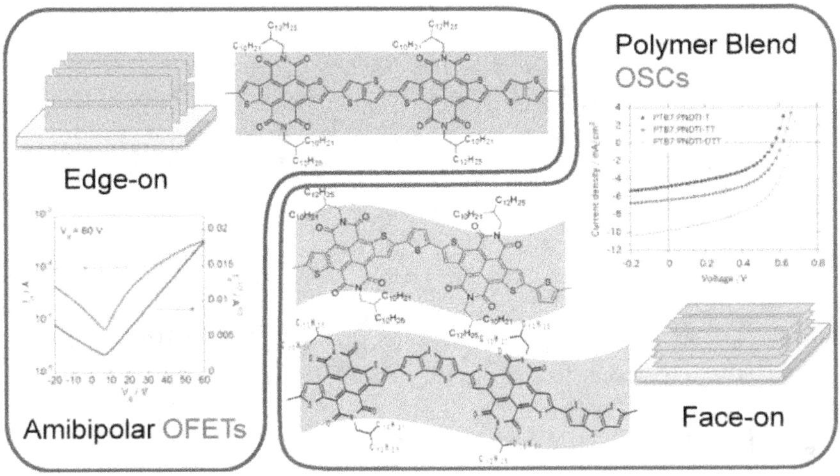

Figure 5.6 Ambipolar semiconductors in field-effect transistors and electron acceptors in polymer blend solar cells. Reproduced from Reference 68 with permission from American Chemical Society.

and bithiophene (PNDTI-BT-DT) copolymers integrated with poly[[4,8-bis[(2-ethylhexyl)oxy]benzo[1,2-b:4,5-b]-bithiophene-2,6-diyl][3-fluoro-2-[(2-ethylhexyl)carbonyl]-thieno[3,4-b]thiophenediyl]] (PTB7) as an electron donor polymer which has been utilized for many polymer/fullerene polymer solar cells [49] and other polymer solar cells [71]. The authors reported that higher photo-response was exhibited by all polymer solar cells in the near infrared region, as compared to visible region. As shown in Figure 5.7, strong absorption band was exhibited by PTB7 films in the 500-750 nm region and PNDTI-BT-DT films exhibited absorption band in the region of 600-900 nm. More sunlight could be utilized by this absorption behavior which was comparatively more beneficial than polymer-fullerene system. Under the irradiation of solar light, maximum PCE was recorded as 2.59%. Crystalline structure was formed by PNDTI-BT-DT and PTB7 blend films and was almost similar to the structure of the pristine films.

Marchiori *et al.* [72] reported the application of poly[2,7-(9,9-bis(2-ethylhexyl)-dibenzosilole)-alt-4,7-bis(thiophen-2-yl)benzo-2,1,3-thiadiazole] copolymer (PSiF-DBT) for photovoltaic devices which exhibited PCE of 4%. The PSiF-DBT copolymer was utilized as active layer in bilayer solar cell with C60, exposed to thermal annealing at 100 °C. However, it was observed that even without annealing exposure, PCE of 2% was exhibited by PSiF-DBT-based devices. This

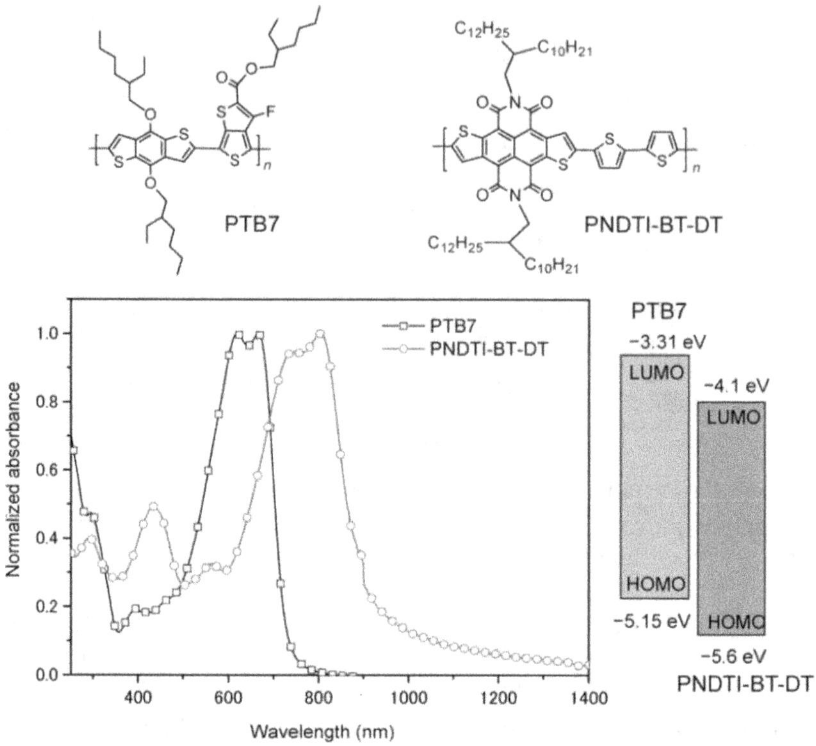

Figure 5.7 Chemical structures of PTB7 and PNDTI-BT-DT, along with absorption spectra and energy level diagrams. Reproduced from Reference 70 with permission from American Chemical Society.

was due to the reason that the cast film of PSiF-DBT contained well stacked morphology after deposition which reduced the defects concentration, thus, behaving as charge carrier traps. Consequently, there was not much improvement achieved by thermal annealing in the transportation of holes of films of PSiF-DBT. The power conversion efficiency was basically induced by the increment in roughness of the polymer surface which contributed for more efficacious interface for dissociating exciton at the heterojunction of PSiF-DBT-fullerene. Kim *et al.* [73] also prepared different π-conjugated alternating copolymers by employing cross-coupling polymerization between units of benzotriazole (electron acceptor) and benzodithiophene (electron donator). For improving the resultant polymer solubility, linear n-octyl or branched 1-octylnonyl groups were subjected to the benzotriazole part for generating 2-octyl-2H-benzo[1,2,3]triazole (LBTz) or 2-(heptadecan-9-yl)-2H-benzo[1,2,3]triazole (YBTz). High solubility was exhibited by the copolymers in organic solvents due to the presence of benzotriazole alkyl side chains. Narrow band gaps were also exhibited by the developed copolymers. It was also observed that the optical band gaps of these polymers were within the range of 1.90-2.03 eV depending on the structure of the polymer. Lower HOMO energy levels were exhibited by the as-prepared polymers as compared to poly(3-hexylthiophene) due to the presence of strong electron-accepting units of benzotriazole. It was suggested that the as-prepared polymers exhibited higher open-circuit voltages and better oxidative stability for the application in photovoltaic devices.

In another study, Saeki *et al.* [74] reported benzobisthiazole and benzothiadiazole based alternating copolymer. The copolymer contained solubilizing linear and branched alkyl chains. The authors explored the inverted cell structure where zinc oxide and MoOx/Ag were utilized as cathode buffer layer and anode respectively for improving PCE of photovoltaic devices. Enhancement of power conversion efficiency was observed from 3.8% to 6.5% after the introduction of inverted cell structure (ZnO/polymer:fullerene/MoOx/Ag). Furthermore, usage of optimized film thickness and $PC_{71}BM$ exhibited better interaction. Guo *et al.* [75] applied microphase-separated conjugated block copolymers (poly(3-hexylthiophene)–block–poly-((9,9-dioctylfluorene)-2,7-diyl-alt-[4,7-bis(thiophen-5-yl)-2,1,3-benzothiadiazole]-2′,2″-diyl) (P3HT-b-PFTBT)) for controlling donor-acceptor heterojunctions (Figure 5.8). As compared to blends of homopolymers, the use of block copolymers as active layer of cells led to higher photo-conversion. Furthermore, it was revealed by X-ray scattering that block copolymer based solar cells exhibited better performance due to the existence of self-assembly into meso-scale lamellar morphologies. Therefore, it was concluded that performance could be enhanced in excitonic solar cells through control of donor–acceptor interfaces by applying conjugated block copolymers.

For the purpose of harvesting solar energy, cost effective and flexible photovoltaic devices are required which can be prepared by employing conjugated polymers [43,76-78]. The generation of electron-donor polymers was taken as an inspiration for preparing polymer-fullerene polymer solar cells which gained significant enhancement in respect to their power conversion efficiencies recently [24,79-83]. Donor-acceptor (D-A) copolymers could provide properties of charge transportation, absorption bands with fine-tuning, HOMO/LUMO levels through the tuning of donor-acceptor interactions and π-electron

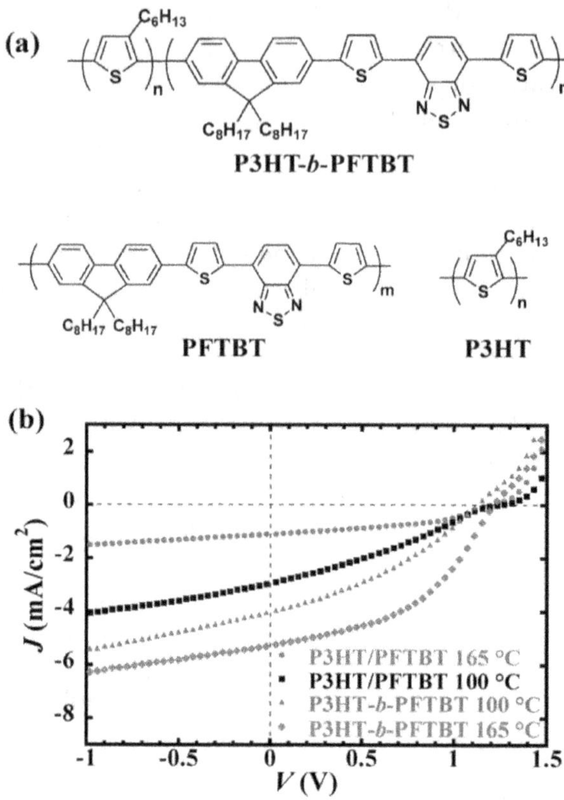

Figure 5.8 Performances of block copolymer P3HT-b-PFTBT and polymer blend P3HT/PFTBT solar cells: (a) chemical structures of P3HT-b- PFTBT, PFTBT and P3HT; (b) current–voltage characteristics of P3HT/PFTBT and P3HT-b-PFTBT photovoltaic devices. Reproduced from Reference 75 with permission from American Chemical Society.

delocalization. Researchers have also studied various electron-withdrawing units such as quinoxaline, thieno-[3,4-c]pyrrole-4,6-dione (TPD), 2,1,3-benzothiadiazole (BT) and isoindigo [28,84-90]. Tao *et al.* [91] prepared and characterized a variety of D–A1–D–A2 copolymers for generating effective polymer based solar cells. The copolymers (Figure 5.9) contained donor segments and were synthesized depending on quinoxaline-fluorinated quinoxaline and isoindigo with different electron-donating segment lengths. For the purpose of gaining increased open circuit voltage and low-lying HOMO level, fluorine atoms were subjected onto the weaker quinoxaline acceptor. It was observed that the introduction of fluorine in the qunioxaline unit

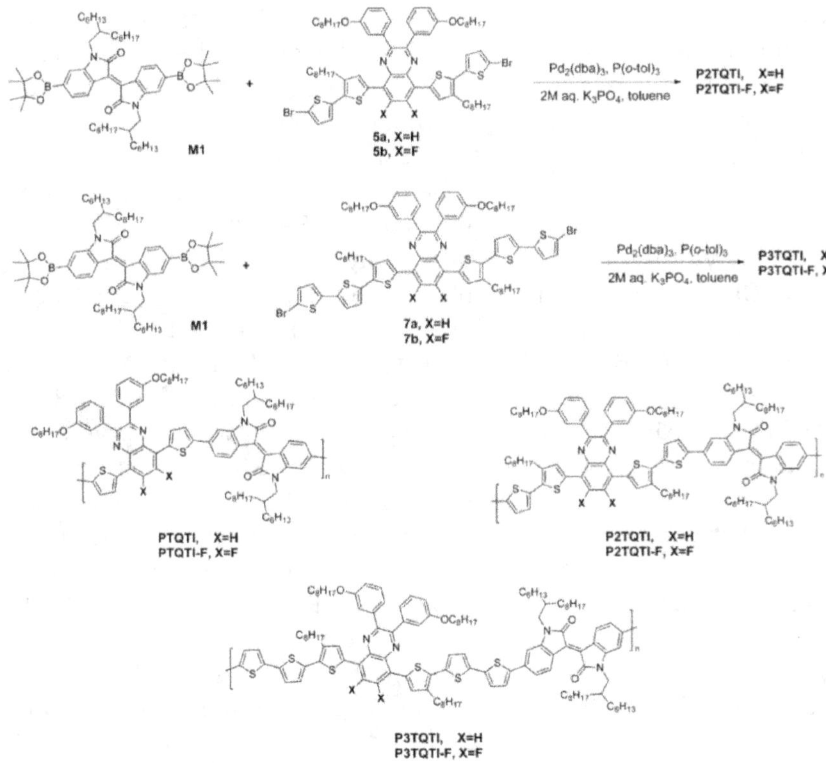

Figure 5.9 Synthetic routes of P2TQTI(F) and P3TQTI(F). Reproduced from Reference 91 with permission from American Chemical Society.

resulted in enhanced absorption coefficients, leading to high short-circuit current density and reduced HOMO levels, thus, exhibiting higher open-circuit voltage. It was also noticed that lower open-circuit voltage was exhibited by polymer solar cells which were based on P2TQTI(F) and P3TQTI(F) polymers containing extended donor segments. In contrast, polymer solar cells based on PTQTI(F) polymer containing short donor segments exhibited higher open-circuit voltage. PCE of 7% and J_{sc} of 15.5 mAcm^{-2} were exhibited by P3TQTI-F polymer containing high absorption efficient and an optimized nanostructure. In another study, Liu *et al.* [92] prepared a variety of benzodithiophene-thienothiophene copolymers containing varying fluorine content. It was observed that deep HOMO energy levels could be achieved with incorporation of fluorine in the units of thienothiophene with no absolute influence on polymer light harvesting properties and band gap. Consequently, an enhanced open-circuit voltage was gained in photovoltaic devices and high short-circuit currents were maintained at the same time. The charge transport properties and morphology significantly influenced the fill factors of photovoltaic devices in the bulk-heterojunction cell films.

Kim *et al.* [93] reported the synthesis and characterization of conjugated copolymers with low band gap, containing alternating difluorinated benzoselenadiazole (DFDTBSe) and ethylhexyloxy (EH)- or octyldodecyloxy (OD) substituted benzo[1,2-b:4,5-b]-dithiophene (BDT). The two conjugated copolymers were prepared by applying Stille cross-coupling polymerization and named as PEHBDT-DFDTBSe and PODBDT-DFDTBSe, respectively. Almost similar band gap energies of 1.66 eV and 1.69 eV were exhibited by PEHBDT-DFDTBSe and PODBDT-DFDTBSe, respectively. It was observed that molecular ordering in the $PC_{71}BM$ blended active layer films could be influenced by alkyloxy side chains present in the polymers. The 'face-on' crystal structure having a short intermolecular $\pi-\pi$ stacking distance and an 'edge-on' structure were exhibited by PEHBDT-DFDTBSe:$PC_{71}BM$ and PODBDT-DFDTBSe:$PC_{71}BM$, respectively. In case of inverted bulk heterojunction organic photovoltaic cells, power conversion efficiency of 5.74% was achieved by fabricating PEHBDT-DFDTBSe:$PC_{71}BM$, while in tandem organic photovoltaic cells, PCE of 7.15% was achieved by fabricating PEHBDT-DFDTBSe:$PC_{71}BM$ and P3HT:IC60BA (poly(3-hexylthiophene):indene-C60-bisadduct) as the top cell and bottom cell, respectively. Figure 5.10 also shows the schematic morphological models of these materials.

Chen *et al.* [94] also reported the preparation of donor-π-acceptor copolymer which consisted of alkylthienylbenzodithiophene (BDTT) (electron-donating moiety) and fluorinated thienyl-quinoxaline (TTFQ) (electron-accepting segment) containing units of thiophene π-bridge. The authors applied Stille coupling polymerization technique for generating the copolymers and subsequently utilized them as p-type semiconducting polymer. A wide visible-near-infrared absorption band with optical band gap of 1.67 eV was exhibited by PBDTT-TTFQ copolymer and possessed a HOMO level at 5.34 eV. Additionally, high dense fibrillar nanostructure was exhibited by the films of PBDTT-TTFQ which suggested self-assembly of the segments of PBDTT and TTFQ at nanoscale. The V_{oc}, J_{sc} and FF of the bulk-heterojunction polymer solar cell were recorded as 0.75 V, 14.6 mA.cm^2 and 56.1%, respectively. The polymer solar cell comprised of blend of PBDTT-TTFQ : PC71BM (1 : 1) and exhibited the power conversion

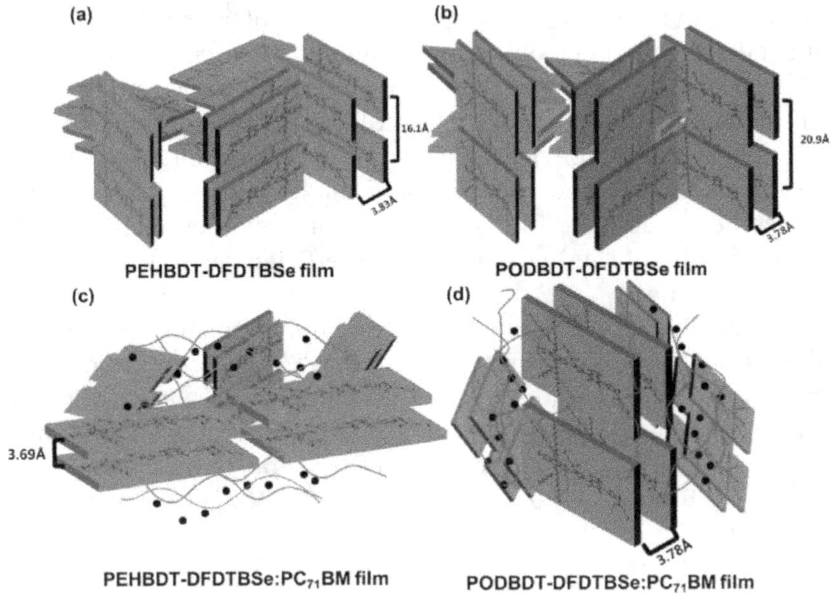

Figure 5.10 Schematic morphological models of (a, b) pure PEHBDT-DFDTBSe and PODBDT-DFDTBSe and (c, d) PEHBDT-DFDTBSe and PODBDT-DFDTBSe:PC71BM blend films. Reproduced from Reference 93 with permission from American Chemical Society.

efficiency of 6.1%. The authors concluded that the copolymers could be utilized for next-generation solar cells. Also, TTFQ unit was observed to be an ideal building block for developing polymer solar cells with enhanced performance. In another study, Dang *et al.* [80] reported the preparation of different copolymers containing a donor unit of indacenodithiophene (IDT), which is considered to be one of the most potential electron rich units because of its high degree of planarity which provides increased delocalization of an electron and therefore, a high charge-carrier mobility [54,95,96]. The copolymers were designed with either meta-hexyl-phenyl or para-hexylphenyl side chains and thiophene-quinoxalinethiophene acceptor unit. The authors utilized different type of acceptors, i.e. fluorinated unit (TQF) and non-fluorinated unit (TQ) where fluorinated copolymer exhibited little lower HOMO level as compared to non-fluorinated copolymers because of fluorine atom, which has electron-withdrawing nature [97,98]. Enhanced photovoltaic performance, higher molecular weight and improved solubility were exhibited by meta-substituted copolymers in both non-fluorinated and fluorinated copolymers. In addition, enhanced V_{oc} of 0.96 V and PCE of around 6.6% were achieved by bulk-heterojunction polymer solar cell based on the blends of fluorinated IDT copolymer (PIDTTQF-m) and $PC_{71}BM$, thereby making the material compatible for the generation of tandem devices. In case of devices based on non-fluorinated PIDTTQ-m: $PC_{71}BM$, enhanced photocurrent and slightly lower photo-voltage were achieved along with high power conversion efficiency (7.8%), which was one of the highest solar cell efficiencies in terms of IDT-based copolymers.

Ye *et al.* [99] reported the synthesis of polymer solar cell based on PBDT-TS1:PC71BM, exhibiting PCE of 9.48%. The authors designed conjugated (BDT) and thieno[3,4-b]thiophene (TT)-based copolymer (PBDTTS1, Figure 5.11a) through the replacement of 2-ethylhexyl with octyl, which was similar to the other reported polymers structures, i.e. PBT7-Th [100] and PBDTT-S-TT [82]. Linear alkylthio

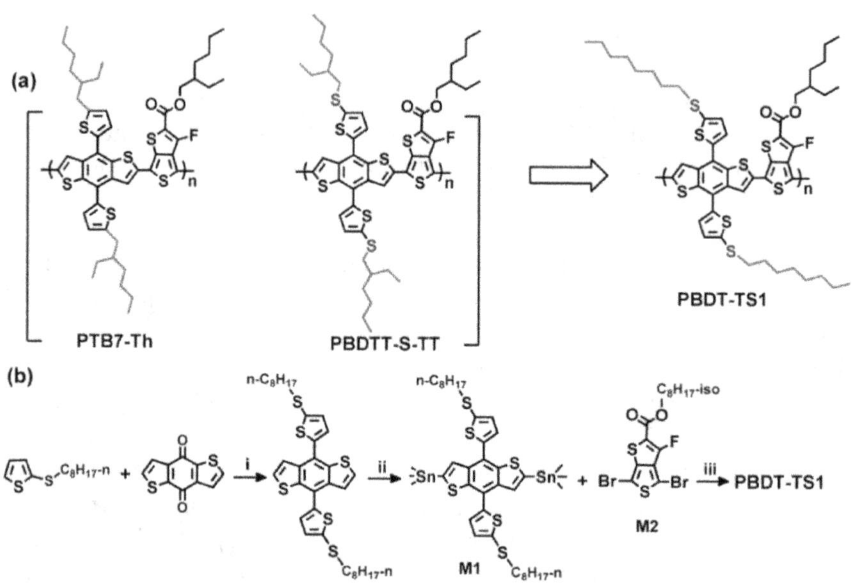

Figure 5.11 (a) Molecular structure of two reference polymers and target polymer; (b) synthesis procedure of PBDT-TS1. Reproduced from Reference 99 with permission from American Chemical Society.

chains were introduced for synthesizing PBDT-TS1 and its preparation route is demonstrated in Figure 5.11b. It is well known that alkyl chains are utilized as basic units to design organic semiconductors for obtaining well balanced intermolecular packing, high crystallinity and enhanced performance of device [101-105]. Moreover, since high starting point could be gained in conventional device, power conversion efficiency of 10% or more could be achieved in single-junction polymer solar cells through the integration of effective structures of devices [58,92,100,106,107].

Cui *et al.* [82] also reported the preparation of a low band gap two-dimensional conjugated polymer (PBDTT-S-TT). The authors subjected an alkylthio substituent onto the thiophene side chain in the polymer. The polymer's hole mobility was observed to be increased to 4.08 x $10^{-3}cm^2V^{-1}s^{-1}$ due to the alkylthio substituent and HOMO energy level was shifted downwards by 0.11 eV. It was noticed that improved V_{oc} of 0.84 eV was exhibited by polymer solar cell device based on PBDTT-S-TT which was recorded as highest value among all BDT and TT units based copolymers. Moreover, enhanced power conversion efficiency of around 8.42% was exhibited by PBDTT-S-TT based device. It could be concluded that molecular modification achieved through the introduction of an alkylthio side chain was a beneficial approach for broadening the absorption, down shifting the HOMO energy level and increasing the hole mobility of the two-dimensional conjugated polymer containing low band gap in order to enhance the photovoltaic performance of the polymer solar cells. Zhao *et al.* [108] reported the preparation of the semiconducting conjugated polymer, (BTT-BTz) poly{benzo(1,2-b:3,4-b':5,6-d") trithiophene-alt-4,4'-dihexyl-2,2'-bithiazole} containing a broad band gap and higher open circuit voltage based on benzotrithiophene and bithiazole units. The power conversion efficiency and open circuit voltage exhibited by BTT-BTz copolymer/$PC_{71}BM$ photoactive layer based device were recorded as 2.49% and 0.83 V, respectively. After treatment with 1,8-diiodoctane, appropriate morphology of the blend film was achieved and the PCE value reached 5.06% because of an increment in the values of J_{sc} and FF. So far, the observed efficiency of device has been the highest among all reported BTT or BTz derivatives based conjugated copolymers. It could be concluded that BTT-BTz based semiconducting copolymer can act as one of the potential active layers in the tandem cell structure for capturing various parts of the solar spectrum.

In another study, Keshtov *et al.* [109] applied Stille reaction conditions for the preparation of two donor-acceptor functional copolymers 4,8-bis-(5-bromothiophene-2-yl)-benzo [1,2,5]thiadiazole (P1) and 4,8-dithiophene-2-yl-benzo[1,2-c;4,5-c']-bis-[1,2,5]thiadiazole

(P2) (acceptor segment) with BDT (donor segment). Broad absorption bands were exhibited by the copolymers. With regards to optical band gaps, P1 and P2 exhibited the values of 1.62 eV and 1.16 eV, respectively. For the fabrication of polymer solar cells, P1 and P2 along with fullerene derivatives (PC60BM and PC70BM) were taken as electron donors and acceptor, respectively. The power conversion efficiency of around 3.45%, 1.19%, 2.17% and 0.80% were exhibited by the bulk heterojunction solar cells based on P1:PC_{70}BM, P2:PC_{70}BM, P1:PC_{60}BM and P2:PC_{60}BM blends (casted from THF solvent), respectively. P1 as electron donor exhibited enhanced power conversion efficiency due to its high V_{oc} which was attributed to the deep lying HOMO level, higher short circuit current density attributed to more force for exciton dissociation at the donor–acceptor joint into the layer of the bulk heterojunction solar cell and its relatively high mobility. Subsequently, the authors utilized DIO/THF (1,8-diiodooctane/tetrahydrofuran) solvent for casting blends and improved power conversion efficiencies of P1:PC_{70}BM and P2:PC_{70}BM blends were observed as 5.30% and 1.58%, respectively. The observed increment in power conversion efficiency was due to the larger donor-acceptor interface in the active layer for exciton dissociation, enhanced crystallinity of copolymer phase and balanced transportation of charge in the device.

Gedefaw *et al.* [110] prepared various p-type copolymers such as P(1)-PFQBDT-OR, P(2)-PFQBDT-TROR, P(3)-PFQBDT-TR and P(4)-PFQTT, termed as P(1), P(2), P(3), and P(4), respectively, which involved the fluorinated quinoxaline (FQ) acceptor unit with substituted benzodithiophene (BDT) or an unsubstituted thieno[3,2-b]thiophene (TT). The copolymers (P(1), P(2), P(3), and P(4)) in combination with PC61BM were utilized for the fabrication of blade-coated bulk heterojunction solar cells. The power conversion efficiencies of the devices based on P(1), P(2), P(3) and P(4) were recorded as 4.1%, 3.8%, 5.2% and 3.8%, respectively. Ho *et al.* [111] also reported the preparation of isoindigo-based low band gap copolymers (PnTI) with an expanded thiophene unit in the donor segment of the polymer. Isoindigo, being an acceptor unit, has gained much attention because of its sustainability (as it is extracted from plants) as well as its strong electron-withdrawing capability [112]. Absorption range of polymer was expanded by the increment in the thiophene unit length, while the HOMO and LUMO levels remained within an appropriate range for polymer solar cells. After the introduction of bithiophene into the donor segment with centrosymmetric conformation, the polymers exhibited improved crystallinity and consequently, enhanced hole mobility. Eventually, polymer P6TI exhibited the power conversion efficiency of around 7.25% as well as high J_{sc} of 16.24 mAcm^{-2}. So far, this PCE has been observed to be highest among all other isoindigo-based low band gap copolymers.

In a recent study, Andersen *et al.* [113] demonstrated the synthesis of slot-die coated photovoltaic device based on semi-random copolymer of poly-3-hexylthiophene and diketopyrrolopyrrole (P3HTT-DPP-10%) mixed with phenyl-C61-butyric acid methyl ester ($PC_{60}BM$). The device was fabricated by utilizing small roll-coater (slot-die coating only) and flexographic printing under an appropriate environment onto an indium tin oxide (ITO)-free substrate. For overcoming the low values of J_{sc} and FF achieved for single junction devices, the devices were generated by following tandem geometry, enabling the viability for employing thinner junction films. Single junction geometry based device exhibited power conversion efficiency of 1.31%, whereas the tandem geometry based device exhibited an efficiency of 1.36%. Earmme *et al.* [71] prepared naphthalene diimide-selenophene (PNDIS-HD) copolymer acceptor and a thiazolothiazole-dithienosilole (PSEHTT) copolymer donor based bulk heterojunction solar cells (Figure 5.12). The solar cells exhibited power conversion efficiency of around 3.3%. External quantum efficiency and J_{sc} were recorded as 47% and 7.78 mA cm^{-2} respectively. These parameters could be compared to the performance of similarly examined [6,6]-Phenyl-C60-butyric acid methyl ester (PC60BM)/PSEHTT devices. Appropriate electron and hole transport in the polymer-polymer blend solar cells was attributed to the PNDIS-HD morphology, i.e. lamellar crystalline which confirmed its potential photovoltaic properties.

Yuan *et al.* [114] reported the preparation and characterization of narrow band gap conjugated polymers based on a high mobility template for generating high performance visibly transparent photovoltaic devices. Nguyen *et al.* [81] reported the synthesis of a series of dialkoxyphenylene-BT which contained semi-crystalline low band gap polymers. The polymers were prepared by non-covalent conformational locking for enhancing intermolecular ordering, chain planarity and thermal stability without disturbing solution processability. Uniformly-distributed interpenetrating nano-fibrillar network morphologies with $PC_{70}BM$ were obtained by the polymers, exhibiting properly-balanced electron and hole mobilities. Power conversion efficiency of around 9.39% was exhibited by polymer solar cells

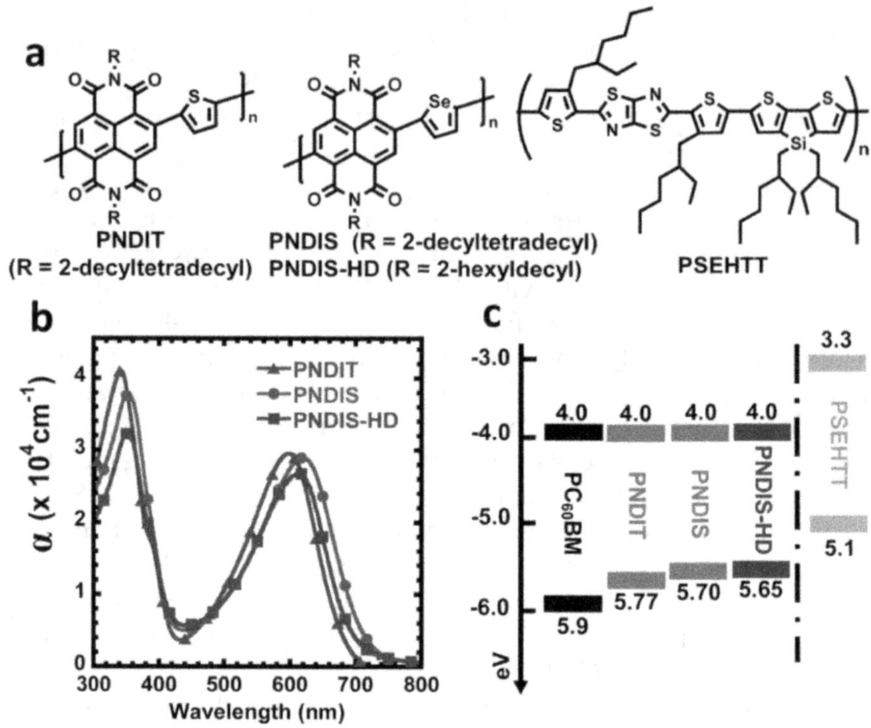

Figure 5.12 (a) Molecular structures of acceptor (PNDIT, PNDIS and PNDIS-HD) and donor (PSEHTT) polymers, (b) UV-vis absorption spectra of PNDIT, PNDIS and PNDIS-HD, (c) LUMO/HOMO energy levels of PNDIT, PNDIS, PNDIS-HD, PC60BM and PSEHTT. Reproduced from Reference 71 with permission from American Chemical Society.

based on these polymers in a structure of 290 nm thick single-cell device without using an extra interfacial layer. In another study, Liao *et al.* [115] employed dual doped zinc oxide nano-film as the cathode interlayer for generating single junction inverted polymer solar cell which exhibited PCE of around 10.31%. Figure 5.13 also demonstrates the chemical structures of the materials and energy level diagrams.

Chen *et al.* [116] reported another single-junction polymer solar cell which exceeded PCE of around 10%. Huang *et al.* [117] also employed Stille coupling reaction for the preparation of two conjugated copolymers (named as PSBDTeFBT and PSBDTeTID) based on asymmetric 4-(2-ethylhexyloxy)-8-(2-ethylhexylthio)benzo(1,2-b:4,5-b')dithiophene (SBDT) unit and different conjugated side chain group

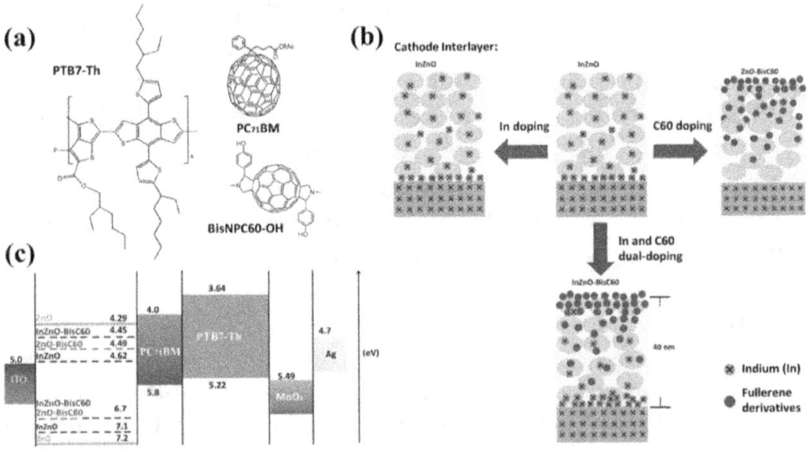

Figure 5.13 (a) Chemical structures of the materials, (b) illustration of the proposed cathode interlayer from XPS depth profile, (c) energy levels diagram for ZnO, InZnO, ZnO-BisC60 and InZnO-BisC60. Reproduced from Reference 115 with permission from Springer Nature.

DTffBT and TID (5,6-difluoro-4,7-bis(4-(2-ethylhexyl)thiophen-2-yl)benzo[c][1,2,5]thiadiazole and 6-(thiophen-2-yl)-di(2-ethylhexy)Isoindigo). The authors characterized the polymers by using cyclic voltammetry, UV-Vis absorption spectroscopy and thermal gravimetric analysis for analyzing electrochemical, optical and thermal properties, respectively. Good solubility in common organic solvents and enhanced thermal stability were exhibited by the copolymers. The values of deep HOMO energy levels for 5,6-difluoro-benzodiathiazole and isoindigo-based copolymers were observed as -5.56 eV and -5.48 eV, respectively. The power conversion efficiency of 5% and high open circuit voltage of 0.92 V were exhibited by the photovoltaic device based on 5,6-difluoro-benzodiathiazole-based polymer/phenyl-C71-butyric acid methyl ester.

Li *et al.* [118] reported the synthesis of hybrid rod-coil diblock copolymers with coordinatively attached nanocrystals of cadmium selenide (CdSe) and studied their optical properties and applicability for photovoltaic devices. Lu *et al.* [119] reported ternary blend polymer solar cell having poly-3-oxothieno[3,4-d]isothiazole-1,1-dioxide/benzodithiophene (PID2), polythieno[3,4-b]-thiophene/benzodithiophene (PTB7) as donors and [6,6]-phenyl C71 butyric acid methyl ester (PC71BM) as an acceptor. The values of FF, PCE, V_{oc} and J_{sc} of the as prepared ternary solar cell were observed as 68.7%, 8.22%,

0.72 V and 16.8 mA cm^{-2}, respectively. In another study, Najari *et al.* [120] employed direct heteroarylation polymerization for the synthesis of a series of push-pull conjugated polymers based on thieno[3,4- c]pyrrole-4,6-dione (TPD), furo[3,4- c]pyrrole-4,6-dione (FPD) and selenopheno[3,4-c]-pyrrole-4,6-dione (SePD). Squeo *et al.* [121] employed Stille cross-coupling technique for the preparation of ultra-low band gap a,b-unsubstituted 4,4-(difluoro-4-bora-3a,4a-diaza-s-indacene)-based conjugated polymer.

5.5 Conclusions

In the past few years, organic photovoltaics have exhibited immense development with respect to improvements in both lifetime and efficiency. In this chapter, the necessity of low band gap polymers as well as other requirements for the fabrication of polymer solar cells have been discussed. The review presents thorough compilation of recent studies reporting the design, synthesis and property evaluation of various polymers and copolymers for utilization in photovoltaic applications.

References

1. Brabec, C. J. (2004) Organic photovoltaics: technology and market. *Solar Energy Materials and Solar Cells*, **83**, 273-292.
2. Krebs, F. C., and Spanggaard, H. (2005) Significant improvement of polymer solar cell stability. *Chemistry of Materials*, **17**, 5235-5237.
3. Reyes-Reyes, M., Kim, K., Dewald, J., Lopez-Sandoval, R., Avadhanula, A., Curran, S., and Carroll, D. L. (2005) Meso-structure formation for enhanced organic photovoltaic cells. *Organic Letters*, **7**, 5749-5752.
4. Ma, W., Yang, C., Gong, X., Lee, K., and Heeger, A. J. (2005) Thermally stable, efficient polymer solar cells with nanoscale control of the interpenetrating network morphology. *Advanced Functional Materials*, **15**, 1617-1622.
5. Hoppe, H., and Sariciftci, N. S. (2006) Morphology of polymer/fullerene bulk heterojunction solar cells. *Journal of Materials Chemistry*, **16**, 45-61.
6. Yu, G., Gao, J., Hummelen, J. C., Wudl, F., and Heeger, A. J. (1995) Polymer photovoltiac cells: Enhanced efficiencies via a network of internal donor-acceptor heterojunctions. *Science*, **270**, 1789-1791.
7. Arias, A. C., MacKenzie, J. D., McCulloch, I., Rivnay, J., and Salleo, A. (2010) Materials and applications for large area electronics: soluti-

on-based approaches. *Chemical Reviews*, **110**, 3-24.

8. Chen, J., and Cao, Y. (2009) Development of novel conjugated donor polymers for high-efficiency bulk-heterojunction photovoltaic devices. *Accounts of Chemical Research*, **42**, 1709-1718.
9. Li, Y. (2012) Molecular design of photovoltaic materials for polymer solar cells: toward suitable electronic energy levels and broad absorption. *Accounts of Chemical Research*, **45**, 723-733.
10. Su, Y.-W., Lan, S.-C., and Wei, K.-H. (2012) Organic photovoltaics. *Materials Today*, **15**, 554-562.
11. Po, R., Carbonera, C., Bernardi, A., and Camaioni, N. (2011) The role of buffer layers in polymer solar cells. *Energy & Environmental Science*, **4**, 285-310.
12. Kim, J.-H., Song, C.E., Kang, I.-N., Shin, W.S., and Hwang, D.-H. (2013) A highly crystalline low band-gap polymer consisting of perylene and diketopyrrolopyrrole for organic photovoltaic cells. *Chemical Communications*, **49**, 3248-3250.
13. Kim, J.-H., Song, C.E., Kim, H. U., Grimsdale, A. C., Moon, S.-J., Shin, W. S., Choi, S. K., and Hwang, D.-H. (2013) High open circuit voltage solution-processed tandem organic photovoltaic cells employing a bottom cell using a new medium band gap semiconducting polymer. *Chemistry of Materials*, **25**, 2722-2732.
14. Wienk, M. M., Kroon, J. M., Verhees, W. J., Knol, J., Hummelen, J. C., van Hal, P. A., and Janssen, R. A. (2003) Efficient methano [70] fullerene/MDMO-PPV bulk heterojunction photovoltaic cells. *Angewandte Chemie, International Edition*, **115**, 3493-3497.
15. Li, G., Zhu, R., and Yang, Y. (2012) Polymer solar cells. *Nature Photonics*, **6**, 153-161.
16. Cheng, Y.-J., Yang, S.-H., and Hsu, C.-S. (2009) Synthesis of conjugated polymers for organic solar cell applications. *Chemical Reviews*, **109**, 5868-5923.
17. Li, Y., and Zou, Y. (2008) Conjugated polymer photovoltaic materials with broad absorption band and high charge carrier mobility. *Advanced Materials*, **20**, 2952-2958.
18. Gendron, D., and Leclerc, M. (2011) New conjugated polymers for plastic solar cells. *Energy & Environmental Science*, **4**, 1225-1237.
19. Zhou, H., Yang, L., and You, W. (2012) Rational design of high performance conjugated polymers for organic solar cells. *Macromolecules*, **45**, 607-632.
20. Mihailetchi, V. D., van Duren, J. K., Blom, P. W., Hummelen, J. C., Janssen, R. A., Kroon, J. M., Rispens, M. T., Verhees, W. J. H., and Wienk, M. M. (2003) Electron transport in a methanofullerene. *Advanced Functional Materials*, **13**, 43-46.
21. Dou, L., You, J., Hong, Z., Xu, Z., Li, G., Street, R. A., and Yang, Y. (2013) A decade of organic/polymeric photovoltaic research. *Advanced Materials*, **25**, 6642-6671.

22. Dhanabalan, A., Van Duren, J., Van Hal, P., Van Dongen, J., and Janssen, R. (2001) Synthesis and characterization of a low bandgap conjugated polymer for bulk heterojunction photovoltaic cells. *Advanced Functional Materials*, **11**, 255-262.
23. Li, W., Furlan, A., Hendriks, K. H., Wienk, M. M., and Janssen, R. A. (2013) Efficient tandem and triple-junction polymer solar cells. *Journal of the American Chemical Society*, **135**, 5529-5532.
24. He, Z., Zhong, C., Su, S., Xu, M., Wu, H., and Cao, Y. (2012) Enhanced power-conversion efficiency in polymer solar cells using an inverted device structure. *Nature Photonics*, **6**, 591-595.
25. Liu, S., Zhang, K., Lu, J., Zhang, J., Yip, H.-L., Huang, F., and Cao, Y. (2013) High-efficiency polymer solar cells via the incorporation of an amino-functionalized conjugated metallopolymer as a cathode interlayer. *Journal of the American Chemical Society*, **135**, 15326-15329.
26. Dou, L., You, J., Yang, J., Chen, C.-C., He, Y., Murase, S., Moriarty, T., Emery, K., Li, G., and Yang, Y. (2012) Tandem polymer solar cells featuring a spectrally matched low-bandgap polymer. *Nature Photonics*, **6**, 180-185.
27. You, J., Dou, L., Yoshimura, K., Kato, T., Ohya, K., Moriarty, T., Emery, K., Chen, C.-C., Gao, J., and Li, G. (2013) A polymer tandem solar cell with 10.6% power conversion efficiency. *Nature Communications*, **4**, 1446.
28. Wang, E., Ma, Z., Zhang, Z., Vandewal, K., Henriksson, P., Inganas, O., Zhang, F., and Andersson, M. R. (2011) An easily accessible isoindigo-based polymer for high-performance polymer solar cells. *Journal of the American Chemical Society*, **133**, 14244-14247.
29. Guo, X., Zhang, M., Tan, J., Zhang, S., Huo, L., Hu, W., Li, Y., and Hou, J. (2012) Influence of D/A ratio on photovoltaic performance of a highly efficient polymer solar cell system. *Advanced Materials*, **24**, 6536-6541.
30. Zhang, M., Guo, X., Zhang, S., and Hou, J. (2014) Synergistic effect of fluorination on molecular energy level modulation in highly efficient photovoltaic polymers. *Advanced Materials*, **26**, 1118-1123.
31. Brabec, C. J., Sariciftci, N. S., and Hummelen, J. C. (2001) Plastic solar cells. *Advanced Functional Materials*, **11**, 15-26.
32. He, Y., and Li, Y. (2011) Fullerene derivative acceptors for high performance polymer solar cells. *Physical Chemistry and Chemical Physics*, **13**, 1970-1983.
33. He, Y., Chen, H.-Y., Hou, J., and Li, Y. (2010) Indene–C60 bisadduct: a new acceptor for high-performance polymer solar cells. *Journal of the American Chemical Society*, **132**, 1377-1382.
34. Guo, X., Cui, C., Zhang, M., Huo, L., Huang, Y., Hou, J., and Li, Y. (2012) High efficiency polymer solar cells based on poly (3-hexylthiophene)/indene-C 70 bisadduct with solvent additive. *Energy & Envi-*

ronmental Science, **5**, 7943-7949.

35. Shockley, W., and Queisser, H. J. (1961) Detailed balance limit of efficiency of p-n junction solar cells. *Journal of Applied Physics*, **32**, 510-519.
36. Hadipour, A., de Boer, B., Wildeman, J., Kooistra, F. B., Hummelen, J. C., Turbiez, M. G., Wienk, M. M., Janssen, R. A., and Blom, P. W. (2006) Solution-Processed Organic Tandem Solar Cells. *Advanced Functional Materials*, **16**, 1897-1903.
37. Kim, J. Y., Lee, K., Coates, N. E., Moses, D., Nguyen, T.-Q., Dante, M., and Heeger, A. J. (2007) Efficient tandem polymer solar cells fabricated by all-solution processing. *Science*, **317**, 222-225.
38. Chou, C. H., Kwan, W. L., Hong, Z., Chen, L. M., and Yang, Y. (2011) A metal-oxide interconnection layer for polymer tandem solar cells with an inverted architecture. *Advanced Materials*, **23**, 1282-1286.
39. Gevaerts, V. S., Furlan, A., Wienk, M. M., Turbiez, M., and Janssen, R. A. (2012) Solution processed polymer tandem solar cell using efficient small and wide bandgap polymer: fullerene blends. *Advanced Materials*, **24**, 2130-2134.
40. Bailey-Salzman, R. F., Rand, B. P., and Forrest, S. R. (2006) Semitransparent organic photovoltaic cells. *Applied Physics Letters*, **88**, 233502.
41. Ng, G.-M., Kietzke, E. L., Kietzke, T., Tan, L.-W., Liew, P.-K., and Zhu, F. (2007) Optical enhancement in semitransparent polymer photovoltaic cells. *Applied Physics Letters*, **90**, 103505.
42. Meiss, J., Holzmueller, F., Gresser, R., Leo, K., and Riede, M. (2011) Near-infrared absorbing semitransparent organic solar cells. *Applied Physics Letters*, **99**, 193307.
43. Chen, C.-C., Dou, L., Zhu, R., Chung, C.-H., Song, T.-B., Zheng, Y. B., Hawks, S., Li, G., Weiss, P. S., and Yang, Y. (2012) Visibly transparent polymer solar cells produced by solution processing. *ACS Nano*, **6**, 7185-7190.
44. Thompson, B. C., and Frechet, J. M. (2008) Polymer–fullerene composite solar cells. *Angewandte Chemie, International Edition*, **47**, 58-77.
45. Boudreault, P.-L. T., Najari, A., and Leclerc, M. (2010) Processable low-bandgap polymers for photovoltaic applications. *Chemistry of Materials*, **23**, 456-469.
46. Beaujuge, P. M., and Frechet, J. M. (2011) Molecular design and ordering effects in π-functional materials for transistor and solar cell applications. *Journal of the American Chemical Society*, **133**, 20009-20029.
47. Li, G., Shrotriya, V., Huang, J., Yao, Y., Moriarty, T., Emery, K., and Yang, Y. (2005) High-efficiency solution processable polymer photovoltaic cells by self-organization of polymer blends. *Nature Materials*, **4**, 864-868.

48. Chen, H.-Y., Hou, J., Zhang, S., Liang, Y., Yang, G., Yang, Y., Yu, L., Wu, Y., and Li, G. (2009) Polymer solar cells with enhanced open-circuit voltage and efficiency. *Nature Photonics*, **3**, 649-653.
49. Liang, Y., Xu, Z., Xia, J., Tsai, S.T., Wu, Y., Li, G., Ray, C., and Yu, L. (2010) For the bright future-bulk heterojunction polymer solar cells with power conversion efficiency of 7.4%. *Advanced Materials*, **22**(20), E135-E138.
50. He, Z., Zhong, C., Huang, X., Wong, W.Y., Wu, H., Chen, L., Su, S., and Cao, Y. (2011) Simultaneous enhancement of open-circuit voltage, short-circuit current density, and fill factor in polymer solar cells. *Advanced Materials*, **23**, 4636-4643.
51. Amb, C. M., Chen, S., Graham, K. R., Subbiah, J., Small, C. E., So, F., and Reynolds, J. R. (2011) Dithienogermole as a fused electron donor in bulk heterojunction solar cells. *Journal of the American Chemical Society*, **133**, 10062-10065.
52. Chu, T.-Y., Lu, J., Beaupre, S., Zhang, Y., Pouliot, J.-R., Wakim, S., Zhou, J., Leclerc, M., Li, Z., and Ding, J. (2011) Bulk heterojunction solar cells using thieno [3, 4-c] pyrrole-4, 6-dione and dithieno [3, 2-b: 2′, 3′-d] silole copolymer with a power conversion efficiency of 7.3%. *Journal of the American Chemical Society*, **133**, 4250-4253.
53. Huo, L., Zhang, S., Guo, X., Xu, F., Li, Y., and Hou, J. (2011) Replacing alkoxy groups with alkylthienyl groups: a feasible approach to improve the properties of photovoltaic polymers. *Angewandte Chemie, International Edition*, **123**, 9871-9876.
54. Zhang, Y., Zou, J., Yip, H.-L., Chen, K.-S., Zeigler, D.F., Sun, Y., and Jen, A. K.-Y. (2011) Indacenodithiophene and quinoxaline-based conjugated polymers for highly efficient polymer solar cells. *Chemistry of Materials*, **23**, 2289-2291.
55. Ye, L., Jiao, X., Zhang, H., Li, S., Yao, H., Ade, H., and Hou, J. (2015) 2D-conjugated benzodithiophene-based polymer acceptor: design, synthesis, nanomorphology, and photovoltaic performance. *Macromolecules*, **48**, 7156-7163.
56. Zhang, M., Guo, X., Ma, W., Ade, H., and Hou, J. (2015) A Large-Bandgap Conjugated Polymer for Versatile Photovoltaic Applications with High Performance. *Advanced Materials*, **27**, 4655-4660.
57. Liu, D., Gu, C., Xiao, M., Qiu, M., Sun, M., and Yang, R. (2015) A new highly conjugated crossed benzodithiophene and its donor–acceptor copolymers for high open circuit voltages polymer solar cells. *Polymer Chemistry*, **6**, 3398-3406.
58. Dou, L., Chang, W. H., Gao, J., Chen, C. C., You, J., and Yang, Y. (2013) A selenium-substituted low-bandgap polymer with versatile photovoltaic applications. *Advanced Materials*, **25**, 825-831.
59. Dou, L., Gao, J., Richard, E., You, J., Chen, C.-C., Cha, K. C., He, Y., Li, G., and Yang, Y. (2012) Systematic investigation of benzodithiophene- and diketopyrrolopyrrole-based low-bandgap polymers designed

for single junction and tandem polymer solar cells. *Journal of the American Chemical Society*, **134**, 10071-10079.

60. Kronemeijer, A. J., Gili, E., Shahid, M., Rivnay, J., Salleo, A., Heeney, M., and Sirringhaus, H. (2012) A selenophene-based low-bandgap donor–acceptor polymer leading to fast ambipolar logic. *Advanced Materials*, **24**, 1558-1565.
61. Shahid, M., McCarthy-Ward, T., Labram, J., Rossbauer, S., Domingo, E. B., Watkins, S. E., Stingelin, N., Anthopoulos, T. D., and Heeney, M. (2012) Low band gap selenophene–diketopyrrolopyrrole polymers exhibiting high and balanced ambipolar performance in bottom-gate transistors. *Chemical Science*, **3**, 181-185.
62. Ballantyne, A. M., Chen, L., Nelson, J., Bradley, D. D., Astuti, Y., Maurano, A., Shuttle, C. G., Durrant, J. R., Heeney, M., and Duffy, W. (2007) Studies of highly regioregular poly (3-hexylselenophene) for photovoltaic applications. *Advanced Materials*, **19**, 4544-4547.
63. Klein, M. F., Pfaff, M., Muller, E., Czolk, J., Reinhard, M., Valouch, S., Lemmer, U., Colsmann, A., and Gerthsen, D. (2012) Poly (3-hexylselenophene) solar cells: Correlating the optoelectronic device performance and nanomorphology imaged by low-energy scanning Transmission electron microscopy. *Journal of Polymer Science, Part B: Polymer Physics*, **50**, 198-206.
64. Zhang, Z., Lin, F., Chen, H.-C., Wu, H.-C., Chung, C.-L., Lu, C., Liu, S.-H., Tung, S.-H., Chen, W.-C., and Wong, K.-T. (2015) A silole copolymer containing a ladder-type heptacylic arene and naphthobisoxadiazole moieties for highly efficient polymer solar cells. *Energy & Environmental Science*, **8**, 552-557.
65. Liu, D., Zhao, W., Zhang, S., Ye, L., Zheng, Z., Cui, Y., Chen, Y., and Hou, J. (2015) Highly efficient photovoltaic polymers based on benzodithiophene and quinoxaline with deeper HOMO levels. *Macromolecules*, **48**, 5172-5178.
66. Lu, L., Zheng, T., Wu, Q., Schneider, A. M., Zhao, D., and Yu, L. (2015) Recent advances in bulk heterojunction polymer solar cells. *Chemical Reviews*, **115**, 12666-12731.
67. Duan, C., Gao, K., van Franeker, J. J., Liu, F., Wienk, M. M., and Janssen, R. A. (2016) Toward practical useful polymers for highly efficient solar cells via a random copolymer approach. *Journal of the American Chemical Society*, **138**, 10782-10785.
68. Nakano, K., Nakano, M., Xiao, B., Zhou, E., Suzuki, K., Osaka, I., Takimiya, K., and Tajima, K. (2016) Naphthodithiophene diimide-based copolymers: Ambipolar semiconductors in field-effect transistors and electron acceptors with near-infrared response in polymer blend solar cells. *Macromolecules*, **49**, 1752-1760.
69. Liu, T., Pan, X., Meng, X., Liu, Y., Wei, D., Ma, W., Huo, L., Sun, X., Lee, T. H., and Huang, M. (2017) Alkyl side-chain engineering in wide-bandgap copolymers leading to power conversion efficiencies over

10%. *Advanced Materials*, **29**(6), 1604251.

70. Zhou, E., Nakano, M., Izawa, S., Cong, J., Osaka, I., Takimiya, K., and Tajima, K. (2014) All-polymer solar cell with high near-infrared response based on a Naphthodithiophene Diimide (NDTI) copolymer. *ACS Macro Letters*, **3**, 872-875.
71. Earmme, T., Hwang, Y.-J., Murari, N.M., Subramaniyan, S., and Jenekhe, S. A. (2013) All-polymer solar cells with 3.3% efficiency based on naphthalene diimide-selenophene copolymer acceptor. *Journal of the American Chemical Society*, **135**, 14960-14963.
72. Marchiori, C. F., Yamamoto, N. A., Matos, C. F., Kujala, J., Macedo, A. G., Tuomisto, F., Zarbin, A. J., Koehler, M., and Roman, L. S. (2015) Annealing effect on donor-acceptor interface and its impact on the performance of organic photovoltaic devices based on PSiF-DBT copolymer and C60. *Applied Physics Letters*, **106**, 133301.
73. Kim, J.-H., Kim, H. U., Song, C. E., Kang, I.-N., Lee, J.-K., Shin, W. S., and Hwang, D.-H. (2013) Benzotriazole-based donor–acceptor type semiconducting polymers with different alkyl side chains for photovoltaic devices. *Solar Energy Materials and Solar Cells*, **108**, 113-125.
74. Saeki, A., Tsuji, M., Yoshikawa, S., Gopal, A., and Seki, S. (2014) Boosting photovoltaic performance of a benzobisthiazole based copolymer: a device approach using a zinc oxide electron transport layer. *Journal of Materials Chemistry A*, **2**, 6075-6080.
75. Guo, C., Lin, Y.-H., Witman, M. D., Smith, K. A., Wang, C., Hexemer, A., Strzalka, J., Gomez, E. D., and Verduzco, R. (2013) Conjugated block copolymer photovoltaics with near 3% efficiency through microphase separation. *Nano Letters*, **13**, 2957-2963.
76. Krebs, F. C., Espinosa, N., Hosel, M., Sondergaard, R. R., and Jorgensen, M. (2014) Rise to power-OPV-based solar parks. *Advanced Materials*, **26**, 29-39.
77. Espinosa, N., Hosel, M., Jorgensen, M., and Krebs, F. C. (2014) Large scale deployment of polymer solar cells on land, on sea and in the air. *Energy & Environmental Science*, **7**, 855-866.
78. Zhan, X., and Zhu, D. (2010) Conjugated polymers for high-efficiency organic photovoltaics. *Polymer Chemistry*, **1**, 409-419.
79. Dou, L., Chen, C.-C., Yoshimura, K., Ohya, K., Chang, W.-H., Gao, J., Liu, Y., Richard, E., and Yang, Y. (2013) Synthesis of 5 H-dithieno [3, 2-b: 2′, 3′-d] pyran as an electron-rich building block for donor-acceptor type low-bandgap polymers. *Macromolecules*, **46**, 3384-3390.
80. Dang, D., Chen, W., Himmelberger, S., Tao, Q., Lundin, A., Yang, R., Zhu, W., Salleo, A., Muller, C., and Wang, E. (2014) Enhanced photovoltaic performance of indacenodithiophene-quinoxaline copolymers by side-chain modulation. *Advanced Energy Materials*, **4**(15), 1400680.
81. Nguyen, T. L., Choi, H., Ko, S.-J., Uddin, M. A., Walker, B., Yum, S.,

Jeong, J.-E., Yun, M., Shin, T., and Hwang, S. (2014) Semi-crystalline photovoltaic polymers with efficiency exceeding 9% in a ~300 nm thick conventional single-cell device. *Energy & Environmental Science*, **7**, 3040-3051.

82. Cui, C., Wong, W.-Y., and Li, Y. (2014) Improvement of open-circuit voltage and photovoltaic properties of 2D-conjugated polymers by alkylthio substitution. *Energy & Environmental Science*, **7**, 2276-2284.
83. Chen, Z., Cai, P., Chen, J., Liu, X., Zhang, L., Lan, L., Peng, J., Ma, Y., and Cao, Y. (2014) Low band-gap conjugated polymers with strong interchain aggregation and very high hole mobility towards highly efficient thick-film polymer solar cells. *Advanced Materials*, **26**, 2586-2591.
84. Stalder, R., Mei, J., and Reynolds, J.R. (2010) Isoindigo-based donor-acceptor conjugated polymers. *Macromolecules*, **43**, 8348-8352.
85. Lei, T., Wang, J.-Y., and Pei, J. (2014) Design, synthesis, and structure–property relationships of isoindigo-based conjugated polymers. *Accounts of Chemical Research*, **47**, 1117-1126.
86. Robb, M., Ku, S., Brunetti, F., and Hawker, C. (2013) A renaissance of color: New structures and building blocks for organic electronics. *Journal of Polymer Science, Part A: Polymer Chemistry*, **51**(6), 1263-1271.
87. Chen, M. S., Niskala, J. R., Unruh, D. A., Chu, C. K., Lee, O. P., and Frechet, J. M. (2013) Control of polymer-packing orientation in thin films through synthetic tailoring of backbone coplanarity. *Chemistry of Materials*, **25**, 4088-4096.
88. Wang, E., Mammo, W., and Andersson, M. R. (2014) Isoindigo-based polymers and small molecules for bulk heterojunction solar cells and field effect transistors. *Advanced Materials*, **26**, 1801-1826.
89. Wang, E., Ma, Z., Zhang, Z., Henriksson, P., Inganas, O., Zhang, F., and Andersson, M. R. (2011) An isoindigo-based low band gap polymer for efficient polymer solar cells with high photo-voltage. *Chemical Communications*, **47**, 4908-4910.
90. Ma, Z., Wang, E., Vandewal, K., Andersson, M. R., and Zhang, F. (2011) Enhance performance of organic solar cells based on an isoindigo-based copolymer by balancing absorption and miscibility of electron acceptor. *Applied Physics Letters*, **99**, 143302.
91. Tao, Q., Xia, Y., Xu, X., Hedstrom, S., Backe, O., James, D. I., Persson, P., Olsson, E., Inganas, O., and Hou, L. (2015) D–A1–D–A2 copolymers with extended donor segments for efficient polymer solar cells. *Macromolecules*, **48**, 1009-1016.
92. Liu, P., Zhang, K., Liu, F., Jin, Y., Liu, S., Russell, T. P., Yip, H.-L., Huang, F., and Cao, Y. (2014) Effect of fluorine content in thienothiophene-benzodithiophene copolymers on the morphology and performance of polymer solar cells. *Chemistry of Materials*, **26**, 3009-3017.

93. Kim, J.-H., Shin, S. A., Park, J. B., Song, C. E., Shin, W. S., Yang, H., Li, Y., and Hwang, D.-H. (2014) Fluorinated benzoselenadiazole-based low-band-gap polymers for high efficiency inverted single and tandem organic photovoltaic cells. *Macromolecules*, **47**, 1613-1622.
94. Chen, H.-C., Chen, Y.-H., Liu, C.-H., Hsu, Y.-H., Chien, Y.-C., Chuang, W.-T., Cheng, C.-Y., Liu, C.-L., Chou, S.-W., and Tung, S.-H. (2013) Fluorinated thienyl-quinoxaline-based D–π–A-type copolymer toward efficient polymer solar cells: synthesis, characterization, and photovoltaic properties. *Polymer Chemistry*, **4**, 3411-3418.
95. McCulloch, I., Ashraf, R. S., Biniek, L., Bronstein, H., Combe, C., Donaghey, J. E., James, D. I., Nielsen, C. B., Schroeder, B. C., and Zhang, W. (2012) Design of semiconducting indacenodithiophene polymers for high performance transistors and solar cells. *Accounts of Chemical Research*, **45**, 714-722.
96. Chen, Y.-C., Yu, C.-Y., Fan, Y.-L., Hung, L.-I., Chen, C.-P., and Ting, C. (2010) Low-bandgap conjugated polymer for high efficient photovoltaic applications. *Chemical Communications*, **46**, 6503-6505.
97. Dang, D., Chen, W., Yang, R., Zhu, W., Mammo, W., and Wang, E. (2013) Fluorine substitution enhanced photovoltaic performance of a D–A 1–D–A 2 copolymer. *Chemical Communications*, **49**, 9335-9337.
98. Albrecht, S., Janietz, S., Schindler, W., Frisch, J., Kurpiers, J., Kniepert, J., Inal, S., Pingel, P., Fostiropoulos, K., and Koch, N. (2012) Fluorinated copolymer PCPDTBT with enhanced open-circuit voltage and reduced recombination for highly efficient polymer solar cells. *Journal of the American Chemical Society*, **134**, 14932-14944.
99. Ye, L., Zhang, S., Zhao, W., Yao, H., and Hou, J. (2014) Highly efficient 2D-conjugated benzodithiophene-based photovoltaic polymer with linear alkylthio side chain. *Chemistry of Materials*, **26**, 3603-3605.
100. Liao, S. H., Jhuo, H. J., Cheng, Y. S., and Chen, S. A. (2013) Fullerene derivative-doped zinc oxide nanofilm as the cathode of inverted polymer solar cells with low-bandgap polymer (PTB7-Th) for high performance. *Advanced Materials*, **25**, 4766-4771.
101. Mei, J., and Bao, Z. (2013) Side chain engineering in solution-processable conjugated polymers. *Chemistry of Materials*, **26**, 604-615.
102. Chen, M. S., Lee, O. P., Niskala, J. R., Yiu, A.T., Tassone, C. J., Schmidt, K., Beaujuge, P. M., Onishi, S. S., Toney, M. F., and Zettl, A. (2013) Enhanced solid-state order and field-effect hole mobility through control of nanoscale polymer aggregation. *Journal of the American Chemical Society*, **135**, 19229-19236.
103. Zhang, F., Hu, Y., Schuettfort, T., Di, C.-a., Gao, X., McNeill, C. R., Thomsen, L., Mannsfeld, S. C., Yuan, W., and Sirringhaus, H. (2013) Critical role of alkyl chain branching of organic semiconductors in enabling solution-processed n-channel organic thin-film transistors with mobility of up to 3.50 $cm^2 V^{-1} s^{-1}$. *Journal of the American Chemical*

Society, **135**, 2338-2349.
104. Lei, T., Dou, J. H., and Pei, J. (2012) Influence of alkyl chain branching positions on the hole mobilities of polymer thin-film transistors. *Advanced Materials*, **24**, 6457-6461.
105. Meager, I., Ashraf, R. S., Mollinger, S., Schroeder, B. C., Bronstein, H., Beatrup, D., Vezie, M. S., Kirchartz, T., Salleo, A., and Nelson, J. (2013) Photocurrent enhancement from diketopyrrolopyrrole polymer solar cells through alkyl-chain branching point manipulation. *Journal of the American Chemical Society*, **135**, 11537-11540.
106. Guo, X., Zhang, M., Ma, W., Ye, L., Zhang, S., Liu, S., Ade, H., Huang, F., and Hou, J. (2014) Enhanced photovoltaic performance by modulating surface composition in bulk heterojunction polymer solar cells based on PBDTTT-C-T/PC71BM. *Advanced Materials*, **26**, 4043-4049.
107. Chen, K. S., Yip, H. L., Salinas, J. F., Xu, Y. X., Chueh, C. C., and Jen, A. K. Y. (2014) Strong photocurrent enhancements in highly efficient flexible organic solar cells by adopting a microcavity configuration. *Advanced Materials*, **26**, 3349-3354.
108. Zhao, X., Yang, D., Lv, H., Yin, L., and Yang, X. (2013) New benzotrithiophene derivative with a broad band gap for high performance polymer solar cells. *Polymer Chemistry*, **4**, 57-60.
109. Keshtov, M., Marochkin, D., Kochurov, V., Khokhlov, A., Koukaras, E., and Sharma, G. (2014) New conjugated alternating benzodithiophene-containing copolymers with different acceptor units: synthesis and photovoltaic application. *Journal of Materials Chemistry A*, **2**, 155-171.
110. Gedefaw, D., Tessarolo, M., Prosa, M., Bolognesi, M., Henriksson, P., Zhuang, W., Seri, M., Muccini, M., and Andersson, M. R. (2016) Induced photodegradation of quinoxaline based copolymers for photovoltaic applications. *Solar Energy Materials and Solar Cells*, **144**, 150-158.
111. Ho, C.-C., Chen, C.-A., Chang, C.-Y., Darling, S. B., and Su, W.-F. (2014) Isoindigo-based copolymers for polymer solar cells with efficiency over 7%. *Journal of Materials Chemistry A*, **2**, 8026-8032.
112. Puchalska, M., Polec-Pawlak, K., Zadrozna, I., Hryszko, H., and Jarosz, M. (2004) Identification of indigoid dyes in natural organic pigments used in historical art objects by high-performance liquid chromatography coupled to electrospray ionization mass spectrometry. *Journal of Mass Spectrometry*, **39**, 1441-1449.
113. Andersen, T. R., Dam, H. F., Burkhart, B., Angmo, D., Corazza, M., Thompson, B. C., and Krebs, F. C. (2014) Medium area, flexible single and tandem junction solar cells based on roll coated semi-random copolymers. *Journal of Materials Chemistry C*, **2**, 9412-9415.
114. Yuan, J., Ford, M., Ding, G., Dong, H., Wang, M., Han, L., Li, Y., Bazan, G. C., and Ma, W. (2016) Narrow bandgap conjugated polymers

based on a high-mobility polymer template for visibly transparent photovoltaic devices. *Journal of Materials Chemistry A*, **4**, 17333-17343.

115. Liao, S.-H., Jhuo, H.-J., Yeh, P.-N., Cheng, Y.-S., Li, Y.-L., Lee, Y.-H., Sharma, S., and Chen, S.-A. (2014) Single junction inverted polymer solar cell reaching power conversion efficiency 10.31% by employing dual-doped zinc oxide nano-film as cathode interlayer. *Scientific Reports*, **4**, 6813.
116. Chen, J. D., Cui, C., Li, Y.Q., Zhou, L., Ou, Q. D., Li, C., Li, Y., and Tang, J. X. (2015) Single-junction polymer solar cells exceeding 10% power conversion efficiency. *Advanced Materials*, **27**, 1035-1041.
117. Huang, Y., Wu, F., Zhang, M., Mei, S., Shen, P., and Tan, S. (2015) Synthesis and photovoltaic properties of conjugated polymers with an asymmetric 4-(2-ethylhexyloxy)-8-(2-ethylhexylthio) benzo [1, 2-b: 4, 5-b′] dithiophene unit. *Dyes and Pigments*, **115**, 58-66.
118. Li, S., Li, Y., Wisner, C. A., Jin, L., Leventis, N., and Peng, Z. (2014) Synthesis, optical properties and photovoltaic applications of hybrid rod–coil diblock copolymers with coordinatively attached CdSe nanocrystals. *RSC Advances*, **4**, 35823-35832.
119. Lu, L., Xu, T., Chen, W., Landry, E. S., and Yu, L. (2014) Ternary blend polymer solar cells with enhanced power conversion efficiency. *Nature Photonics*, **8**, 716-722.
120. Najari, A., Beaupre, S., Allard, N., Ouattara, M., Pouliot, J. R., Charest, P., Besner, S., Simoneau, M., and Leclerc, M. (2015) Thieno, furo, and selenopheno [3, 4-c] pyrrole-4, 6-dione copolymers: Air-processed polymer solar cells with power conversion efficiency up to 7.1%. *Advanced Energy Materials*, **5**(23), 1501213.
121. Squeo, B. M., Gasparini, N., Ameri, T., Palma-Cando, A., Allard, S., Gregoriou, V. G., Brabec, C. J., Scherf, U., and Chochos, C. L. (2015) Ultra low band gap α, β-unsubstituted BODIPY-based copolymer synthesized by palladium catalyzed cross-coupling polymerization for near infrared organic photovoltaics. *Journal of Materials Chemistry A*, **3**, 16279-16286.

6

Polymer and Polymer Nanocomposite Nanofoams: Synthesis, Properties and Applications

6.1 Introduction

The challenges associated with the use of polymers in different sectors have resulted in the development of novel lightweight polymer nanocomposites by merging foaming with the incorporation of small amounts of nanosized particles with functional features. The combination of these results in materials known as multifunctional polymer foams.

Carbon based nanomaterials such as graphene, carbon nanotubes and other nanomaterials have attracted significant research interest as reinforcing nanoparticles because of their inherently extraordinary mechanical and transport properties. For instance, the addition of carbon based nanoparticles has been observed as a promising strategy to tune the electronic transport through polymers due to their conducting/semiconducting structure-dependent electronic behavior. The developed lightweight materials can find applications from electrostatic painting, electrostatic discharge ESD (packaging materials for ESD sensitive items, fuel system components, etc.) to electromagnetic interference (EMI) shielding (gaskets for electronic devices, fuel cells, etc.) [1]. For instance, Krause [2] and Wessling *et al.* [3] developed foams which could be used for membrane separations. Efforts are also largely focused on creating foams with poor dielectric behaviors for use as insulating materials [4,5]. A large segment of the energy used for heating of houses and utility buildings is wasted from these structures and the use of nanofoam materials can help to increase the thermal insulation [6].

The current chapter reviews recent studies describing the synthesis and properties of polymer nanofoams as well as polymer nanocomposite nanofoams for use in a wide variety of advanced applications.

Swati Singh and Vikas Mittal, The Petroleum Institute (part of Khalifa University of Science and Technology), Abu Dhabi, UAE*
**Current address: Bletchington, Wellington County, Australia*

6.2 Polymer Nanofoams

One of the useful characteristics of polymeric materials is low density, relative to metals and ceramics. Moreover, to further enhance this advantage, bulk polymer materials can be simply processed into polymer foams. Undeniably, polymer foams make significant contributions in our daily life and in a variety of industrial fields in which impact resistance, weight reduction and heat or sound insulation are of importance [7]. At the same time, the controllable mass transferring property and high surface ratio of polymer foams provide them new promising applications in supporting catalysts, gas capture and storage, bio scaffolding [8,9], separation, [10,11] and drug release [12]. For many of these applications, different cellular matrices based on polystyrene and polyurethane are commonly used. These materials exhibit conductivities similar to that of insulating gases [13]. Also, the Knudsen effect and heat conduction in nanoporous foams is limited further due to their low thermal conductivity [14]. Poly(methyl methacrylate) (PMMA) based foams have also been utilized for noise and electrical insulation [15-17].

Comparing with micro-cellular foams, nanocellular foams have better insulation properties owing to the reduction in density [18] as well as the pores with nano-dimensions limiting the gaseous phase conduction [19-20]. The reduction in solid phase ratio can also have a significant effect on the solid phase conduction owing to the decrease in density. Theoretical studies and analytical models have been developed to exhibit the dependence of conductivity on the size and density of cells in polymer foams [21]. It is proposed that the foam should possess density between 0.1 and 0.2 $g.cm^{-3}$ and a cell size lower than 100 nm. Effective methods for synthesizing such challenging systems need to be developed, in addition to the traditionally used chemical processes and etching via chemical route [22,23]. Use of CO_2 for polymer foaming has also shown convincing results.

6.2.1 Common Polymer Foams Depending on the Foaming Processes

Table 6.1 briefly mentions a few common polymer foams depending on the conventional foaming processes. Foams generated from polymers such as polystyrene, polyurethane and polyolefins have been mentioned, whereas processes like extrusion, injection and compression molding, among others, are listed.

Table 6.1 Common polymer foams based on foaming processes

Polystyrene foams	• Extrusion physical foaming foams (XPS) • Expanded bead molding foams (EPS) • Expanded loose-fill foams • Injection-molding structural foams	
Polyurethane foams	• Flexible PU foams (FPF)	➢ Slab stock foams ➢ Molded foams
	• Rigid PU foams	➢ Foamed blocks and sheets (continuous/discontinuous processes) ➢ Poured-in-place foams ➢ Molded foams ➢ Sprayed foams
Polyolefin foams	• Continuous/semi-continuous foaming processes (chemical or physical foaming)	➢ Foams obtained by extrusion ➢ Foams obtained by injection-molding
	• Batch foaming processes	➢ Compression-molded foams ➢ Foams obtained by gas dissolution ➢ Expandable bead foams (EPP and EPE)

6.2.2 Synthesis of Polymer Nanofoams

Use of Blowing Agents

The polymer foams (porous polymers) are one of the most appealing areas of materials science [24-29]. The blowing agent is vital to process polymer foams as it discharges a gaseous substance under specific conditions. Polymer blowing agents can operate based on physical or chemical mechanisms. Azo-dicarbonamide (AC), 1,1-dichloro-1-fluoroethane (HCFC-141b) and hydrogen-containing polysiloxane are common blowing agents. These organic blowing agents have a negative impact on environment as these are generated from fossil based raw materials and are ultimately released to the atmosphere. Replacement of these blowing agents with environmentally friendly chemicals is of great advantage for sustainable expansion of polymer foaming industry. CO_2 exemplifies a beneficial candidate as a blowing agent due to its significant abundance as well as human and environmentally favorable characteristics, though direct use of CO_2 as a blowing agent is quite demanding since it is in gaseous state under normal conditions. Simple process of CO_2 foaming can be divided into two different stages. During the first stage, the polymer is drenched in gas in a pressure autoclave. The cell growth and nucleation is activated by thermodynamic instability. Control on the temperature and pressure can help in elongating the sorption stage. The generation of foams can be controlled by regulating the coalescence phenomenon, which is an important factor in determining cell growth. Several attempts have been made to improve cell density and reduce cell size [30]. The pore size can be reduced by maintaining high cell density and by using high pressure during the sorption stage (more than 20 MPa). It helps in increasing pressure drop rate and gas uptake. Nucleation is also prompted by sudden depressurization.

Yokoyama and Sugiyama [31] used a saturation pressure of 30 MPa for the formation of foams of copolymers of polystyrene–perfluorooctylethyl methacrylate (PFMA). The use of high pressure increased the intake of gas in the block of PFMA, which was observed to have high CO_2-philicity than polystyrene and confirmed the creation of nanopores within fluorinated zones [32]. Figure 6.1 also shows the mechanism of expanding nanocells in block copolymer nanodomains and elastic forces acting on the cells. Forest *et al.* [33] used CO_2 foaming process for PMMA and displayed that the PMMA/CO_2 system indicated backward vitrification below a 5 MPa saturation pressure. For the same pressure, the system had two glass transitions, in a broad array of temperatures [34]. Due to the plasticization effect of CO_2 on the PMMA matrix, the first glass transition occurred near to that of neat PMMA. Certainly, the CO_2 sorption inside polymers causes drop in glass transition temperature by abating the pull among polymer molecules. However, far from the first glass transition temperature, in case the saturation stage was achieved at a lower temperature, the CO_2 solubility in PMMA could have become high enough to once more reach the PMMA's rubbery state [35].

Ren *et al.* [36] prepared epoxy foam by optimal choice of amine for the synthesis of carbamate as latent blowing and curing agent. Some

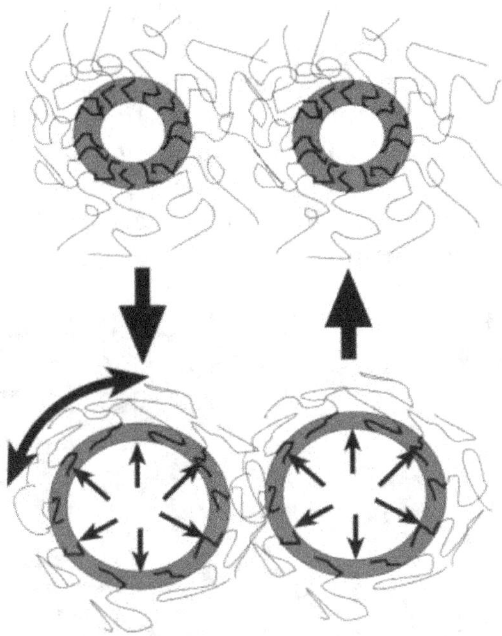

Figure 6.1 Schematic illustration of expanding nanocells in block copolymer nanodomains and elastic forces acting on the cells. Reproduced from Reference 31 with permission from American Chemical Society.

liquid amines used as curing agents for epoxy resins can react with gaseous CO_2 during bubbling to form solid ammonium carbamates. The carbamates are CO_2-blocked amines, which can decompose and liberate CO_2 at high temperatures, and simultaneously free the amines. The carbamates and epoxy resin diglycidyl ether of bisphenol A can be premixed to form the foaming formulation and are stable under ambient conditions. This is known as a novel single-pack technology for epoxy foaming. At high temperature, in the foaming process, the freed amines act as the curing agent and the released gaseous CO_2 acts as the blowing agent. The usage of volatile organic compound (VOC) blowing agents is, thus, avoided. However, in this technology, the coordination of blowing and curing actions is important. The curing rate of the freed amine and the decomposition rate of the blocked amine with epoxy needs to be coordinated within a small gap between their onset temperatures. Figure 6.2 demonstrates the morphology of the epoxy foams using different blocked amines. In another study, Forest *et al.* used gaseous CO_2 foaming process to produce acrylonitrile butadiene styrene (ABS) nano-cellular foams [37].

By carrying out the sorption process at low temperature, acrylonitrile/styrene could be made into a rubbery form in a rigid phase with enhanced CO_2 uptake. Lowest foam densities could be achieved by keeping foaming temperature between 90-100 °C. Cell nucleation and growth was limited by using low foaming temperature, thereby, forming denser foams. High foaming temperature increased rate of nucleation, cell growth, gas diffusion, however, reduced the CO_2 uptake by cell growth control.

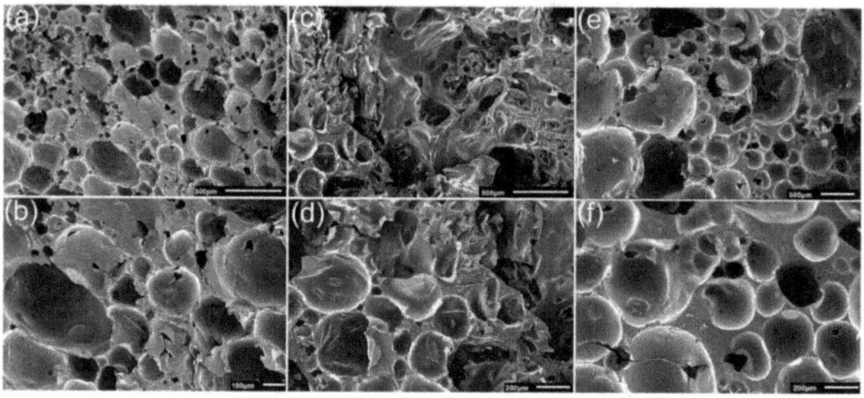

Figure 6.2 Foam morphology of the epoxy foams prepared from different blocked amines. Reproduced from Reference 36 with permission from American Chemical Society.

Thermal Decomposition

The thermal decomposition process can be carried out as two different approaches. First approach focuses on the thermal treatment of polymers with high glass transition blended with thermally labile organic components. On the other hand, the second approach comprises of degrading the graft or block copolymers thermally which consist of thermally labile and stable blocks [38]. This approach is restricted to polymers having high stability like polyamide, polyimide, etc., since the polymer matrix must stay firm when it goes through the thermal degradation of the labile component. Furthermore, polymer matrix can be contaminated if the added porogen is not completely degraded. Thermal decomposition technique is useful for the introduction of nanopores in the polymers for specific use in electronic applications requiring polymers with low dielectric constant

[39,40]. The amount of added porogen determines the precise pore size [41,42].

Phase Separation

In this method, the solvent is mixed with binary or multicomponent polymer system along with additive mixtures. The dissolved polymer in the solvent is encouraged to disperse in a controlled way and separate from the solvent by either thermally or chemically. In this way, the solvent present in the isolated places of polymer matrix forms two different phases or the components form two bi-continuous phases. Continuous removal of the organic solvent results in the ultimate structure of foam. It is crucial to have a closed cycle of the process streams owing to the use of environmentally unfriendly solvents. Also, due to the cross-linking in the polymer, it is difficult to dissolve such polymers in the solvent, thus, this technique cannot be used for network polymers.

By monitoring the process environments, open or closed morphologies of the cell having pore sizes ranging from 0.1 nm to 100 μm can be formed [43-48]. Ling *et al.* [47] exhibited the formation of PEI/graphene nanocomposite foam using water vapor induced phase separation method (Figure 6.3). In another study, Pavia *et al.* [43] generated blend of two types of poly(lactic acid) (PLA) with disparate average molecular weights (50,000 and 175,000 g/mol) in different ratios (90/10, 80/20 and 70/30). Thermally induced phase separation using ternary systems was to obtain microporous material, where water was the non-solvent and dioxane was the solvent. Fabrication of foams with average cell size in the range of 40-70 μm and open porosity of 87-92% by altering factors like temperature of de-mixing was reported. Such foams with microporosity are reported to have applications in tissue engineering as biocompatible and biodegradable scaffolds.

Leaching

In this method, selective leaching of inorganic salts like KCl or NaCl, which are also water soluble, is carried out. The process also involves an organic compound such as gelatin microspheres and glucose [49] which acts as porogen to create pores in the polymer. The method of particulate leaching or solvent casting mainly consists of the dispersion of solvent and polymer where solvent is highly volatile in nature.

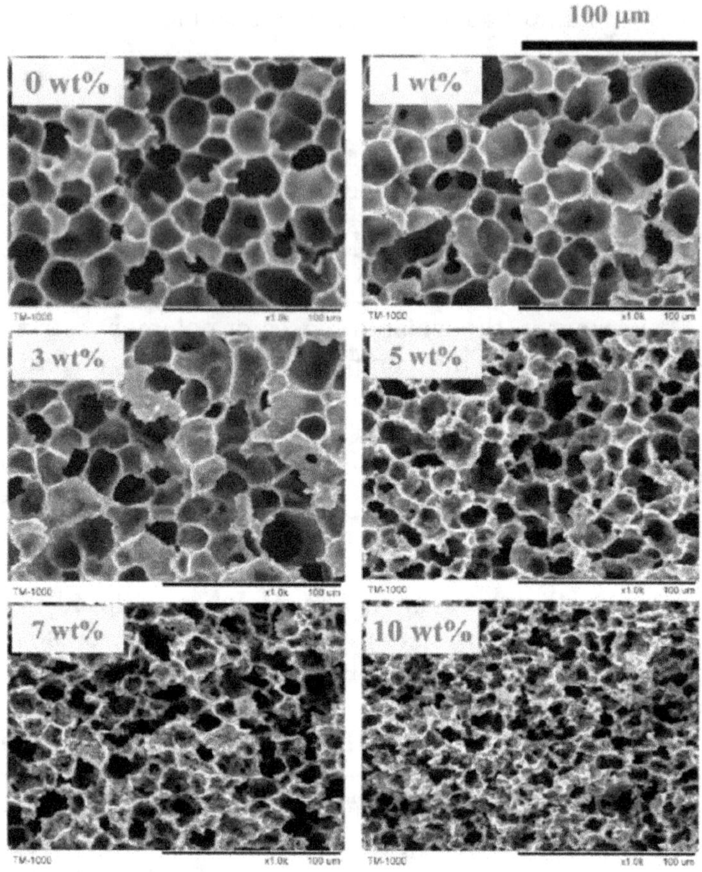

Figure 6.3 Scanning electron micrographs of PEI/graphene nanofoams as a function of graphene content. Reproduced from Reference 47 with permission from American Chemical Society.

The solution is subsequently casted in a mold having a solid porogen. After the solvent gets evaporated, the porogen is washed out to generated the polymer foam [50,51]. The major concern of this system is the possibility of unsolicited impurities in the porous polymeric structures like traces of salts and solvents [52]. Various studies have suggested to use particulate leaching method or melt molding process to overcome such challenges [53-55]. Briefly, the melt molding process includes premixing solid porogen particulates with polymer matrix followed by combined hot-pressing of both. The material is, thus, brought to the solid porogen leaching stage, same as that for the solvent-cast samples. By this method, control of various properties like pore size, pore structure and porosity can be achieved with the help of shape, size and concentration of leachable particles. Reignier and Huneault [52] used melt processing technique to prepare porous scaffolds by combining co-continuous polymer and particulate leaching. By using the method of selective removal of polyethylene oxide and salt, the authors prepared a permeable polycaprolactone scaffold. As a result, porous supports with various pore sizes and interlocked pores were obtained. Furthermore, it was possible to control the porosity by changing the concentrations of polymers and salt.

Etching

Dentures are created on the surface of polymers using scissions made by high energy radiation [56,57]. Loose polymer part can be removed using acid or alkaline solution phase etching. In addition, this method can also be used to remove inorganic particles, which are present in the composites to form porous structures [58]. Radiation and etching durations influence the shape and size of the pores. Track etched membranes of polycarbonate and polyester foams have been developed using this method and commercially used by various industries.

Other Techniques

For the development of polymer foams, several other techniques have also been utilized. For example, stretching can be used for introducing porosity in the structure of a polymer. Controlled stretching of polymeric membranes which have a dry semi-crystalline nature introduces porosity. It can also be achieved in the presence of appropriate solvents [59,60]. Repetition of stretching reduces the amorphous areas in the vicinity of crystallite sites. The pores produced using this method have a size range from 20 to 250 nm. However, this technique can be used only for polymers like polypropylene and polyethylene which are semi-crystalline in nature and can be stretched relatively easily [61,62].

Sintering is another technique used for the fabrication of polymer foams. At high temperature, applying extremely high stress mechanically on a large quantity of polymer is known as sintering [63-65]. Using this method, molded materials with homogeneity and pore size ranging from 0.1 to10 μm can be achieved. Thermoplastic polymers and non-melt-processable polymers can be manipulated using this technique.

6.2.3 Properties of Polymer Nanofoams

Many studies have reported the structure property correlations for polymer nanofoams [66-71]. As also mentioned earlier, polyurethane (PU) shows exceptional performance owing to the characteristics like high elasticity, abrasion resistance, damping properties, good impact strength, etc. [72,73]. In many studies, PU was prepared in the presence of dimethylformamide (DMF) by the chemical reaction of poly(tetramethylene glycol) (PTMG) with 4,4′-diphenylmethanediisocyanate (MDI) and 1,3-propylenediamine. By subjecting PU films to supercritical CO_2 (sCO_2) under particular conditions of pressure and temperature, porous structure was obtained. The impact of processing parameters, such as pressure (9-18 MPa), soaking time (10-200 min) and temperature (0-60 °C), on the morphology of foams was examined. It was observed that according to Gaussian distribution, the average cell size reduced and became more uniform with growing pressure from 9.0-18.0 MPa. Moreover, it was reported that the cell density increased with a rise in soaking pressure due to high CO_2 pressure. High cell density was achieved as high pressure resulted in an enhanced absorption of CO_2 molecules. However, increment in soaking temperature affected the density of the cells negatively and the cell size positively. Formation of fewer nuclei because of the decrease in CO_2 molecules by increasing the temperature was cited as the reason for this observation. Soaking time change didn't affect the structure of the foams, however, a small change in the dimensions was observed. Nanoporous silica-PU hybrid aerogels were also fabricated by the combination of super critical drying and sol-gel processing [74]. Tetramethyl orthosilicate (TMOSI) was attached to the organic functional groups in polymeric methylene diphenyl diisocyanate for obtaining silica aerogel. 1,4-clioxane and 2-dimethylaminoethanol acted as diluent and tertiary amine catalyst respectively in the process. The solvent could be removed in super critical conditions for obtaining the material from gel. After the completion of gelation procedure, aged wet hybrid aerogel was dried using sCO_2 at low temperature. D-fructose was also used as the porogen in PLLA gel [75]. After drying with sCO_2, the porogens were washed away using water. Dioxane acted as solvent while ethanol acted as the non-solvent. Utilizing the affinity of sCO_2 towards organic solvents, further elimination of any organic residues was possible. This can also be used to remove porogen molecules trapped inside the gel. Depending on the size of porogen and pores, porous structures with nano, micro and macro dimensions were fabricated. The structures were also characterized with a porosity of about 97%. This approach can be used for generating scaffolds with a predetermined size and shape for use in tissue engineering applications. In another study, Leong *et al.* [76] processed nanoporous structures for tissue engineering applications. Solvent cast method was used for the fabrication of the first scaffold with dichloromethane (DCM) as solvent in the presence of 1,1,1,3,3,3-hexafluoroisopropanol (HFIP). Electrospun method was utilized for the creation of second scaffold. Due to little miscibility between DCM and water, during electrospinning, evaporation of DCM took place. This resulted in phase separation as evaporation resulted in cooling and water vapor condensation on the fiber surface. Third support was synthesized by electrospinning in the presence of HFIP. This resulted in the formation of dense fiber porous support with pore size around 1-5 μm. Solvent cast method was able to synthesize materials with higher strength compared to the electrospinning technique. The lower mechanical strength was a result of pores acting as defects which acted as stress concentration points. However, the electrospun scaffolds exhibited fibrous nature with pores in nano-dimension and enhanced surface area to volume ratio. This resulted in higher uptake of protein and an improved interaction between the scaffolding material and protein.

For preparing nanoporous polystyrene, Mao and Hillmeyer [77] synthesized polystyrene-poly(ethylene oxide) (PS-PEO) diblock copolymer. Cylindrical PEO had a volume fraction of 0.3 and diameter of ~17 nm. PEO was etched from block copolymer using chemical etching method resulting in the synthesis of nanoporous polymer. Aqueous hydrogen iodide (HI) solution acted as an etching agent for dissociation of aliphatic ether linkages of PEO. The study revealed that the pores size was approximately 15±1 nm and diameter of PEO in the original block copolymer was maintained. Figure 6.4 shows the scanning electron image representing the formation of nanoporous structure. Nanoporous polyimides were also synthesized by drying polyimide gels using sCO_2 [78]. Initially, the authors prepared polyimide gels by the reaction of 4,4′-oxydiphthalic anhydride (ODPA) and pyromellitic dianhydride (PMDA) with 4, 4′-oxydianiline (ODA) and p-phenylenediamine (PDA) in the presence of 1,3,5-tris(4-aminophenyl)benzene (TAPB) trifunctional end cross-linker. The wet polyimide gel was dried under vacuum ($P=10^{-1/4}$ 4 mm Hg and T¼ 250 °C) or sCO_2 (P¼16 MPa and T¼ 80 °C) to obtain final nanoporous polyimides. The authors observed that the gels generated from the flexible

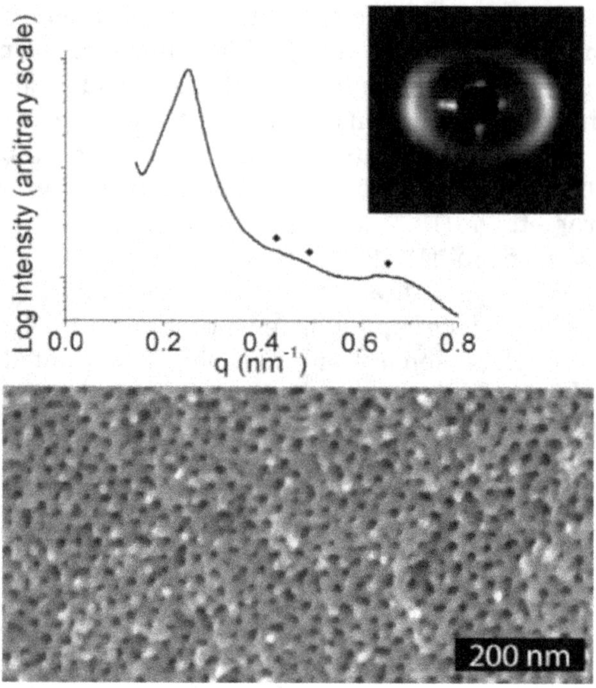

Figure 6.4 Small angle X-ray scattering and scanning electron micrograph of the PS-PEO sample after HI treatment. Reproduced from reference 77 with permission from American Chemical Society.

oligomers such as ODPA-ODA were transparent, whereas foams containing rigid oligomers like PMDA-ODA, PMDA, and PMDA-PDA were opaque. It was concluded that in the presence of sCO_2, porosity was ~90% and the pores size were around 50-800 nm. Further, the use of sCO_2 as dryer as an alternative to vacuum formed the porous structure without collapsing. The authors mentioned that the pore morphology could be adjusted since the progress in crystallization and liquid-liquid phase separation by crosslinking and imidization while gelation were competitive. In another study, Yen *et al.* [79] also generated nanoporous polycaprolactone membranes using non-solvent phase separation and thermal phase separation together. 2-methoxyethanol and 1,4-dioxane acted as non-solvent and solvent respectively in preparing polymer solutions. Evaporation of both liquids was minimized by keeping the end product in a chamber at 5 °C chamber and by dipping it in water at 5 °C for the removal of organic residual materials. The polymer concentration could be changed in order to attain structure with size 55-90 nm, well connected porous structure and enhanced porosity.

6.3 Polymer Nanocomposites Nanofoams and Their Properties

The role of nanoparticles in improving the properties of polymer foams can be summarized as: control of morphology and improvement in the structural properties of foams [80,81]. In the foaming process, increased gas retention can be achieved by attaining improved cellular structures and rheological properties which are introduced by strain hardening [80-84]. Better physical properties are shown by foams which have homogenous and narrower cell distribution [85-87]. The nanometer dimension of nanoparticles can work well with strengthening the cell walls of the foams which are of micron or sub-micron dimensions [80]. For instance, Figure 6.5 shows the cell structure evolution of linear homopolypropylene/clay (HPPC) nanocomposite foams with clay content. With increasing clay content in the composite, the cell morphology of HPPC foams was observed to improve continuously with smaller cell sizes, higher cell densities, higher close cell contents and uniform cell distribution [83]. The other important aspects of having nanoparticles in the matrix are increment in thermal stability and change in the structure of the polymer resulting in enhanced mechanical properties [87-91]. Moreover, the process of nanocomposite foaming can also extend the spectrum of properties like enhancing the acoustic or thermal insulation, in addition to the clear advantage of density reduction [92]. Thus, such polymer foams have higher potential of their use as lightweight materials in transportation, aerospace or construction [93,94]. Polymers are well recognized materials for being naturally electrically and thermally insulating [95,96]. As foaming improves the insulating characteristic of polymers by including a vital void fraction, thus, polymer nanocomposite foams are significant materials to be studied in respect to heat transfer, assisted by the large number of applications and usage that these materials contribute as thermal insulators [97,98]. On the other hand, several industries profit from the usage of polymers with improved thermal conductivity, as in the case of electrical wires for heat dissipation [99] or when heat drops in electronic or electric systems [100]. It has been observed that addition of thermally conductive fillers like carbon fibers, ceramic particles, carbon black, metal particles, graphite, etc., boosts the thermal conductivity of polymers. This method is superior as compared to the use of thermally conducting polymers as these polymers are occasionally expensive along with poor mechanical performance and thermal stability [101-103].

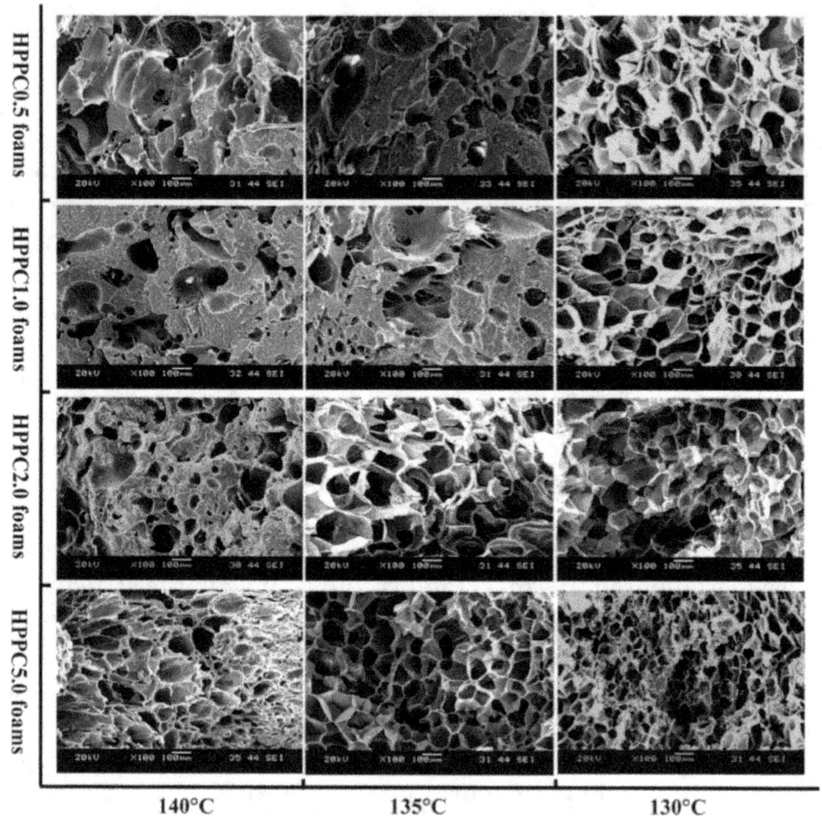

Figure 6.5 Cell structure evolution of HPPC nanocomposite foams with die temperature and clay content. Reproduced from Reference 83 with permission from American Chemical Society.

6.3.1 Mechanical Properties

Conventionally, polymer foams are utilized for packaging and muffling of delicate devices such as shock absorption materials and body protection apparels [104]. Owing to rising concerns about global warming along with increase in oil prices, the requirement of fuel effective transportation vehicles stays an active area of research. Thus, in order to decrease fuel consumption, light-weight materials are progressively put into use in vehicles. Polymer composites, low density metal alloys and fiber composites are generally used for these applications. Owing to their low density, another candidate for this application is polymer foams. However, due to density reduction, the mechanical characteristics in polymer foams are nearly always decreased [105], which considerably limits their use as light-weight structured materials. In research on polymer foams, a combination of high strength and light-weight is, thus, a foremost research aim. The mechanical characteristics of polymer foams are dependent on the innate characteristics of the polymer, foam's microstructure, density and cell anisotropy [106], cell size [107] and cell orientation [108]. Generally, strength and elastic modulus of polymer foams increases with increasing relative density, which is the ratio of the foam density to the solid polymer density [109-114]. By improving the polymer matrix characteristics by keeping the relative density constant, mechanical properties of polymer foams can be enhanced. Generally, nanofillers are utilized to improve polymer foam's mechanical characteristics by microstructure optimization and matrix enhancement [114-118].

Ahmadzadeh *et al.* [119] reported cellulose nanocomposite foams reinforced with surface-modified montmorillonite (SM-MMT) fabricated using high shear homogenization process. A strain hardening behavior was observed from the stress–strain curves of nanocomposite foams. However, pure cellulose foams indicated a flat area in stress–strain curves and acted as elastic-plastic materials capable of absorbing large amount of energy. The nanocomposite foams exhibited greater Young's modulus and tensile strength with respect to the pure cellulose foam. The foam reinforced with 5% of SM-MMT had the highest tensile strength (13.7 MPa). Antunes *et al.* [120] demonstrated an improvement in the mechanical properties by the addition of graphene nano-platelets (GnP) into polymers. GnP induced the formation of finer cellular structures and, on the other side, foaming increased GnP dispersion throughout the polymer, thus, maximizing GnP's strengthening effect. Thus, the microstructure modifications facilitated a way to achieve foams with mechanical performance evidently higher than unfilled foams and in some circumstances even similar to unfilled solid material. Mrlik *et al.* [121] examined the mechanical properties in the static and dynamic mode and reported that samples irradiated with gamma radiations at lower doses (under 5 kGy) exhibited enhancement in properties due to the cross-linking of the polymer structure. On the other hand, it was observed that the samples irradiated at higher doses also displayed the cross-linking behavior, but were significantly affected by scission, thus, resulting in poor mechanical properties.

Gutierrez *et al.* [122] analyzed mechanical properties of solid and foamed nanocomposites by carrying out compression tests. With respect to the mechanical properties of pure low density polyethylene (LDPE), the results exhibited that silica particles led to an overall enhancement in mechanical response of the studied samples. Svagan *et al.* [123] also studied starch-based biofoam. The authors reported that at 50% relative humidity and similar relative density, the foam exhibited comparable mechanical properties (E = 32 MPa, compressive yield strength, 630 kPa) to expanded polystyrene (EPS). Figure 6.6 compares the microstructure of pure and composite nanofoams.

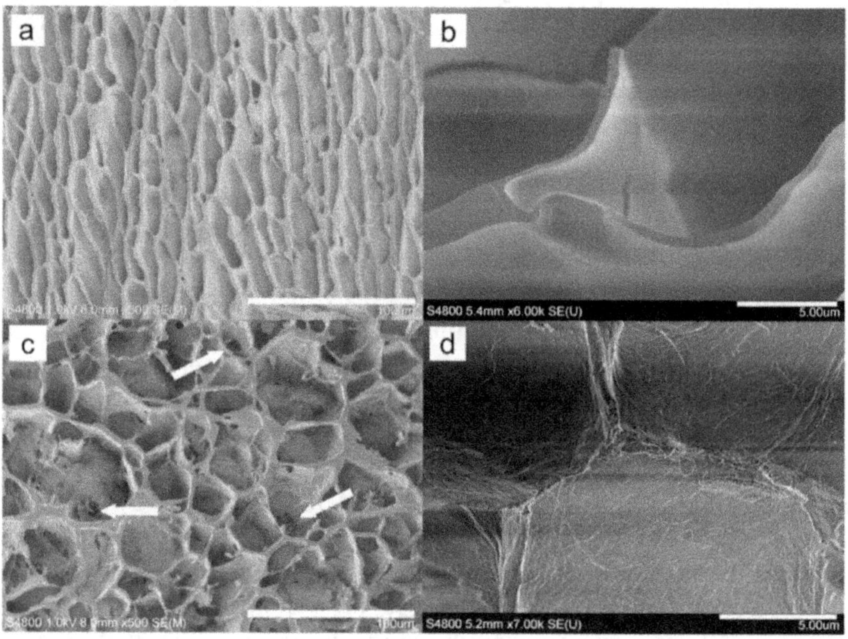

Figure 6.6 Scanning electron images of (a,b) neat amylopectin starch foam and (c,d) a composite foam. Reproduced from Reference 123 with permission from American Chemical Society.

Han *et al.* [124] reported exfoliated and intercalated polystyrene/nano-clay composites foams using in-situ polymerization and mechanical blending. At die temperature of 200 °C and screw rotation speed of 10 rpm, the intercalated composite with a small amount (i.e., 5 wt%) of nano-clay exhibited increased cell density from 2.7×10^7 to 2.8×10^8 cells/cm^3 and significantly reduced cell size from 25.3 to 11.1 μm. The exfoliated nanocomposite displayed smallest cell size (4.9 μm) and the maximum cell density (1.5×10^9 cells/cm^3) for the same particle concentration. The nanocomposite foams exhibited greater tensile modulus, compared with polystyrene foams, with improved barrier properties and fire retardance. Merging the extrusion foaming process and nanocomposite generation offers a new method for the regulation of cell structure in micro-cellular foams. In another study, the influence of long chain branching (LCB) and clay inclusion on the cellular structure of PLA was investigated by Najafi *et al.* [125]. Melt extrusion method was used to prepare linear PLA and LCB-PLA nanocomposites with different clay loadings and were subsequently foamed in an injection molding machine. It was reported that the mechanical properties of the foamed samples were considerably influenced by the cellular structure. At 0.5 wt% clay loading, a significant enhancement of the mechanical properties was attained.

Gultom *et al.* [126] also prepared light-weight polyurethane nanocomposite foams by using natural zeolites as the filler material. Ball milling method was applied to prepare the zeolite. The nanocomposite foams were fabricated by in-situ method by reaction of toluene diisocyanate and polypropylene glycol. It was reported that polyurethane nanocomposites comprising 20% of zeolite had superior mechanical properties as compared to unfilled polyurethane.

6.3.2 Electrical Properties

The development of several conducting nanomaterials such as carbon nanotubes, carbon nanofibers [127] and graphene [128] has contributed to the progress of electrically conductive polymer foams [129,130]. These materials can be broadly employed for electromagnetic interference shielding effectiveness (EMI SE), electrostatic discharge protection [131,132] and lightening-protection panels. For instance, Figure 6.7 demonstrates the EMI shielding effectiveness of carbon nanotube-polystyrene (CNT-PS) foam composites [131]. EMI SE of 19 dB was achieved in such composite foams.

An enhancement in the electrical conductivity of polymer foams is resulted when an electrically conductive filler forms an electron transport path in the foam structure by the use of cell walls and cell structure [133]. Microstructure of the foams, electric conductivity of

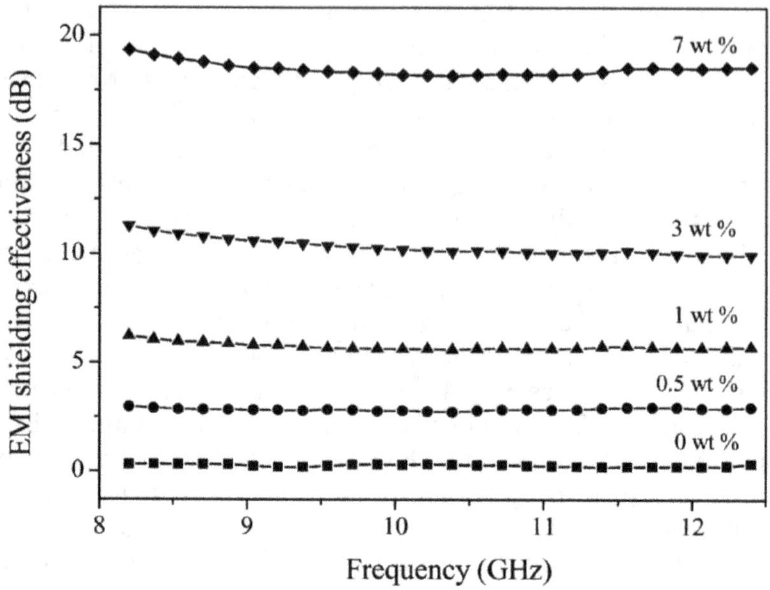

Figure 6.7 EMI shielding effectiveness (EMI SE) as a function of CNT-PS foam composites. Reproduced from Reference 131 with permission from American Chemical Society.

the matrices and density of the foams affect the electrical conductivity of such polymer foams.

In a recent study, simple, efficient, green and low cost method of preparation of light-weight poly(L-lactic acid)/multi-walled carbon nanotube nanocomposites with high strength and electrical conductivity was developed by Kuang *et al.* [134]. The method involved solid-state sCO_2 foaming and pressure-induced flow technique. In comparison with other traditional materials with similar densities and thickness, the foam had improved equivalent average specific EMI SE. The developed material represents a promising candidate for using biologically degrading materials in EMI shielding. Shen *et al.* [135] also reported the synthesis of graphene coated polymer foams with ultralow density for electromagnetic interference shielding with an absorption-dominant mechanism, possibly due to both conductive dissipation (Figure 6.8).

Zhang *et al.* [136] reported the preparation of functional PMMA/graphene nanocomposite micro-cellular foams generated by foaming the PMMA and graphene blend with subcritical CO_2. The incorporation of graphene provided high electrical conductivity as well

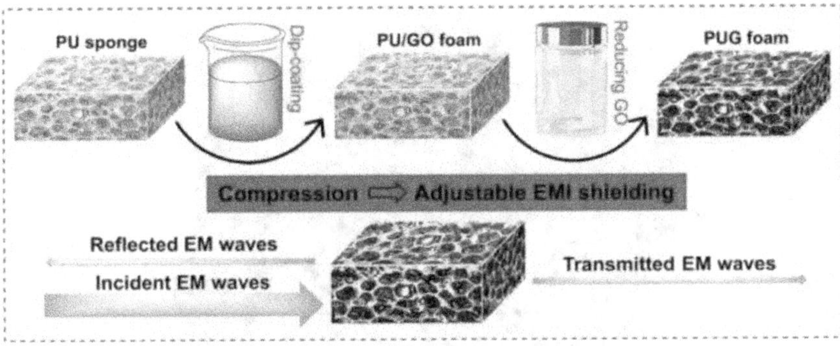

Figure 6.8 Schematic of the generation of graphene coated polymer foams and their principle of operation. Reproduced from Reference 135 with permission from American Chemical Society.

as enhanced electromagnetic interference shielding efficiency. The shielding efficiency was attributed to the absorption phenomenon, rather than reflection in the measured frequency range. The foam exhibited significantly higher ductility and toughness as compared to the bulk polymer owing to the presence of a large number of microcellular cells. Figure 6.9 demonstrates the scanning electron analysis of the cross-section of PMMA nanocomposite microcellular foam as well as transmission electron analysis exhibiting well-dispersed graphene sheets in the cell wall of the foam.

Polypropylene (PP)/stainless-steel fiber (SSF) composite foams were also prepared using injection molding by Ameli *et al.* [137] (Figure 6.10) with EMI SE of 48 dB. The authors also reported that the physical foaming significantly affected the morphology of the composites due to the plasticizing effect of dissolved gas in the melt, which resulted in the reduction in fiber breakage as well as attenuated the preferential orientation of the fibers. The study confirmed that foaming in injection molding process is useful to generate light-weight conducting polymer composites with low filler content and the generated materials have potential of application in many sectors such as electronics, aerospace and automotive industries.

As mentioned earlier, Antunes *et al.* [120] studied the incorporation of graphene nanoplatelets in polymer foams and confirmed that it can lead to significant enhancement in electrical conductivity depending upon the composite microstructure. As foaming enhanced GnP dispersion and simultaneously decreased the volume fraction of solid, a reduction of distance between nanoplatelets was observed which reached the critical distance for achieving tunnel-like electrical

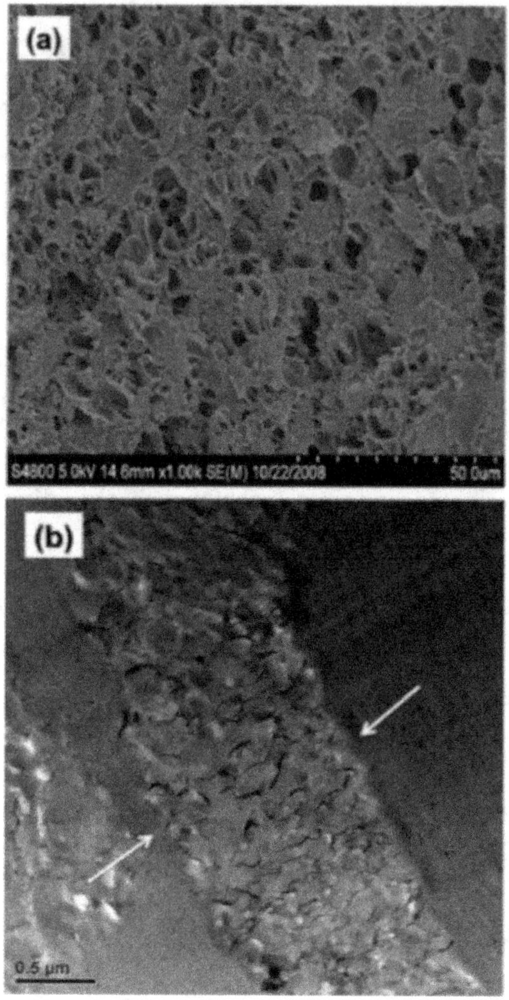

Figure 6.9 (a) Scanning electron image of the cross-section of PMMA nanocomposite microcellular foam, (b) transmission electron image of well- dispersed graphene sheets in the cell wall of the foam. Reproduced from Reference 136 with permission from American Chemical Society.

conduction. It was suggested that the foams may show higher electrical conductivities compared to that of the respective unfoamed composites. This, in combination with elevated mechanical performance, confirmed the novel application of the developed materials as EMI shielding, ESD or piezoelectric rigid components. Menner *et al.* [138] also reported nanocomposite foams generated using polymerization

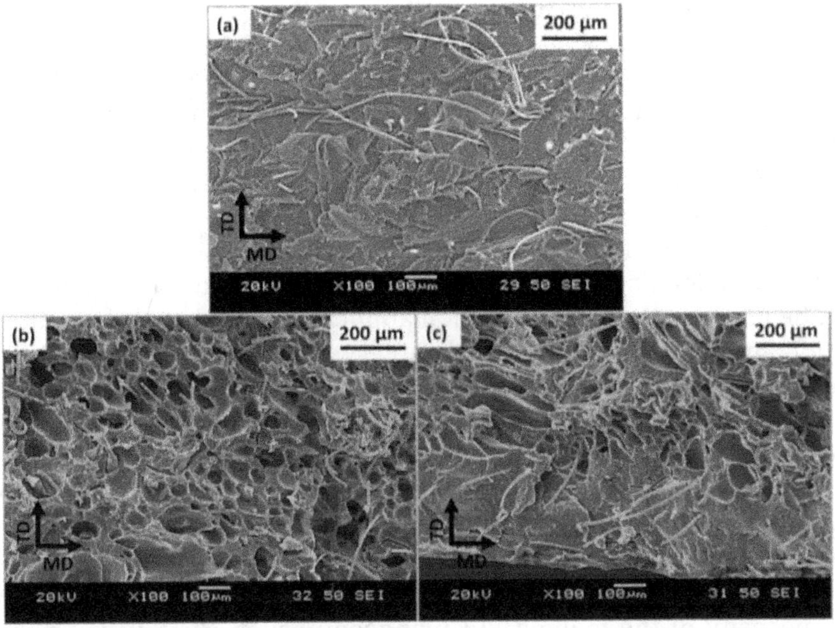

Figure 6.10 Scanning electron micrographs of (a) solid injection molded, (b) the core of foam injection molded, and (c) the skin of foam injection molded PP-0.5 vol % SSF composites. MD and TD denote the machine and thickness directions, respectively. Reproduced form Reference 137 with permission from American Chemical Society.

of high internal phase emulsions containing titania nanoparticles in the organic phase or carbon nanotubes in aqueous phase. The resulting foams exhibited an interconnected pore structure. The presence of titania led to an enhanced compressive modulus and crush strength of the foam. In another study, multi walled carbon nanotubes reinforced polycaprolactone (PCL/MWCNT) nanofoams were developed by Thomassin *et al.* [139]. With a very low volume percentage of 0.25%, a high shielding efficiency of 60-80 dB was attained with low reflectivity. Nano/or micro-cellular polypropylene/MWCNT (PP/MWCNT) composites were fabricated by Ameli *et al.* [140] and were confirmed to be efficient in EMI shielding applications. Conductivity of the composite structure was increased to a range of 10^{-3}-10^{-2} S cm^{-1} with 0.4 vol% concentration of MWCNT. Polycarbonate/CNT (PC/CNT) composite structures were also fabricated by Monnereau *et al.* [141] for use as EMI absorbents. The gradient foams exhibited enhanced efficiency as compared to the homogenous foams.

6.3.3 Thermal Properties

Lorenzetti *et al.* [142] reported polyurethane-graphene nanocomposite foams with superior thermal insulation. It was observed that graphene at very low filler fraction reduced the radiative contribution of the initial thermal conductivity by decreasing the cell size and increasing the extinction coefficient. In another study, Jiang *et al.* [143] developed foamed polymer nanocomposite system using nanoclay, polyetherimide and a thermoplastic elastomer Kraton (PCK), which can be used for thermal protection applications. Under various saturation and foaming conditions, samples with different compositions were foamed using a solid-state foaming method. It was revealed that the fire-retardance ability of the foamed PCK samples improved with increase in nanoclay loading. On the other hand, the fire retardancy decreased with increase in Kraton loading. Nevertheless, foaming reduced the mass loss in the ablation test, when the Kraton loading was high, signifying that the behavior of the material system could be improved by modifying the loading of nanoclay and Kraton, along with the porosity of the foamed PCK nanocomposites. By governing the foaming method parameters, porous structures with altered porosity were achieved. In this study, the resulted foams had a graded structure with a general porosity of 20-60%. The generated materials have the potential to be used in a range of light-weight thermal protection applications, such as thermal insulation for firefighting equipment, payload protection during an atmospheric entry of a spacecraft and automobiles. Moreover, the gas concentration model used in this work can be further advanced to guide future process growth for attaining different graded porous structures in polymer nanocomposites. In another study, relative density, cell structure and graphene incorporation were used as factors for studying thermal conductivity in polycarbonate-graphene nanocomposite foams [144]. sCO_2 foaming of two different types was used to obtain foams with different relative density and cellular morphology. The thermal conductivity decreased to 0.03 W/ (m K) on foaming. The thermal conductivity was also observed to improve two-fold by adding 5 wt% graphene.

6.4 Specific Applications in EMI Shielding

As mentioned earlier, due to their electrical insulating behavior, it is generally difficult for most of the polymers to stop electromagnetic waves from propagating. In order to overcome this limitation, one of the common routes is to incorporate electrically conductive fillers [145]. Thus, the electrically conductive polymer composites (foamed and unfoamed) have gained significant research attention for EMI shielding applications [145-159] owing to light-weight, low cost, preparation flexibility and processability as compared to the conventional metal-based materials. The EMI shielding effectiveness of a polymer composite is influenced critically by the content of the conductive filler, intrinsic electrical conductivity, magnetic permeability, dielectric constant and aspect ratio.

Micro-cellular epoxy/MWCNT (EP/MWCNT) composite was generated by Li *et al.* [160] using batch foaming process with sCO_2. Specific EMI SE was observed to be improved to 21.3 dB by using 1 wt% CNT. The combined effect of MWCNTs and micro-cellular structure was confirmed to improve EMI shielding performance. An improvement in EMI SE by increasing the content of graphene in rGO-polyimide (PI) composite foam was also reported [161]. Stepwise increase in graphene concentration improved the shielding efficiency gradually. The results concluded that the increase in efficiency was due to absorption rather than reflection. Increasing GO content, however, affected the absorption ability negatively. The thermal stability of the composite also exhibited significant improvement owing to the increase in filler content. A weight loss of 5% was observed till 508 °C. For composites with 16% filler loading, same loss in weight was observed till 581 °C. Thus, it was concluded that the increased filler content improved the heat stability of the foam by 80 °C.

Other Applications

Chavan *et al.* [162] described the production and application of elastomeric polymer nanocomposite foams for the efficient removal of Hg^{2+} and Pb^{2+} heavy metal ions from polluted water (Figure 6.11). The polydimethylsiloxane (PDMS) foams were modified so that these became hydrophilic and allowed the polluted water to enter inside their structure. The ZnSe colloidal nanocrystals (NCs), decorated on the surface of the foams, acted as active components capable to capture the metal ions. In this manner, once the nanocomposite foams were dipped in water polluted with Pb^{2+} or Hg^{2+}, a cation exchange reaction took place, thus, resulting in the successful removal of the heavy metal ions. In the considered concentration region of 20-40 ppm, the removal capacity for the Pb^{2+} ions was above 98%, whereas the removal of Hg^{2+} ions reached almost 100%. Also, the reaction was

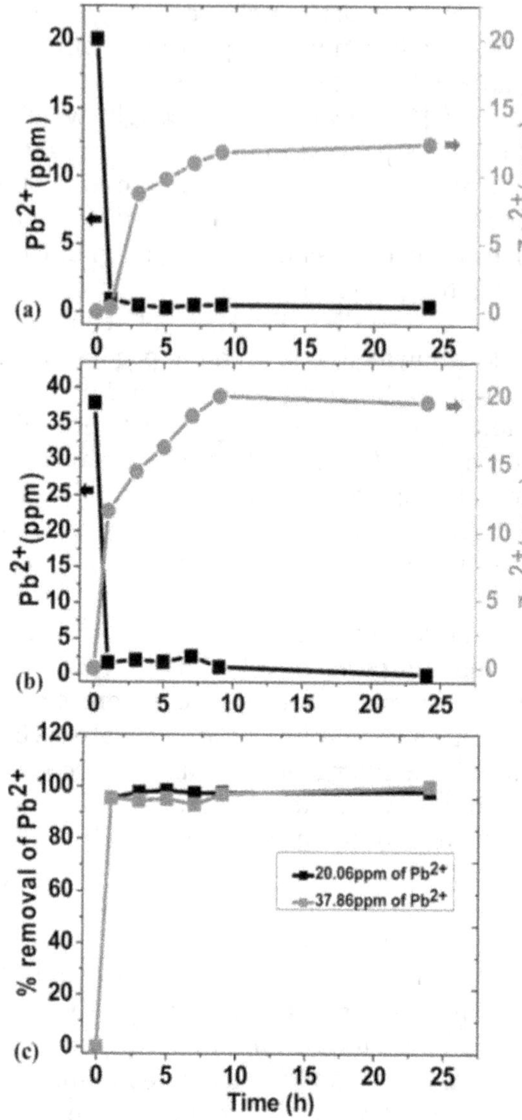

Figure 6.11 Kinetics of ion removal (a) 20.06 ppm of Pb^{2+} and (b) 37.86 ppm of Pb^{2+} ion concentration, (c) removal percentage of Pb^{2+} ions for both concentrations. Reproduced from Reference 162 with permission from American Chemical Society.

completed after 24 h, however, 95% concentration of both the metal ions was removed after the first hour, thus, underlining the robust performance of the functional foams. After the entrapment of heavy metal ions, the color of the foams altered, thus, providing a simple detection. Considering that the prepared foams provided good resistance to heat and contained good elastic properties, these can, thus, be utilized in diverse conditions of water remediation.

The synthesis of CuS nanoparticles to functionalize MWCNTs-polymer nanocomposite foams was reported by Altay *et al.* [163]. Uniform size distribution of particles, large surface area, thermal stability and economical method of preparation were attained by using sonication during the synthesis process. Van der Waals interaction between CNTs were overcame by ultrsasonic dispersion of polymer matrix. This material exhibited an optical band gap of 2.28 eV.

In another study, Hussein *et al.* [164] designed a novel nanocomposite material for arsenic removal from drinking water, in ppb concentrations, through introducing 15-20 nm sized iron oxide nanoparticles (IONPs) into a porous polyurethane foam (PU) having an open cell structure. The arsenic removal efficiency increased by altering size and wt% of IONPs as well as the composition ratio of polypropylene glycol (PPG) and toluene di-isocyanate (TDI). It was observed that the composition ratio of PPG:TDI (1:2) with 12% IONPs (15-20 nm) resulted in the highest affinity towards arsenic species with respect to other possible combinations, with a removal capacity of 40% using single stage batch analysis. Thus, for water filtration applications, the developed polyurethane nanocomposite foam facilitated low-cost operation with high potential and versatility.

6.5 Conclusions

Polymer nanofoams and polymer nanocomposite foams are functional materials comprising of unique configuration and characteristics. In the nanocomposite foams, the nanomaterials of various sizes, shapes and surface chemistries are utilized to govern the structure, properties and functionality of the foams. Many synthetic strategies have been developed in recent years which provide accurate control on the various characteristics such as porosity, cell size, density, conductivity, mechanical properties, etc. Due to such control on the structure, morphology and properties of the nanofoams, these foams find applications in a large number of commercial areas for specialty uses such as electromagnetic shielding, insulation, packaging, etc. With continued research focus on these materials, it is envisaged that the functional uses of such materials would further expand in near future, thereby, replacing many traditional metal based materials.

References

1. Antunes, M., and Velasco, J. I. (2014) Multifunctional polymer foams with carbon nanoparticles. *Progress in Polymer Science*, **39**, 486-509.
2. Krause, B. (2001) *Polymer Nanofoams*, Thesis, University of Twente, Netherlands.
3. Wessling, M., Borneman, Z., Boomgaard, Den, A. V., and Smolders, C. A. (1994) Selectivity as a function of membrane thickness: Gas separation and pervaporation. *Journal of Applied Polymer Science*, **53**(12), 1639-1651.
4. Kumar, V., (1993) Semi-continuous production of solid state PET foams. *Cellular Polymers*, **12**, 207-223.
5. Hedrick, J. L., Carter, K. R., Cha, H. J., Haker C.J., Pietro, R. A. D., Labadie, J. W., Miller, R. D., Russell, T. P., Sanchez, M. I., Volksen, W., Yoon, D. Y., Mecerreyes, D., Jerome, R., and Mcgrath, J. L. (1996) High-temperature polyimide nanofoams for microelectronic applications. *Reactive & Functional Polymers*, **30**, 43-53.
6. Hofland, G. W., Kersch, C., and Woerlee, G. F. (2003) Polyetherimide Nanofoams using Carbon dioxide. *6th International Symposium on Supercritical Fluids*, France. Online: http://www.isasf.net/fileadmin/files/Docs/Versailles/Papers/PMp15.pdf (assessed 21st May 2018).
7. Klempner, D., and Sendijarevic, V. (2004) *Handbook of Polymeric Foams and Foam Technology*, Hanser Publishers, Germany.
8. Nofar, M., and Park, C. B. (2014) Poly (lactic acid) foaming. *Progress in Polymer Science*, **39**, 1721-1741.
9. Darder, M., Aranda, P., Ferrer, M. L., Gutierrez, M. C., Monte, F., and Hitzky, E. R. (2011) Progress in bio nanocomposite and bioinspired foams. *Advanced Materials*, **23**, 5262-5267.
10. Dawson, R., Cooper, A. I., and Adams, D. J. (2012) Nanoporous organic polymer networks. *Progress in Polymer Science*, **37**, 530-563.
11. Zhang, W., Liu, M., Liu, Y., Liu, R., Wei, F., Xiao, R., and Liu, H. (2015) 3D porous poly(l-lactic acid) foams composed of nanofibers, nanofibrous micro sheaves and microspheres and their application in oil-water separation. *Journal of Material Chemistry A*, **3**, 14054-14062.
12. Ahmed, A., Hearn. J., Abdelmagid, W., and Zhang, H. (2012) Dual-tuned drug release by nanofibrous scaffolds of chitosan and mesoporous silica microspheres. *Journal of Materials Chemistry*, **22**, 25027-25035.
13. Danowska, M., Piszczyk, L., Strankowski, M., Gazda, M., and Haponiuk, J. T. J. (2013) Rigid polyurethane foams modified with selected layered silicate nanofillers. *Journal of Applied Polymer Science*, **130**, 2272-2281.

14. Lu, X., Caps, R., Fricke, J., Alviso, C. T., and Pekala, R. W. (1995) Correlation between structure and thermal conductivity of organic aerogels. *Journal of Non-Crystalline Solids*, **188**, 226-34.
15. Atalla, Y., Fu, J., Atalla, N., and Naguib, H. E. (2010) Study of the effects of processing parameters on the sound absorption of open-cell microcellular polymeric foams. *Noise Control Engineering Journal*, **58**, 18-26.
16. Yeh, J. M., Chang, K. C., Peng, C. W., Chand, B. G., Chiou, S. C., and Huang, H. H. (2009) Preparation and insulation property studies of thermoplastic PMMA-silica nanocomposite foams. *Polymer Composites*, **30**, 715-722.
17. Manninen, A. R., Naguib, H. E., Nawaby, A. V., and Day, M. (2005) CO_2 sorption and diffusion in polymethyl methacrylate-clay nanocomposites. *Polymer Engineering and Science*, **45**, 904-914.
18. Ferkl, P., Pokorny, R., Bobak, M., and Kosek, J. (2013) Heat transfer in one-dimensional micro and nano-cellular foams. *Chemical Engineering Science*, **97**, 50-58.
19. Schmidt, D., Raman, V. I., Egger, C., Fresne, C., and Schadler, V. (2007) Templated crosslinking reactions for designing nanoporous materials. *Materials Science and Engineering C*, **27**, 1487-1490.
20. Forest, C., Chaumont, P., Cassagnau, P., Swoboda, B., and Sonntag, P. (2015) Polymer nanofoams for insulating applications prepared from CO_2 foaming. *Progress in Polymer Science*, **41**, 122-145.
21. Seo, M., Amendt, M. A., and Hillmyer, M. A. (2011) Cross-linked nanoporous materials from reactive and multifunctional block polymers. *Macromolecules*, **44**, 9310-9318.
22. Olson, D. A., Chen, L., and Hillmyer, M. A. (2007) Templating nanoporous polymers with ordered block copolymers. *Chemistry of Materials*, **20**, 869-890.
23. Wu, D., Xu, F., Sun, B., Fu, R., He, H., and Matyjaszewski, K. (2012) Design and preparation of porous polymers. *Chemical Reviews*, **112**, 3959-4015.
24. Costeux, S. (2014) CO_2-Blown nano cellular foams. *Journal of Applied Polymer Science*, **131**(23), doi: 10.1002/app.41293.
25. Yoda, S., Ohara, M., Takebayashi, Y., Sue, K., Hakuta, Y., Furuya, T., Yamada, M., and Otake, K. (2013) A porous polymer-silica composite with a "vespula-like" structure for thermal insulating materials prepared via high pressure phase control. *Journal of Material Chemistry A*, **1**, 9620-9623.
26. Wu, Q., Andersson, R. L., Holgate, T., Johansson, E., Gedde, U. W., Olsson, R. T., and Hedenqvist, M. S. (2014) Highly porous flame-retardant and sustainable biofoams based on wheat gluten and in situ polymerized silica. *Journal of Material Chemistry A*, **2**, 20996-21009.
27. Huang, S., Li, X., Jiao, Y., and Shi, J. (2015) Fabrication of a superhydrophobic, fire-Resistant, and mechanical robust sponge upon poly-

phenol chemistry for efficiently absorbing oils/organic solvents. *Industrial & Engineering Chemistry Research*, **54**, 1842-1848.

28. Liu, L., Fu, M., and Wang, Z. (2015) Synthesis of boron-containing toughening agents and their application in phenolic foams. *Industrial & Engineering Chemistry Research*, **54**, 1962-1970.
29. Li, Y., Chen, Z., and Zeng, C. (2014) Poly(methyl methacylate) (PMMA) nanocomposite foams. In: *Polymer Nanocomposite Foams*, Mittal, V. (ed.), CRC Press, USA, pp. 1-33.
30. Shinkai, T., Ito, M., Sugiyama, K., Ito, K., and Yokoyama, H. (2012) Ordered and foam structures of semi-fluorinated block copolymers in supercritical carbon dioxide. *Soft Matter*, **8**, 5811-5817.
31. Yokoyama, H., and Sugiyama, K. (2005) Nanocellular structures in block copolymers with CO_2-philic blocks using CO_2 as a blowing agent: crossover from micro- to nanocellular structures with depressurisation temperature. *Macromolecules*, **38**, 10516-10522.
32. Zhang, R., and Yokoyama, H. (2009) Fabrication of nanoporous structures in block copolymer using selective solvent assisted with compressed carbon dioxide. *Macromolecules*, **42**, 3559-3564.
33. Forest, C., Chaumont, P., Cassagnau, P., Swoboda, B., and Sonntag, P. (2015) CO_2 nano-foaming of nanostructured PMMA, *Polymer*, **58**, 76-87.
34. Tsioptsias, C., and Panayiotou, C. (2011) Simultaneous determination of sorption, heat of sorption, diffusion coefficient and glass transition depression in polymer CO_2 systems. *Thermochimica Acta*, **521**, 98-106.
35. Hu, X., Liu, M., Wouterson, E. M., and Zhang, L. (2014) Nanotoughening and microtoughening of polymer syntactic foams. In: *Polymer Nanocomposite Foams*, Mittal, V. (2014), CRC Press, USA, pp. 35-59.
36. Ren, Q., Xu, H., Yu, Q., and Zhu, S. (2015) Development of epoxy foaming with CO_2 as latent blowing agent and principle in selection of amine curing agent. *Industrial & Engineering Chemistry Research*, **54**, 11056-11064.
37. Forest, C., Chaumont, P., Cassagnau, P., Swoboda, B., and Sonntag, P. (2015) Generation of nano cellular foams from ABS terpolymers. *European Polymer Journal*, **65**, 209-220.
38. Aram, E., and Ataei, S. M. (2015) A review on the micro- and nanoporous polymeric foams: Preparation and properties. *International Journal of Polymeric Materials and Polymeric Biomaterials*, **65**, 358-375.
39. Lee, Y. J., Huang, J. M., Kuo, S. W., and Chang, F. C. (2005) Low-dielectric, nanoporous polyimide films prepared from PEO–POSS nanoparticles. *Polymer*, **46**, 10056-10065.
40. Aram, E., and Ataei, S. M. (2013) Preparation of thermally stable, low dielectric constant, pyridine-based polyimide and related nanofoams. *Journal of Applied Polymer Science*, **128**, 4387-4394.

41. Ataei, S. M., and Aram, E. (2014) New polyimide and nanoporous structures with low dielectric constant. *Advances in Polymer Technology*, **33**, 21407.
42. Ataei, S. M., and Saidi, S. (2007) Preparation and characterization of sulfone -based polyimide nanofoams grafted with poly propylene glycol. *E-polymers.* **112**, 1-11.
43. Pavia, F. C., Carrubba, V. L., and Brucato, V. (2012) Morphology and thermal properties of foams prepared via thermally induced phase separation based on polylactic acid blends. *Journal of Cellular Plastics*, **48**(5), 399-407.
44. Guan, J., Fujimoto, K. L., Sacks, M. S., and Wagner, W. R. (2005) Preparation and characterization of highly porous, biodegradable polyurethane scaffolds for soft tissue applications. *Biomaterials*, **26**, 3961-3971.
45. Cai, Q., Yang, J., Bei, J., and Wang, S. (2002) A novel porous cells scaffold made of polylactide–dextran blend by combining phase-separation and particle-leaching techniques. *Biomaterials*, **23**, 4483-4492.
46. Matsuyamaa, H., Yano, H., Makia, T., Teramoto, M., Mishima, K., and Matsuyama, K. (2001) Formation of porous flat membrane by phase separation with supercritical CO_2. *Journal of Membrane Science*, **194**, 157-163.
47. Ling, J., Zhai, W., Feng, W., Shen, B., Zhang, J., and Zheng, W. G. (2013) Facile preparation of lightweight microcellular polyetherimide/graphene composite foams for electromagnetic interference shielding. ACS *Applied Materials and Interfaces*, **5**, 2677-2684.
48. Liu, J., Morykwas, M. J., Argenta, L. C., and Wagner, W. D. (2011) Development of a biodegradable foam for use in negative pressure wound therapy. *Journal of Biomedical Materials Research Part B: Applied Biomaterials*, **98**, 316-322.
49. Holy, C. E., Dang, S. M., Davies, J. E., Shiochet, M.S. (1999) In vitro degradation of a novel poly(lactide-co-glycolide) 75/25 foam. *Biomaterials*, **20**, 1177-1185.
50. Hou, Q., Grijpma, D. W., and Feijen, J. (2003) Porous polymeric structures for tissue engineering prepared by a coagulation, compression moulding and salt leaching technique. *Biomaterials*, **24**, 1937-1947.
51. Sedghi, S. M., Rodrigue, D., Brisson, J., and Iliuta, M. C. (2013) Highly hydrophobic microporous low-density polyethylene hollow fiber membranes by melt-extrusion coupled with salt-leaching technique. *Polymers for Advanced Technologies*, **24**, 584-592.
52. Reignier, J., and Huneault, M. A. (2006) Preparation of interconnected poly (ε-caprolactone) porous scaffolds by a combination of polymer and salt particulate leaching. *Polymer*, **47**, 4703-4717.
53. Oh, S. H., Kang, S. G., and Lee, J. H. (2006) Degradation behavior of

hydrophilized PLGA scaffolds prepared by melt-molding particulate-leaching method: comparison with control hydrophobic one. *Journal of Materials Science: Materials in Medicine*, **17**, 131-137.

54. Mi, H. Y., Jing, X., Salick, M. R., Turng, L. S., and Peng, X. F. (2014) Fabrication of thermoplastic polyurethane tissue engineering scaffold by combining microcellular injection molding and particle leaching. *Journal of Materials Research*, **29**, 911-922.
55. Oh, S. H., Kang, S. G., Kim, E. S., Cho, S. H., and Lee, J. H. (2003) Fabrication and characterization of hydrophilic poly (lactic-co-glycolic acid)/poly (vinyl alcohol) blend cell scaffolds by melt-molding particulate-leaching method. *Biomaterials*, **24**, 4011-4021.
56. Baker, R. W. (2004) *Membrane Technology and Applications*, Wiley, USA.
57. Zeman, L. J., and Zydney, A. L. (1996) *Microfiltration and Ultrafiltration: Principles and Application*, Marcel Dekker, USA.
58. Jiang, L., Liu, J., Wu, D., Li, H., and Jin, R. (2006) A methodology for the preparation of nanoporous polyimide films with low dielectric constants. *Thin Solid Films*. **510**, 241-246.
59. Baker, R. W. (2000) *Membrane Technology and Application*, McGraw Hill, USA.
60. Volynskii, A. L., and Bakeev, N. F. (1995) *Solvent Crazing of Polymers*, Elsevier, The Netherlands.
61. Gore, R. W., (1973) Process for Producing Porous Products, US patent 3953566.
62. Druin, M. L., Loft, J. T., and Plovan, S. G., (1972) Novel Open-Celled Microporous Film, US patent 3801404.
63. Taboas, J. M., Maddox, R. D., Krebsbach, P. H., and Hollister S. (2003) Indirect solid free form fabrication of local and global porous, biomimetic and composite 3D polymer-ceramic scaffolds. *Journal of Biomaterials*, **24**, 181-194.
64. Nor, M. A., Hong, L. C., Ahmad, Z. A., and Akil, H. M. (2008) Preparation and characterization of ceramic foam produced via polymeric foam replication method. *Journal of Materials Processing Technology*, **207**, 235-239.
65. Wang, X., Fan, H., Xiao, Y., and Zhang, X. (2006) Fabrication and characterization of porous hydroxyapatite/β-tricalcium phosphate ceramics by microwave sintering. *Material Letters*, **60**, 455-458.
66. Chiellini. E., Cinelli, P., Ilieva, V. I., Imam, S. H., and Lawton, J. L. (2009) Environmentally compatible foamed articles based on potato starch, corn fiber, and poly(vinyl alcohol). *Journal of Cellular Plastics*, **45**, 17-32.
67. Liu, H., Xie, F., Yu, L., Chen, L., and Li, L. (2009) Thermal processing of starch-based polymers. *Progress in Polymer Science*, **34**, 1348-1368.
68. Lee, S. Y., and Hanna, M. A. (2008) Tapioca starch-poly(lactic acid)-

Cloisite 30B nanocomposite foams. *Polymer Composites*, **30**, 665-672.

69. Lawton, J. W., Shogren, R. L., and Tiefenbacher, K. F. (2004) Aspen fiber addition improves the mechanical properties of baked corn-starch foams. *Industrial Crops and Products*, **19**, 41-48.
70. Shorgen, R. L., Lawton, J. W., Doanne, W. M., and Tiefenbacher, F. K. (1998) Structure and morphology of baked starch foams. *Polymer*, **39**, 6649-6655.
71. Zhou, J., Song, J., and Parker, R. (2006) Structure and properties of starch-based foams prepared by microwave heating from extruded pellets. *Carbohydrate Polymers*, **63**, 466-475.
72. Rokicki, G., Parzuchowski, P. G., and Mazurek, M. (2015) Non-isocyanate polyurethanes: synthesis, properties, and applications. *Polymers for Advanced Technologies*, **26**, 707-761.
73. Mahajan, N., and Gupta, P. (2015) New insights into the microbial degradation of polyurethanes. *RSC Advances*, **5**, 41839-41854.
74. Yim, T. J., Kim, S. Y., and Yoo, K. P. (2002) Fabrication and thermophysical characterization of nano-porous silica-polyurethane hybrid aerogel by sol-gel processing and supercritical solvent drying technique. *Korean Journal of Chemical Engineering*, **19**, 159-166.
75. Reverchon, E., Cardea, S., and Rapuano, C. (2008) A new supercritical fluid-based process to produce scaffolds for tissue replacement. *The Journal of Supercritical Fluids*, **45**, 365-373.
76. Leong, M. F., Chian, K. S., Mhaisalkar, P. S., Ong, W. F., and Ratne, B. D. (2009) Effect of electrospun poly(D,L-lactide) fibrous scaffold with nanoporous surface on attachment of porcine esophageal epithelial cells and protein adsorption. *Journal of Biomedical Materials Research Part A*, **89A**(4), 1040-1048.
77. Mao, H., and Hillmyer, M. A. (2005) Nanoporous polystyrene by chemical etching of poly(ethylene oxide) from ordered block copolymers. *Macromolecules*, **38**, 4038-4039.
78. Kawagishi, K., Saito, H., Furukawa, H., and Horie, K. (2007) Superior nanoporous polyimides via supercritical CO2 drying of jungle-gym-type polyimide gels. *Macromolecular Rapid Communications*, **28**, 96-100.
79. Yen, C., He, H., Lee, L. J., and Ho, W. S. W. (2009) Synthesis and characterization of nanoporous polycaprolactone membranes via thermally- and nonsolvent-induced phase separations for biomedical device application. *Journal of Membrane Science*, **343**, 180-188.
80. Ibeh, C. C., and Bubacz, M. (2008) Current trends in nanocomposite foams. *Journal of Cellular Plastics*, **44**, 493-515.
81. Lee, L. J., Zeng, C., Cao, X., Han, X., Shen, J., and Xu, G. (2005) Polymer nanocomposite foams. *Composites Science and Technology*, **65**, 2344-2363.
82. Zhai, W., Park, C. B., and Kontopoulou, M. (2011) Nano silica additi-

on dramatically improves the cell morphology and expansion ratio of polypropylene hetero phasic copolymer foams blown in continuous extrusion. *Industrial & Engineering Chemistry Research*, **50**, 7282-7289.

83. Zhai, W., Kuboki, T., Wang, L., Park, C. B., Lee, E. K., and Naguib, H. E. (2010) Cell structure evolution and the crystallization behavior of polypropylene/clay nanocomposites foams blown in continuous extrusion. *Industrial & Engineering Chemistry Research*, **49**, 9834-9845.
84. Gibson, L. J., and Ashby, M. F. (1997) *Cellular Solids: Structure and Properties*. Cambridge University Press, UK.
85. Eaves, D. (2004) *Handbook of Polymer Foams*. Rapra Technology Limited, UK.
86. Anadao, P. (2014) The use of montmorillonite clay in polymer nanocomposite foams. In: *Polymer Nanocomposite Foams*, Mittal, V. (ed.), CRC Press, USA, pp. 149-168.
87. Fu, S. Y., Feng, X. Q., Lauke, B., and Mai, Y. W. (2008) Effects of particle size, particle/matrix interface adhesion and particle loading on mechanical properties of particulate-polymer composites. *Composites Part B*, **39**, 933-961.
88. Tjong, S. C. (2006) Structural and mechanical properties of polymer nanocomposites. *Materials Science and Engineering: R: Reports*, **53**, 79-197.
89. Zou, H., Wu, S., and Shen, J. (2008) Polymer/sílica nanocomposites: preparation, characterization, properties and applications. *Chemical Reviews*, **108**, 3893-3957.
90. Ray, S. S., and Okamoto, M. (2003) Polymer/layered silicate nanocomposites: a review from preparation to processing. *Progress in Polymer Science*, **28**, 1539-1641.
91. Pavlidou, S., and Papaspyrides, C. D. (2008) A review on polymer-layered silicate nanocomposites. *Progress in Polymer Science*, **33**, 1119-1198.
92. Kumar, V., and Weller, J. (1994) Production of microcellular polycarbonate using carbon dioxide for bubble nucleation. *Journal of Manufacturing Science and Engineering*, **116**, 413-420.
93. Weller, J., and Kumar, V. (2010) Solid-state microcellular polycarbonate foams. II. The effect of cell size on tensile properties. *Polymer Engineering & Science*, **50**, 2170-2175.
94. Mascia, L., Re, G. D., Ponti, P. P., Bologna, S., Giacomo, G. D., and Haworth, B. (2006) Crystallization effects on autoclave foaming of polycarbonate using supercritical carbon dioxide. *Advances in Polymer Technology*, **25**, 225-235.
95. Agari, Y., and Uno, T. (1985) Thermal conductivity of polymer filled with carbon materials: effect of conductive particle chains on thermal conductivity. *Journal of Applied Polymer Science*, **30**(5), 2225-

2235

96. King, J. A., Tucker, K. W., Meyers, J. D., Weber, E. H., Clingerman, M. L., and Ambrosius, K. R. (2001) Factorial design approach applied to electrically and thermally conductive nylon 6,6. *Polymer Composites*, **22**, 142-154.
97. Antunes, M., Velasco, J. I., Solorzano, E., and Perez, M. A. R. (2011) Heat transfer in polyolefin foams. In: Heat Transfer in Multi-Phase Materials, Oechsner, A., Murch, G. E. (eds.), Springer-Verlag, Germany, pp. 131-162.
98. Wicklein, B., Kocjan, A., Alvarez, G. S., Carosio, F., Camino, G., and Antonietti, M. (2015) Thermally insulating and fire-retardant lightweight anisotropic foams based on nanocellulose and graphene oxide. *Nature Nanotechnology*, **10**, 277-283.
99. Verdejo, R., Bujans, F. B., Perez, M. A. R., Saja, J. A., and Manchado, M. A. L. (2008) Functionalized graphene sheet filled silicone foam nanocomposites. *Journal of Material Chemistry*, **18**, 2221-2226.
100. King, J. A., Tucker, K. W., Vogt, B. D., Weber, E. H., and Quan, C. (1999) Electrically and thermally conductive nylon 6,6. *Polymer Composites*, **20**, 643-654.
101. Wolff, S., and Wang, M. (1993) *Carbon Black Science & Technology*, 2nd edition, CRC Press, USA.
102. Pashayi, K., Fard, H. R., Lai, F., Iruvanti, S., Plawsky, J., and Tasciuc, T. B. (2014) Self-constructed tree-shape high thermal conductivity nanosilver networks in epoxy. *Nanoscale*, **6**, 4292-4296.
103. Han, Z., and Fina, A. (2011) Thermal conductivity of carbon nanotubes and their polymer nanocomposites: a review. *Progress in Polymer Science*, **36**, 914-944.
104. Landro, L. D., Sala, G., and Olivieri, D. (2002) Deformation mechanisms and energy absorption of polystyrene foams for protective helmets. *Polymer Testing*, **21**, 217-228.
105. Michel, F. S., Chazeau, L., Cavaille, J., and Chabert, E. (2006) Mechanical properties of high density polyurethane foams: I. Effect of the density. *Composites Science and Technology*, **66**, 2700-2708.
106. Arora, K., Lesser, A., and McCarthy, T. (1998) Compressive behavior of microcellular polystyrene foams processed in supercritical carbon dioxide. *Polymer Engineering & Science*, **38**, 2055-2062.
107. Weller, J. E., and Kumar, V. (2010) Solid-state microcellular polycarbonate foams. II. The effect of cell size on tensile properties. *Polymer Engineering & Science*, **50**, 2170-2175.
108. Kabir, M., Saha, M., and Jeelani, S. (2006) Tensile and fracture behavior of polymer foams. *Material Science and Engineering A*, **429**, 225-235.
109. Wang, J., Wang, H., Chen, X., and Yu, Y. (2010) Experimental and numerical study of the elastic properties of PMI foams. *Journal of Materials Science*, **45**, 2688-2695.

110. Zhang, Y., Rodrigue, D., and Kadi, A. A. (2003) High density polyethylene foams. III. Tensile properties. *Journal of Applied Polymer Science*, **90**, 2130-2138.
111. Liu, H., and Han, C. (2008) Study on the cell structure and compressive behavior of biodegradable poly(ε-caprolactone) foam. *Polymer Engineering Science*, **48**, 2432-2438.
112. Doroudiani, S., and Kortschot, M. T. (2003) Polystyrene foams. III. Structure–tensile properties relationships. *Journal of Applied Polymer Science*, **90**, 1427-1434.
113. Matuana, L. M., and Park. C. (1998) Cell morphology and property relationships of microcellular foamed PVC/wood-fiber composites. *Polymer Engineering Science*, **38**, 1862-1872.
114. Chen, L., Schadler, L. S., and Ozisik, R. (2011) An experimental and theoretical investigation of the compressive properties of multi-walled carbon nanotube/poly(methyl methacrylate) nanocomposite foams. *Polymer*, **52**, 2899-2909.
115. Jo, C., and Naguib, H. E. (2007) Effect of nanoclay and foaming conditions on the mechanical properties of HDPE-clay nanocomposite foams. *Journal of Cellular Plastics*, **43**, 111-121.
116. Saha, M. C., Kabir, E., and Jeelani, S. (2008) Enhancement in thermal and mechanical properties of polyurethane foam infused with nanoparticles. *Material Science Engineering A*, **479**, 213-222.
117. Nikje, M. A., and Tehrani, Z. (2010) Thermal and mechanical properties of polyurethane rigid foam/modified nanosilica composite. *Polymer Engineering & Science*, **50**, 468-473.
118. Sescousse, R., Gavillon, R., and Budtova, T. (2011) Aerocellulose from cellulose–ionic liquid solutions: preparation, properties and comparison with cellulose–NaOH and cellulose–NMMO routes. *Carbohydrate Polymers*, **83**, 1766-1774.
119. Ahmadzadeh, S., Nasirpour, A., Keramat, J., Hamdami, N., Behzad, T., and Desobry, S. (2015) Nanoporous cellulose nanocomposite foams as high insulated food packaging materials. *Colloids and Surfaces A: Physicochemical and Engineering Aspects*, **468**, 201-210.
120. Antunes, M., Gedler, G., Abbasi, H., and Velasco, J. I. (2016) Graphene nanoplatelets as a multifunctional filler for polymer foams. *Materials Today: Proceedings*, **3**(2), S233-S239.
121. Mrlik, M., and Maadeed, M. A. (2016) Tailoring of the thermal, mechanical and dielectric properties of the polypropylene foams using gamma-irradiation. *Polymer Degradation and Stability*, **133**, 234-242.
122. Gutierrez, E. L., Arroyo, C. S., Velasco, J. I., and Perez, M. A. R., (2016) Low density polyethylene/silica nanocomposite foams. Relationship between chemical composition, particle dispersion, cellular structure and physical properties. *European Polymer Journal*, **81**, 173-185.

123. Svagan, A. J., Berglund, L. A., and Jensen, P. (2011) Cellulose nanocomposite biopolymer foam - hierarchical structure effects on energy absorption. *ACS Applied Materials and Interfaces,* **3,** 1411-1417.
124. Han, X., Zeng, C., Lee, L. J., Koelling, K. W., and Tomasko, D. L. (2003) Extrusion of polystyrene nanocomposite foams with supercritical CO_2. *Polymer Engineering and Science*, **43**, 1261-1275.
125. Najafi, N., Heuzey, M. C., Carreau, P. J., Therriault, D., and Park, C. B. (2015) Mechanical and morphological properties of injection molded linear and branched-polylactide (PLA) nanocomposite foams. *European Polymer Journal*, **73**, 455-465.
126. Gultom, F., Wirjosentono, B., Thamrin, Nainggolan, H., and Eddiyanto, (2016) Preparation and characterization of North Sumatera natural zeolite polyurethane nanocomposite foams for light-weight engineering materials. *Procedia Chemistry*, **19**, 1007-1013.
127. Ajayan, P. M., Stephan, O., Colliex, C., and Trauth, D. (1994) Aligned carbon nanotube arrays formed by cutting a polymer resin-nanotube composite. *Science*, **265**, 1212-1214.
128. Choi, S. M., Jhi, S. H., and Son, Y. W. (2010) Effects of strain on electronic properties of graphene. *Physical Review B: Condensed Matter and Materials Physics*, **81**, 23-26.
129. Yang, Y., Gupta, M. C., Dudley, K. L., and Lawrence, R. W. (2005) Conductive carbon nanofiber-polymer foam structures. *Advanced Materials*, **17**, 1999-2003.
130. Verdejo, R., Bujans, F. B., Perez, M. A. R., Saja, J. A., and Manchado, M. A. L. (2008) Functionalized graphene sheet filled silicone foam nanocomposites. *Journal of Materials Chemistry*, **18**, 2221-2226.
131. Yang, Y., Gupta, M. C., Dudley, K. L., and Lawrence, R. W., (2005) Novel carbon nanotube–polystyrene foam composites for electromagnetic interference shielding. *Nano Letters*, **5**, 2131-2134.
132. Esfahani, A. R. S., Katbab, A. A., Pakdaman, A. R., Dehkhoda, P., Shams, M. H., and Ghorbani, A. (2012) Electrically conductive foamed polyurethane/silicone rubber/graphite nanocomposites as radio frequency wave absorbing material: The role of foam structure. *Polymer Composites*, **33**, 397-403.
133. Xu, X. B., Li, Z. M., Shi, L., Bian, X. C., and Xiang, Z. D. (2007) Ultralight conductive carbon-nanotube-polymer composite. *Small*, **3**, 408-411.
134. Kuang, T., Chang, L., Chen, F., Sheng, Y., Fu, D., and Peng, X. (2016) Facile preparation of lightweight high-strength biodegradable polymer/multi-walled carbon nanotubes nanocomposite foams for electromagnetic interference shielding. *Carbon*, **105**, 305-313.
135. Shen, B., Li, Y., Zhai, W., and Zheng, W. (2016) Compressible graphene-coated polymer foams with ultralow density for adjustable

electromagnetic interference (EMI) shielding. *ACS Applied Materials and Interfaces*, **8**, 8050-8057.

136. Zhang, H.-B., Yan, Q., Zheng, W.-G., He, Z., and Yu, Z.-Z. (2011) Tough Graphene-Polymer Microcellular Foams for Electromagnetic Interference Shielding. *ACS Applied Materials and Interfaces*, **3**, 918-924.
137. Ameli, A., Nofar, M., Wang, S., and Park, C. B. (2014) Lightweight polypropylene/stainless steel fiber composite foams with low percolation for efficient electromagnetic interference shielding. *ACS Applied Materials and Interfaces*, **6**, 11091-11100.
138. Menner, A., Salgueiro, M., Shaffer, M. S. P., and Bismarck, A. (2008) Nanocomposite foams obtained by polymerization of high internal phase emulsions. *Journal of Polymer Science, Part A Polymer Chemistry*, **46**(16), 5708-5714.
139. Thomassin, J. M., Pagnoulle, C., Bednarz, L., Huynen, I., Jerome, R., and Detrembleur, C. (2008) Foams of polycaprolactone/MWNT nanocomposites for efficient EMI reduction. *Journal of Material Chemistry*, **18**, 792-796.
140. Ameli, A., Nofar, M., Park, C. B., Potschke, P., and Rizvi, G. (2014) Polypropylene/carbon nanotube nano/microcellular structures with high dielectric permittivity, low dielectric loss, and low percolation threshold. *Carbon*, **71**, 206-217.
141. Monnereau, L., Urbanczyk, L., Thomassin, J., Pardoen, T., Bailly, C., and Huynen, I. (2015) Gradient foaming of polycarbonate/carbon nanotube based nanocomposites with supercritical carbon dioxide and their EMI shielding performances. *Polymer*, **59**, 117-123.
142. Lorenzetti, A., Roso, M., Bruschetta, A., Boaretti, C., and Modesti, M. (2016) Polyurethane-graphene nanocomposite foams with enhanced thermal insulating properties. *Polymers for Advanced Technologies*, **27**(3), 303-307.
143. Jiang, W., Sundarram, S. S., Wong, D., Koo, J. H., and Li, W. (2014) Polyetherimide nanocomposite foams as an ablative for thermal protection applications. *Composites Part B*, **58**, 559-565.
144. Gedler, G., Antunes, M., Tasciuc, T. B., Velasco, J. I., and Ozisik R. (2016) Effects of graphene concentration, relative density and cellular morphology on the thermal conductivity of polycarbonate–graphene nanocomposite foams. *European Polymer Journal*, **75**, 190-199.
145. Das, N. C., Khastgir, D., Chaki, T. K., and Chakraborty, A. (2000) Electromagnetic interference shielding effectiveness of carbon black and carbon fibre filled EVA and NR based composites. *Composites Part A*, **31**, 1069-1081.
146. Kaynak, A. (1996) Electromagnetic shielding effectiveness of galvanostatically synthesized conducting polypyrrole films in the 300-2000 MHz frequency range. *Materials Research Bulletin*, **31**, 845-860.

147. Cheng, K. B., Ramakrishna, S., and Lee, K. C. (2000) Electromagnetic shielding effectiveness of copper/glass fiber knitted fabric reinforced polypropylene composites. *Composites Part A*, **31**, 1039-1045.
148. Lee, B. O., Woo, W. J., and Kim, M. S. (2001) EMI Shielding Effectiveness of Carbon Nanofiber Filled Poly (vinyl alcohol) Coating Materials. *Macromolecular Materials and Engineering*, **286**, 114-118.
149. Babal, A. S., Gupta, R., Singh, B. P., Singh, V. N., Dhakate, S. R., and Mathur, R. B. (2014) Mechanical and electrical properties of high performance MWCNT/polycarbonate composites prepared by an industrial viable twin screw extruder with back flow channel. *RSC Advances*, **4**, 64649-64658.
150. Wojkiewicz, J. L., Fauveaux, S., and Miane, J. L. (2003) Third-order nonlinear circular dichroism in a liquid of chiral molecules. *Synthetic Metals*, **127**, 135-136.
151. Singh, B. P., Saini, K., Choudhary, V., Teotia, S., Pande, S., Saini, P., and Mathur, R. B. (2014) Effect of length of carbon nanotubes on electromagnetic interference shielding and mechanical properties of their reinforced epoxy composites. *Journal of Nanoparticle Research*, **16**, 1-11.
152. Chen, H. C., Lee, K. C., and Lin, J. H. (2004) Electromagnetic and electrostatic shielding properties of co-weaving-knitting fabrics reinforced composites. *Composites Part A*, **35**, 1249-1256.
153. Singh, B. P., Saini, P., Gupta, T. K., Garg, P., Kumar, G., Pande, I., Pande, S., Seth, R. K., Dhawan, S. K., and Mathur, R. B. (2011) Designing of multi-walled carbon nanotubes reinforced low density polyethylene nanocomposites for suppression of electromagnetic radiation. *Journal of Nanoparticle Research*, **13**, 7065-7074.
154. Tellakula, R. A., Varadan, V. K., Shami, T. C., and Mathur, G. N. (2004) Carbon fiber and nanotube based composites with polypyrrole fabric as electromagnetic absorbers. *Smart Materials and Structures*, **13**, 1040.
155. Chen, D., Yang, J., and Chen, G. (2010) The physical properties of polyurethane/graphite nanosheets/carbon black foaming conducting nanocomposites. *Composites Part A*, **41**, 1636-1638.
156. Krueger, Q. J., and King, J. A. (2003) Synergistic effects of carbon fillers on shielding effectiveness in conductive nylon 6,6 and polycarbonate based resins. *Advances in Polymer Technology*, **22**, 96-111.
157. Gupta, T. K., Singh, B. P., Teotia, S., Katyal, V., Dhakate, S. R., and Mathur, R. B. (2013) Designing of multiwalled carbon nanotubes reinforced polyurethane composites as electromagnetic interference shielding materials. *Journal of Polymer Research*, **20**, 1-7.
158. Yan, D.-X., Dai, K., Xiang, Z.-D., Li, Z.-M., Ji, X., Zhang, W.-Q. (2011) Electrical conductivity and major mechanical and thermal properties of carbon nanotube-filled polyurethane foams. *Journal of Appl-*

ied Polymer Science, **120**(5), 3014-3019.

159. Fletcher, A., Gupta, M. C., Dudley, K. L., and Vedeler, E. (2010) Elastomer foam nanocomposites for electromagnetic dissipation and shielding applications. *Composites Science and Technology*, **70**(6), 953-958.
160. Li, J., Zhang, G., Ma, Z., Fan, X., Qin, J., and Shi, X. (2016) Morphologies and electromagnetic interference shielding performances of microcellular epoxy/multi-wall carbon nanotube nanocomposite foams. *Composites Science and Technology*, **129**, 70-78.
161. Li, Y., Pei, X., Shen, B., Zhai, W., Zhang, L., and Zheng, W. (2015) Polyimide/graphene composite foam sheets with ultrahigh thermostability for electromagnetic interference shielding. *RSC Advances*, **5**, 24342-24351.
162. Chavan, A. A., Li, H., Scarpellini, A., Marras, S., Manna, L., Athanassiou, A., and Fragouli, D. (2015) Elastomeric nanocomposite foams for the removal of heavy metal ions from water. *ACS Applied Materials and Interfaces*, **7**, 14778-14784.
163. Altay, M. C., Malikov, E. Y., Eyvazova, G. M., Muradov, M. B., Akperov, O. H., Puskas, R., Madarasz, D., Konya, Z., and Kukovecz, A. (2015) Facile synthesis of CuS nanoparticles deposited on polymer nanocomposite foam and their effects on microstructural and optical properties. *European Polymer Journal*, **68**, 47-56.
164. Hussein, F. B., and Zahra, N. H. A. (2016) Synthesis, characterization and performance of polyurethane foam nanocomposite for arsenic removal from drinking water. *Journal of Water Process Engineering*, **13**, 1-5.

7

Magnetic Polymer Nanocomposites for Environmental Applications

7.1 Introduction

Needless to say, nanomaterials are one of the most vigorously studied subject areas in the recent years. Nanocomposites are a class of such nanomaterials generated by synergistic combination of the components and have dimension of at least one of the constituents below 100 nm. The combination of the constituent materials in the composite results in improved properties. Thus, there has been a plethora of studies which have been carried out to develop a large variety of nanocomposites for different applications [1].

Polymer nanocomposites are a class of materials where polymer acts as the matrix which is incorporated with different type of nanomaterials, thus, improving the properties of the matrix dramatically. Generally, polymer nanocomposites exhibit enhanced stiffness, strength, gas barrier, flame retardance and thermal resistance as compared to pure polymer matrices. Due to significantly enhanced property profiles, polymer nanocomposites have been developed for various applications in the field of microelectronics, biomedicine, automobiles, etc. [2].

Magnetic nanomaterials are functional materials which exhibit different types of magnetism like ferromagnetism, paramagnetism, superparamagnetism, etc. The pure and alloyed forms of Fe, Ni, Co are the most studied magnetic nanomaterials. These have found uses in various fields like information storage, magnetic refrigeration systems, contrast enhancement in magnetic resonance imaging, magnetic carriers for drug targeting, catalysis, etc. [3]. Due to their nanosize, these materials tend to agglomerate swiftly, thus, leading to a reduction in the active surface area. In order to overcome this limitation, various types of coatings or surface functionalities are incorporated with these materials which make these materials more stable

Liyamol Jacob and Vikas Mittal, The Petroleum Institute (part of Khalifa University of Science and Technology), Abu Dhabi, UAE*
**Current address: Bletchington, Wellington County, Australia*

and attractive for various applications [4]. In addition, in particulate form, these materials do not fit many application requirements, where the use of a continuous material or matrix is required. The nanoparticles, if not stabilized/immobilized through organic surface coatings, are also prone to be released in environment, water streams. Thus, to overcome these challenges, the use of polymers as a matrix or surface modifier for the magnetic nanomaterials has been of significant interest. The polymer matrix loaded with (or coated on) magnetic nanomaterials can respond to external magnetic fields, thus, exhibiting potential for a wide range of applications. Also, interaction between the external and internal magnetic field of the magnetic material can bring about a lot of changes in the host material (or organic coating), like induction of magneto-mechanical forces which can be utilized for changing the shape of the material or inducing its movement. Such properties can be utilized for biomedical applications, actuators, etc. In addition, the magnetic materials can work as nano-heaters responding to external alternating magnetic field. This property can be coupled with thermoresponsive polymers to generate advanced materials [5]. In the recent years, magnetic polymer nanocomposites have been studied for application as drug delivery agents, recovering agents in photocatalysts, negative contrast agents in imaging, etc. Synthesis of magnetic polymer composites can be achieved through various methods like plasma polymerization, melt and solution blending and co-precipitation [6]. This review focuses on the recent advances in synthesis and properties of magnetic polymer nanocomposites, along with their environmental applications.

7.2 Magnetic Polymer Nanocomposites

Magnetic properties shown by nanomaterials are size dependent to a large extent. Especially, the materials which have size comparable to the magnetic domains are of great interest. Unlike other nanomaterials, magnetic nanomaterials are expressed as small identical molecules which have magnetic property. As mentioned earlier, the utility of magnetic nanoparticles can be improved by forming composites with different polymers either by incorporating the particles in the polymers or by modifying the surface of the particles by polymers. There are numerous ways to achieve such hybridization of magnetic materials with polymers, such as ex-situ, in-situ, melt blending, solution blending, etc. [7]. The properties a magnetic material should have for the formation of a composite are same as for regular polymer composites such as: 1) easily availability, 2) cost-effectiveness, 3) reproducibility, 4) feasibility for large scale synthesis, 5) low toxicity, 6) no undesirable reactions with the matrix [8].

7.2.1 Fe based Magnetic Polymer Nanocomposites

Fe based magnetic materials have been commonly utilized for the fabrication of polymer nanocomposites with magnetic properties. Fernandez *et al.* [9] synthesized thermochromic thin films of spin crossover (SCO) polymer $[Fe(NH_2trz)_3](BF_4)$ using different organic polymers as hosts. The formation of these materials could be confirmed by the color changes associated with SCO phenomenon. The hydrophobic polymer hosts were characterized by large particles of SCO material, while hydrophilic polymers exhibited more dispersed nanocrystals with fibrous structures in the polymer matrix, which also affected their colorimetric transitions. The fibrous materials exhibited color transition at room temperature, while the bigger particles showed transitions at lower temperature. This was caused not only by the size of the particles, but also by the nature of solvent and polymer interactions during the fabrication of polymer nanocomposite. The color values vs. temperature exhibited a linear relation, thus, avoiding the uncertainty associated with hysteresis loop [9].

Electrocaloric ferroelectric polymer nanocomposites which can operate at high and average electric fields at room temperature were developed by Zhang *et al.* [10]. A low-cost emulsion polymerization was also utilized for the synthesis of magnetic polymer nanocomposites with hydrophobic and superoleophilic surface [11]. The composites were characterized by properties like unsinkability, thermal stability and corrosion resistance. The polymer nanocomposite was used as an oil adsorbent which could be controlled by applying external magnetic field. Repetitive use of the material for oil removal did not affect the surface properties, thus, confirming the development of a functional material which could be readily used for practical applications without any obvious contamination.

In-situ polymerization technique was used by Kalita *et al.* [12] for the fabrication of shape memory polymer composites using Mesua ferrea L. seed oil based hyperbranched polyurethane (HBPU) and Fe_3O_4. The polymer composite was characterized by good dispersity of the magnetic nanoparticles in the polymer matrix and strong interfacial attraction forces between the matrix and filler particles. The material was observed to exhibit shape memory properties, which also enhanced with Fe_3O_4 content in the composite. Under microwave atmosphere, the polymer had full shape fixity and recoverability.

Biocompatible conductive polyfuran and superparamagnetic Fe_3O_4 particles was used to fabricate Hb immobilized biosensor via in-situ polymerization technique [13]. The composite was used for the detection of H_2O_2 from biological samples. Xiong *et al.* [14] synthesized Fe_3O_4/Ag@nanofibrillated cellulose (NFC) nanocomposites which could change from highly porous aerogel to sheet architecture by actuating using a small magnet. The biocompatible and antibacterial polymer composite was observed to be efficient for the removal of 4-nitrophenol. In the composite, cellulose sheet acted both as a biocompatible substrate and reducing agent for silver ions.

Super paramagnetic iron oxide nanoparticles were generated by using double hydrophilic graft copolymers poly(acrylic acid)-g-poly(N-isopropylacrylamide) (PAA-g-PNIPAM) as a water soluble coating [15]. The preparation of the polymeric coating was realized in three different steps, viz. i) a well-defined hydrophobic PtBA-based backbone, poly(tert-butyl 2-((2-chloropropanoyloxy)-methyl)acrylate)-co-poly(tert-butyl acrylate), ($PtBCPMA_{19}$-co-$PtBA_{18}$), was prepared through RAFT copolymerization of a new trifunctional acrylic monomer, tert-butyl 2-((2-chloropropanoyloxy)methyl)acrylate and tert-butyl acrylate; ii) taking this backbone as a macroinitiator to initiate SET-LRP of N-isopropylacrylamide resulted in well-defined (poly(tert-butyl 2-((2-chloropropanoyloxy)methyl)-acrylate)-co-poly(tert-butyl acrylate))-g-poly(N-isopropylacrylamide) ((PtBCPMA-co-PtBA)-g-PNIPAM) amphiphilic graft copolymers with relatively narrow polydispersities ($M_w/M_n \leq 1.31$); iii) handling (PtBCPMA-co-PtBA)-g-PNIPAM in acidic conditions afforded PAA-g-PNIPAM graft copolymers. The polymer acted as a stabilizer during the preparation of superparamagnetic iron oxide particles, with size controllability.

Oxidative chemical polymerization was utilized for the synthesis of polyaniline (PANI)/cobalt-manganese ferrite, (PANI)/$Co_{0.5}Mn_{0.5}Fe_2O_4$ nanocomposite by Sozeri *et al.* [16]. Aniline monomer was polymerized in presence of ammonium peroxydisulfate (APS). The $CoFe_2O_4$ nanoparticles were synthesized using a microwave assisted method. PANI/$Co_{0.5}Mn_{0.5}Fe_2O_4$ nanocomposite exhibited superparamagnetic behavior with a blocking temperature higher than 300 K. Surface spin disorder resulted in the low magnetic saturation of the composite as compared to $CoFe_2O_4$ particles. Unsaturated magnetization could be obtained by temperature decrease which was characterized by increase in coercivity and antiferromagnetic interactions. In another study, ferromagnetic Fe_3O_4 nanoparticles were synthesized using a surfactant-free sonochemical reaction at room temperature without using any inert atmosphere [17]. The particles were integrated into poly(vinyl alcohol), which, in turn, significantly enhanced the thermal stability and flame retardancy of the polymer.

Removal of toxic Pb^{2+}, $Cd^{2+}$and Ni^{2+} from water sources was realized through the use of carboxymethyl-β-cyclodextrin (CM-β-CD) polymer modified Fe_3O_4 nanoparticles (CDpoly-MNPs) by Zayed *et al.* [18]. The magnetic material exhibited selective adsorption of all types of toxic ions from water with a high affinity towards the adsorption of Pb^{2+}. Chitosan based nanocomposite films were also synthesized on the magnetic α-Fe_2O_3 nanoparticles using hydrothermal method [19]. The magnetic nanoparticles exhibited rhombohedral structures and no phase change was observed by the use of chitosan molecules. The material exhibited excellent saturation magnetization and had ferromagnetic properties.

7.2.2 Ni based Magnetic Polymer Nanocomposites

Ni has been used to synthesize a variety of polymer magnetic nanocomposites with the induction of different types of magnetization into the polymer matrix [20-29]. For instance, Figure 7.1 shows the schematic of a phenolic biosensor prepared from the composite of polydopamine (PDA)-laccase (Lac)-nickel nanoparticle loaded carbon nanofibers (NiCNFs) [20]. Yoonessi *et al.* [30] also used in-situ magnetic field solvent method for the synthesis of oriented hybrid nickel tethered graphene polyimide resin nanocomposites with different degrees of orientation. The magnetization of the composite was observed to be maximum along the direction of magnetic field and minimum in the direction perpendicular to the magnetic field, indicating superparamagnetic orientation of Ni nanoparticles. The DC electrical conductivity of the film could be improved by casting it in the presence of a magnetic field. The nanocomposite exhibited actuation in static magnetic field.

In-situ chemical oxidative copolymerization was utilized to synthesize polystyrene@polypyrrole@nickel (PS@PPy@Ni) with core-shell architecture by Li *et al.* [31]. The Ni nanoparticles were deposited on to the surface of the synthesized polymer sheet using activation-electroless plating methods. The coating of Ni around PS@P(Py-

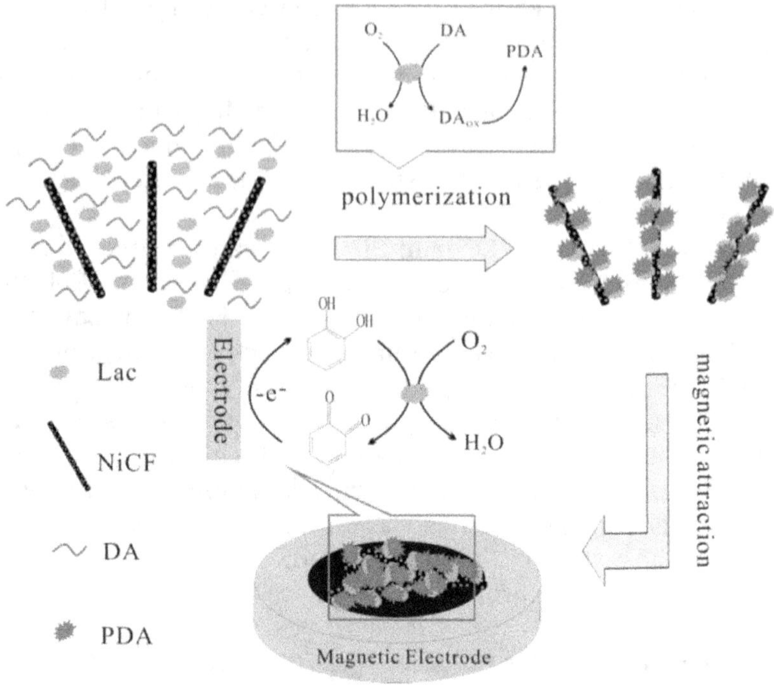

Figure 7.1 Representation of phenolic biosensor prepared from the composite of polydopamine (PDA)-laccase (Lac)-nickel nanoparticle loaded carbon nanofibers (NiCNFs). Reproduced from Reference 20 with permission from American Chemical Society.

PyCOOH) spheres induced ferromagnetism in the material. The material could be used for effective electromagnetic absorption with maximum reflection loss of around -20.06 dB at 10.69 GHz. Figure 7.2 also shows the scanning electron micrographs of the various hybrid materials.

Highly electrically conducting magnetic graphene-Ni polyimide nanocomposites were also synthesized by Yoonessi *et al.* [32]. Increasing the amount of magnetic nanoparticles resulted in an enhancement of saturation magnetization. Nanocomposite exhibited a bimodal nanoparticle distribution throughout the polymer layer and could be actuated in external static magnetic field. In-situ polymerization with the help of cetyltrimethylammonium bromide (CTAB) as a surfactant formed polyaniline-$NiFe_2O_4$ nanofiber composites at water and ionic liquid interphase by Baykal *et al.* [33]. The ionic liquid 1-butyl-3-methyl-imidazolium bromide [BMIM]Br was observed to have significant influence on the structure, morphology, conductivity and the magnetization of the resulting nanocomposite. In general, the introduction of ionic liquids has a positive impact on the conductivity, while a negative impact on the magnetization.

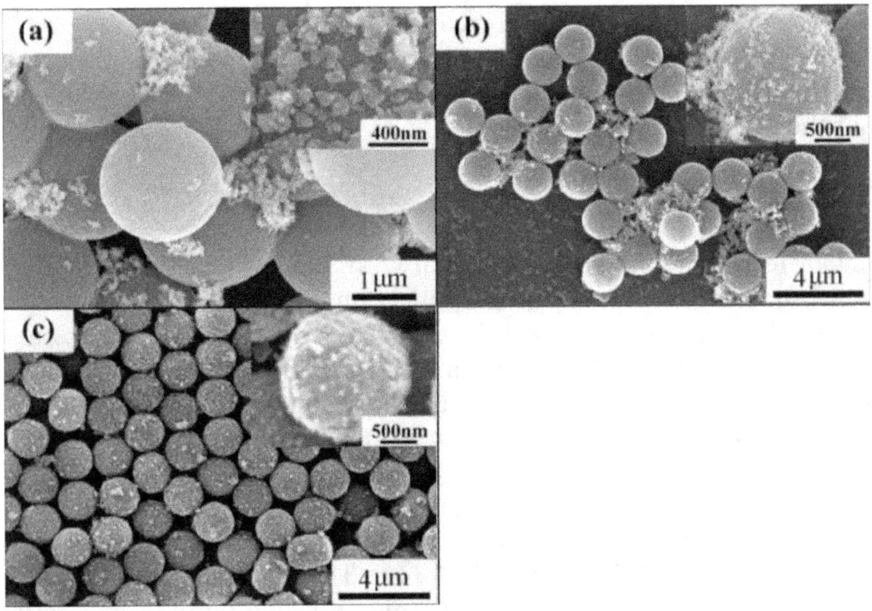

Figure 7.2 SEM images of (a) PS@P(Py-PyCOOH$_0$)@Ni, (b) PS@P(Py- Py-COOH$_{25}$)@Ni and (c) PS@P(Py-PyCOOH$_{50}$)@Ni core/shell microspheres. Reproduced from Reference 31 with permission from American Chemical Society.

Spatially distributed silicon rubber polymer nanocomposites were synthesized with varying the concentration of Ni in the polymer matrix by Boudenne *et al.* [34]. The creation of 1D Nickel structures contributed to higher electric and heat transportability. The spatial distribution did not have any effect on the specific heat capacity of the polymer. The 1D distribution also facilitated an increment of almost 100% thermal conductivity compared to the other orientation of Ni nanoparticles in the polymer matrix. This configuration also facilitated a reduction in percolation threshold from 25% to 2% (vol%) of Ni nanoparticles. In another study, field cooled magnetization of pure Ni ferrite powders and Ni ferrite nanocomposites synthesized using solution casting method exhibited shape memory properties [35]. Increasing particle concentration in the composites resulted in improved shape memory properties. Henkel plots confirmed improvement in the dipolar properties with the increase in particle concentration. However, the increase in cluster size negatively influenced the memory effects. Polydimethylsiloxane (PDMS) magnetic elastomer matrix filled with Ni fiber and Ni particles were also synthesized by speed mixing and curing at ambient temperature conditions [36]. Due to the use of magnetic field during the curing process, the Ni particles and fibers exhibited a distributed alignment throughout the polymer matrix.

7.2.3 Co based Magnetic Polymer Nanocomposites

Cobalt ferrite is a magnetic material with a high Curie temperature, which has found wide ranging applications in biomedical and environmental applications. It has characteristics like high coercivity, medium saturation magnetization, high anisotropy constant, chemical stability, mechanical hardness and electrical insulation [37-39].

The co-assembly of amphiphilic copolymers of PAA20-b-PS280 and oleic acid-stabilized magnetic iron oxide nanoparticles using tetrahydrofuran, N,N-dimethylformamide and water resulted in the synthesis of a well-defined, magnetic shell cross-linked knedel-like nanoparticles (MSCKs) (Figure 7.3) [40]. The material could be used as sequestering agent for hydrocarbons in crude oil with magnetic

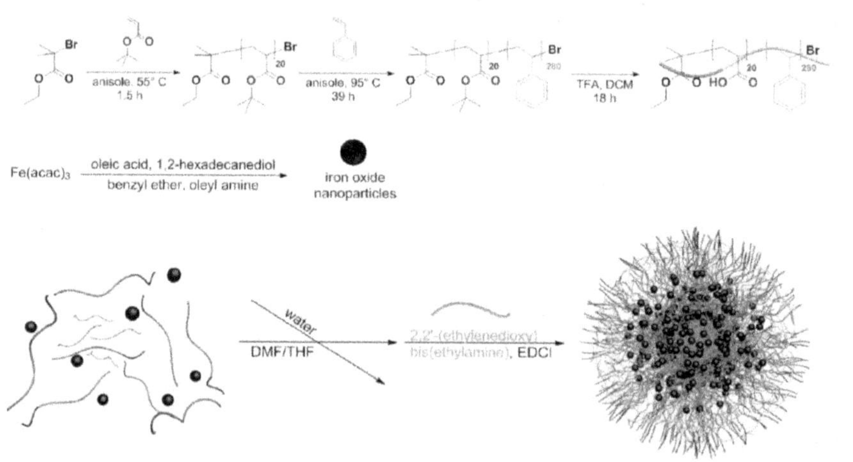

Figure 7.3 Schematic representation of the generation of magnetic shell cross-linked (MSCK) nanoparticles. Reproduced from Reference 40 with permission from American Chemical Society.

responsivity. The generated material in aqueous environments could remove pollutant hydrocarbons from water. The developed material has the potential to be used in the bulk recovery phase at oil spill sites.

The particle agglomeration in solution casting method for polymer magnetic nanomaterial synthesis was overcome by using the nanoparticles in their dry particulate state avoiding the use of solvents [41]. In another study, biocomaptible poly(vinyl alcohol) (PVA) was used for the surface modification of cobalt ferrite nanoparticles ($CoFe_2O_4$ NPs) synthesized via combustion method [42]. The PVA coating transformed the nature of nanoparticles from hydrophobic to hydrophilic. The polymer coated magnetic material exhibited less toxicity and high magnetization with a ferromagnetic behavior compared to the uncoated material.

7.3 Environmental Applications

In the recent years, there has been a significant interest for the use of polymer coated magnetic nanomaterials (or magnetic polymer nanocomposites) for various environmental applications. These materials have been used in areas like pollutant removal from water, oil spill removal, biomedical applications, etc. It is important to gain insights about the interactions of these magnetic nanomaterials with the environment to tune their applications in diverse fields.

7.3.1 Oil Spill Removal

Magnetic polymer nanocomposites have significant potential for use in oil spill removal. These materials are economical, reliable, relatively easy to synthesize and reusable which make them ideal for oil spill removal. The major challenge in the use of materials for the removal of oil is their fast distribution and collection which can be overcome by using magnetic materials as their performance can be efficiently controlled by the application of external magnetic field [43-46]. In a recent study, a well-defined interconnected porous structure cryogel based on resorcinol formaldehyde ($CoFe_2O_4$/graphene/PZTs/R-F) was prepared via freeze drying method, followed by carbonization in order to generate advanced materials to remove crude oil (high viscosity) and vegetable oil (low viscosity) from the oil/water mixture. The sorption capacities of the resultant cryogel for crude oil and vegetable oil were higher than the sorption capacities of $CoFe_2O_4$/graphene/R-F and graphene/R-F/Na_2CO_3 cryogels. The cryogel was responsive to external magnetic field and exhibited good hydrophobic properties with water contact angle of obtuse angle. Figure 7.4 also demonstrates the hydrophobic behavior of the cryogel. From the visual images, it was observed that the oil droplet was quickly sorbed by the cryogel, while the water droplet achieved quasi-spherical shape on the surface of the gel.

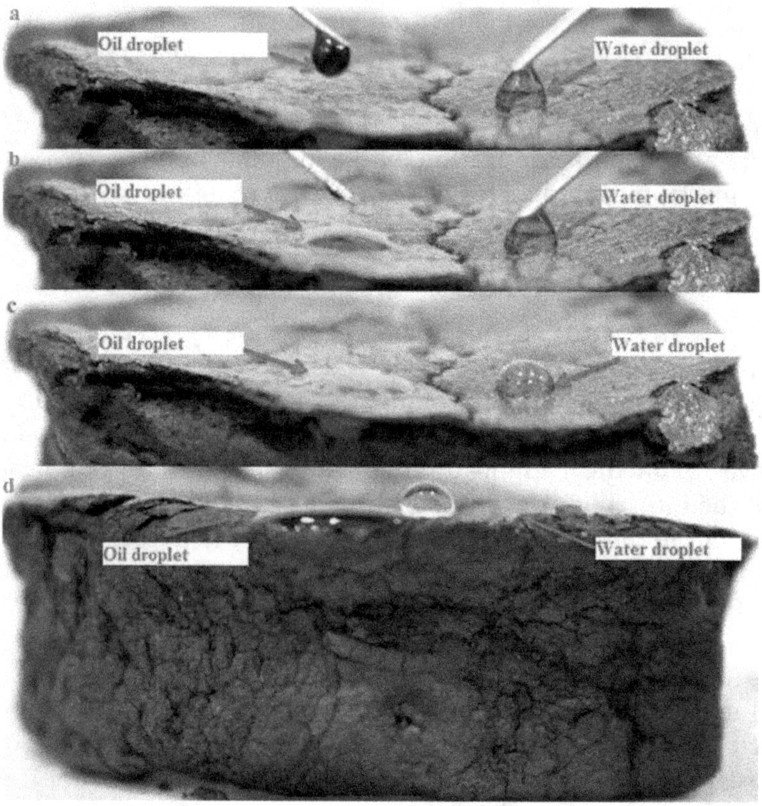

Figure 7.4 Hydrophobic behavior of the $CoFe_2O_4$/graphene/PZTs/R-F cryogel; water droplet gained quasi-spherical shape and oil droplet sorbed quickly onto the top surface of the cryogel. Reproduced from Reference 47.

In another study, superparamagnetic iron oxide nanoparticles functionalized on polyurethane foams were used to efficiently remove oil spills from aqueous environment by Calcagnile *et al.* [48]. The composite was also functionalized with sub-micrometer polytetrafluoroethylene to improve its efficiency. The use of solvent free electrostatic polytetrafluoroethylene particle deposition method rendered the polyurethane foams with water repellent and oil absorbing properties which were otherwise hydrophobic and oleophobic. The use of both types of nanoparticles enhanced the rate of oil absorption using polyurethane foams. In addition to the oil absorbing and water repelling properties, the composite exhibited magnetic responsivity (Figure 7.5). The material has the potential to be used in large area oil spills because the foam easily floats on the surface of water which can be moved using the help of a magnet, thus, cleaning the entire area of water underneath.

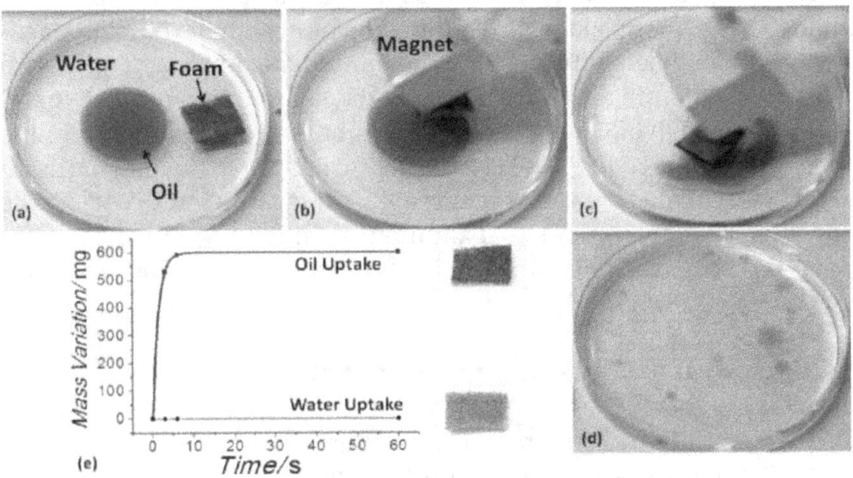

Figure 7.5 (a) Dye-colored oil on the surface of the water, (b,c) removal of the oil by the magnetically driven PU/NPs/PTFE sample, (d) cleaned water surface, (e) graph showing the variation of the weight of the foam with time. Reproduced from Reference 48 with permission from American Chemical Society.

A low-cost emulsion polymerization technique was utilized for the synthesis of superoleophilc and highly hydrophobic magnetic polymer nanocomposites which can remove oil from the surface of water [49]. The developed material had additional properties such as floatability, chemical stability, corrosion resistance and thermal stability, which further reinforced its potential for commercial applications. The material could absorb oil three times its weight with efficiently repelling water and could be retracted from the water surface by using a magnet. It had excellent reusability and the oil from the nanocomposite could be effectively removed using ultrasonication. The protein wastes from the leather industries were also utilized in another study for generating high stability magnetic collagen nanocomposites with superparamagnetic iron oxide nanoparticles [50]. The composite exhibited selective absorption of oils and was trackable by the application of external magnetic field. By heat treatment, the material could be converted into bi-functional graphitic nanocarbon, which proved its environmental sustainability.

Natural rubber's innate ability to resist oil has been utilized for generating oil spill removal materials [51]. Ecofriendly epoxidized natural rubber with magnetite nanoparticles was used for the absorption of gasoline. The epoxidization was used with the aim of retaining the rubber's elasticity and for the adhesion of the nanoparticles. 5% epoxidized natural rubber exhibited maximum stability with absorption capability even after many days of immersion. The material could absorb oil seven times its weight without losing its ability, even after repeated use. The composite formation using magnetic nanoparticles made it easily recoverable using a magnetic field. The magnetic rubber composite exhibited significantly enhanced efficiency as compared to commercially used butyl rubber for oil spill removal.

7.3.2 Removal of Contaminants from Water

Mercury is a highly toxic material for humans and animals even in slightest amounts. Magnetic composite generated from magnetic iron oxide nanoparticles (M-MIONPs) modified with 2-mercaptobenzothiazole were proved to be efficient in the removal of Pb(II) ions from water [52]. The material was observed to be efficient, fast, cost effective and reliable for the removal of toxic mercury ions from aquatic environments. The material was able to remove mercury ions within a short span of 4 min with almost 100% efficiency as compared to non-modified iron oxide nanoparticles.

A nanocomposite composed of multiwalled carbon nanotubes, poly(1-glycidyl-3-methylimidazolium chloride) and ferroferric oxide was used to remove anionic azo dyes orange II, sunset yellow FCF and amaranth [53]. Adsorption kinetics and isotherms followed pseudo second order kinetics and Langmuir isotherm respectively and the adsorption was affected by the pH of the system. The adsorption and separation processes were enhanced by the introduction of the polymer and magnetic nanomaterial together with carbon nanotubes. In another study, radical addition fragmentation chain transfer (RAFT) polymerization using RAFT agent functionalized iron oxide nanoparticles as the chain transfer agent was used to synthesize molecular imprinted core shell magnetic beads [54]. The beads had a polymer film with a thickness of about 22 nm, and exhibited thermal stability, magnetic responsivity and spherical shape. The material was utilized for the selective removal of bisphenol A and could be easily recovered using the application of an external magnetic field. The material also exhibited reusability and highly stable performance without much loss in efficiency up to 5 repeated cycles.

Polyethersulfone (PES) nanocomposite membrane containing polyaniline/iron(II, III) oxide (PANI/Fe_3O_4) nanoparticles synthesized using phase inversion method was utilized for Cu(II) ion removal from aqueous environment [55]. Different nanoparticle loading was tested for optimum removal efficiency. 0.1% loading of nanoparticles was observed to be most efficient due to the nanoparticle occupancy at the superficial pores of the polymer membrane during synthesis. The material exhibited a relatively complex adsorption mechanism with Redlich-Peterson model isotherm. The membrane was also observed to be highly reusable. In another study, in-situ polymerization of polypyrrole monomer with magnetic FE_3O_4 nanoparticles was used to synthesize Fe_3O_4 coated polypyrrole membrane which was efficient in the removal of Cr(VI) ions from water [56]. 100% adsorption was observed for 200 mg/L Cr(VI) aqueous solution at a low pH of 2. The removal was attributed to the ion exchange and reduction on the surface of the nanocomposite. The removal efficiency was negatively affected by an increase in pH of the aqueous solution. The adsorption followed pseudo second order kinetics and it could be fitted with a Langmuir adsorption isotherm. The material exhibited an endothermic reaction during the adsorption, which was spontaneous in nature. The material could be reused for at least two repeated cycles without much loss in the original capacity.

As mentioned earlier, carboxymethyl-β-cyclodextrin (CM-β-CD) polymer modified Fe_3O_4 nanoparticles were utilized for the selective removal of Pb^{2+}, Cd^{2+} and Ni^{2+} ions from aqueous solution [18]. The adsorption capability was dependent on temperature, pH and ionic strength of the aqueous solution. The adsorption exhibited pseudo second order kinetics and could be fitted with Langmuir isotherm. The polymer grafting enhanced the adsorption ability because of the hydroxyl and carboxyl groups from the polymer backbone complexing with the metal ions. The authors also explored the desorption and reusability of the developed materials.

Similar to earlier study, Arora *et al.* [57] investigated the potential application of cobalt ferrite doped graphene/phosphazene nanotubes (PZTs) - resorcinol-formaldehyde (R-F) cryogel nanocomposites as a new magnetic field responsive adsorbent to remove 2-nitrophenol from aqueous solution [47]. Adsorption kinetics and thermodynamics parameters (($\Delta H°$, $\Delta G°$ and $\Delta S°$) were examined from experimental data, which indicated that the adsorption process was spontaneous and endothermic, which followed a pseudo second order model. The maximum adsorption capacity reached 92.37 $mg.g^{-1}$ at initial 2-nitrophenol concentration of 600 $mg.L^{-1}$ and 303 K. The experimental adsorption data was also fitted using Langmuir, Freundlich and Temkin isotherm models and the adsorption data was observed to fit well with Freundlich isotherm model (Figure 7.6).

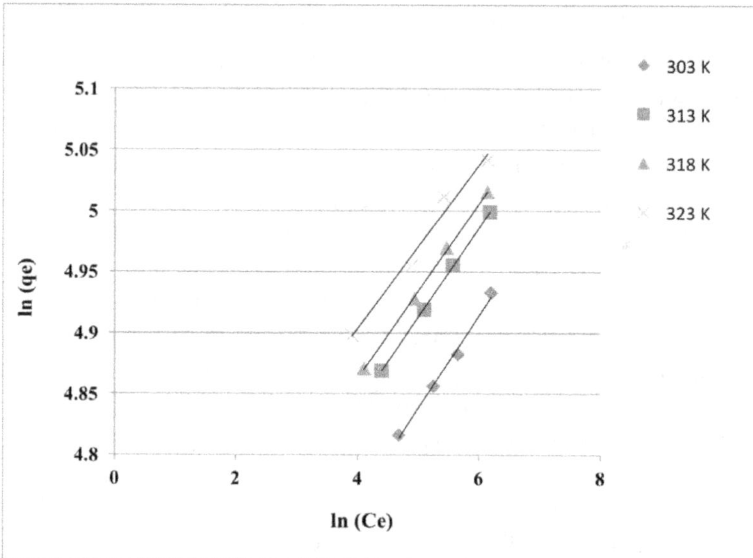

Figure 7.6 Freundlich isotherm for adsorption of 2-nitrophenol (600 $mg.L^{-1}$) onto magnetic adsorbent at 303, 313, 318 and 323 K. Reproduced from Reference 57.

7.3.3 Biocatalysts and Biological Host Systems

Lin *et al.* [58] reported the synthesis of enzyme nanogels with magnetic nanoparticle encapsulation. The synthesis was carried out as follows: i) synthesis of the aminated, acryloylated or carboxylated magnetic nanoparticles, ii) binding the oppositely charged enzymes on to the surface of magnetic nanoparticles using electrostatic interaction, iii) in-situ polymerization using acrylamide and methylenebis(acrylamide) as monomer and crosslinker respectively. The dissociation of enzyme from the polymer surface could be prevented by effectively controlling the pH of the system and its catalytic activity could be modified by covalent linkage. The polyacrylamide gel network was able to provide hydrophilic confinement and improve the thermal stability of the enzyme.

Copolymerization of styrene and maleic anhydride in the presence of magnetite seeds was carried out by Qiu *et al.* [59] to synthesize Fe_3O_4/poly(styrene-co-maleic anhydride) core-shell composite microspheres. The affinity between magnetite nanoparticles and monomers was improvised by encapsulating magnetite nanoparticles with polyethylene glycol. It was observed that the amount of surface anhydride, dimension of microspheres and percentage of magnetite present in the composite was influenced by the comonomer ratio, magnetite particles and solvent. Covalent binding led to the immobilization of free α-amylase to the reactive surfaces containing anhydride groups. The material exhibited a high magnetic immobilized enzyme and binding capacity with an activity recovery of 47.2%. Zhang *et al.* [60] reported the microwave solvothermal method for the rapid synthesis of Fe_3O_4 nanospheres with uniform dimensions. Chemical oxidative polymerization of pyrrole in the presence of magnetic nanospheres were carried out to synthesize Fe_3O_4/polypyrrole (PPy) composite nanospheres with well-defined core/shell structures. The coated shell exhibited an average thickness of 25 nm. Subsequently, gold nanoparticles were immobilized on to the surface of the polymer nanocomposite to generate Fe_3O_4/PPy/Au core/shell/shell structure. The composite exhibited excellent electrocatalytic and biocatalytic activity towards bio-species like ascorbic acid. The magnetic particles were controllable by the application of an external magnetic field and exhibited excellent conductivity. Figure 7.7 shows the micrographs of the generated materials.

β-cyclodextrin-polyurethane polymer coated Fe_3O_4 magnetic nanoparticle composite was prepared as solid-liquid phase transfer catalyst and molecular host system for nucleophilic substitution reactions [61]. The polymeric nanocomposite material was able to catalyze the nucleophilic substitution reaction of benzyl halides with thiocyanate, azide, cyanide and acetate anions in aqueous environment. The recovery and reuse of the developed material was achieved by the application of an external magnetic field. In addition, no toxic by-products were formed during the reaction. The nanoparticles exhibited superparamagnetic properties with a mean diameter of 59 nm.

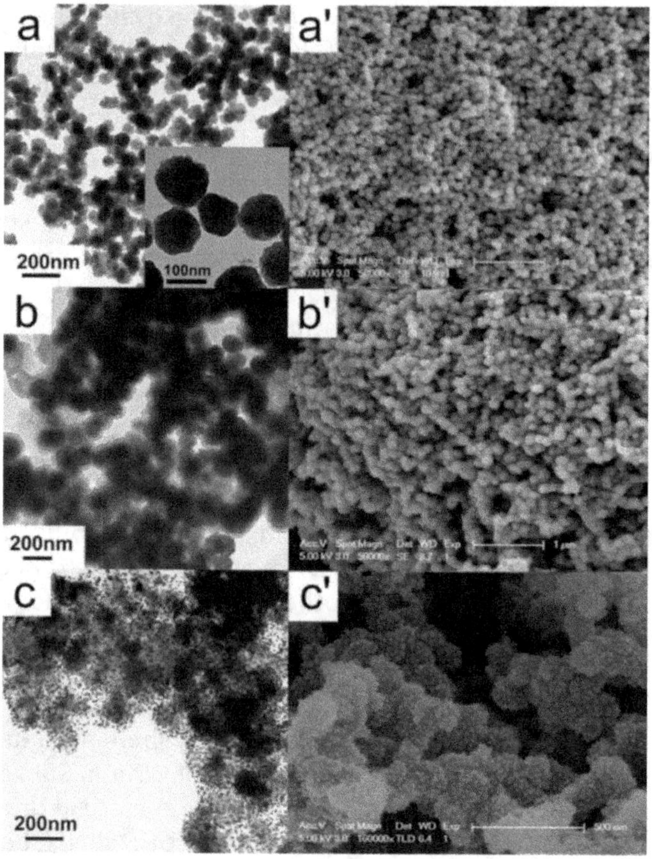

Figure 7.7 (a-c) transmission and (a′-c′) scanning electron images of (a, a′) Fe_3O_4, (b, b′) Fe_3O_4/PPy, and (c, c′) Fe_3O_4/PPy/Au NPs. Reproduced from Reference 60 with permission from American Chemical Society.

In another study, the efficiency of immobilized enzymes could be enhanced by using polymer nanocomposite microspheres (PNCMs) as solid supports, which reduced diffusional limitation and increased the surface area per mass unit [62]. Fe_3O_4/SiO_2-g-poly(PSStNa) was used as a solid support for the immobilization of pectin enzyme by covalent interaction. The material exhibited an enhanced storage capacity and activity over a wide range of temperature and pH. By using

Michaelis-Menten equation, the catalytic activity of the enzyme immobilized polymer magnetic nanocomposite was studied. The separation and recovery could be achieved using an external magnetic field and the method reduced the cost of enzyme immobilization considerably, thus, confirming the potential for a large variety of practical applications.

7.3.4 Gas and Humidity Sensors

Nanocomposites of polypyrrole and iron oxide have been successfully used as humidity sensors. In one such study, the composite was formed using a simultaneous gelation and polymerization process [63,64]. The pellet shaped nanocomposite was utilized for humidity and gas sensing studies. The humidity sensitivity of the as prepared composite was improved by increasing the polypyrrole content. The gas sensitivity studies conducted on CO_2, N_2 and CH_4 gases concluded that the composite exhibited a linear relationship between pressure and sensitivity.

Free standing ultrathin nanocomposite films were also synthesized based on conducting polymers like poly(3,4-ethylenedioxythiophene):polystyrene sulfonate (PEDOT:PSS) with embedded iron oxide nanoparticles [65]. Spin coating was utilized for the synthesis of the nanocomposite films in conjunction with a release technique to detach them from the substrate. The films were tested for their humidity sensing properties and the sensitivity was observed to increase with increase in nanoparticle concentration. Figure 7.8 also demonstrates the change in the resistance of the nanocomposite film during humidity step-change experiments. The response was observed to be fast and effectively matched to the changes in humidity. In addition, the resistance of the sensor was observed to return to the original value after the completion of the test.

In another study, polyaniline/γ-ferric oxide (PANi/γ-Fe_2O_3) nanocomposite films were synthesized by Sen *et al.* [66] for room temperature liquefied petroleum gas (LPG) sensing. The gas sensitivity of the nanocomposite was improved by its large surface area which enhanced the adsorption of gas molecules. The response time was as low as 60 s for the composite and the maximum response was observed for 200 ppm LPG concentration. The sensing mechanism of the nanocomposite was attributed to the change in the depletion region of the p–n junction, which was formed between polymer and magnetic nanoparticles.

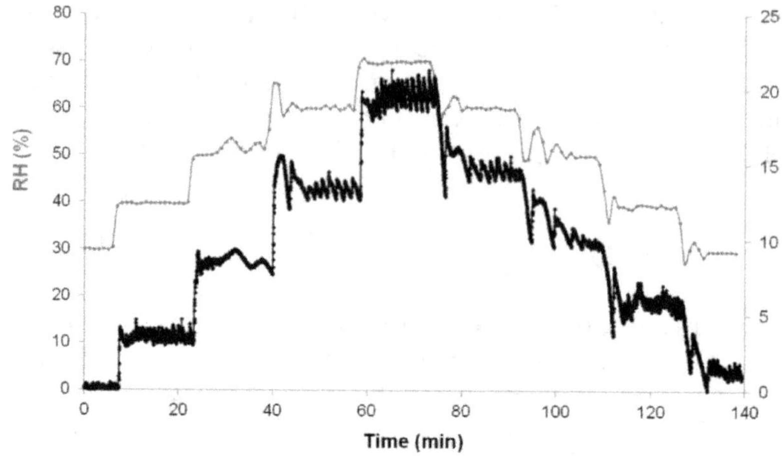

Figure 7.8 The change in the resistance (dark curve) of a PEDOT:PSS/NP nanofilm during humidity step-change experiments (light curve) in the range of 30-70% RH. Reproduced from Reference 65 with permission from American Chemical Society.

7.3.5 Biosensors

Nanocomposites formed by the conjugation of polymers and nanomaterials have found significant applications in biosensing field due to enhanced surface area and tunable surface properties. Many synthesis methods have been utilized for the generation of polymeric nanocomposites, especially magnetic polymer nanocomposites, for biosensing applications [67-69].

Estrone imprinted polymer nanocomposite with Fe_3O_4 coating was synthesized using semi-covalent imprinting method [70]. Firstly, silicone-estrone monomer complex was prepared using 3-(triethoxysilyl)propyl isocyanate and estrone, where the template could be linked to the silica coating on the iron oxide core through a thermally reversible bond. The material exhibited high saturation magnetization and specific recognition. The developed magnetic polymer nanocomposites were successfully utilized for the specific biochemical separation of estrone. In another study, emulsion polymerization was utilized for the synthesis of biocompatible poly(p-phenylenediamine) (PpPDA)-Fe_3O_4 nanocomposite which was further immobilized with hemoglobin (Hb) molecules for sensing of H_2O_2 [71]. The material exhibited good sensitivity and reproducibility towards the detection of H_2O_2. The catalytic reduction current of H_2O_2 was observed to be proportional to its concentration. The high affinity of Hb on PpPDA@Fe_3O_4 nanocomposite was also confirmed by the value of Michaelis–Menten constant (0.088 mM).

Biosensing nanocomposite polymer films based on polypyrrole were also synthesized by immobilizing magnetic nanoparticles on polypyrrole surface under appropriate magnetic field [72]. Streptavidin labeled magnetic particles were used to functionalize NPs assembly with biotin-Fab fragment K47 antibody. The biosensor was used to detect atrazine concentrations. The nanocomposite was observed to be fast, simple and accurate for the detection and exhibited a low concentration detection limit of 5 ng/ml. These properties make these materials attractive candidates for atrazine biosensing for practical and commercial applications. Zhang *et al.* [73] developed Fe_3O_4/polydopamine/Au composite nanoparticles with core shell structure by the polymerization of dopamine on Fe_3O_4 core. The changes in reaction time and reactant concentrations affected the polydopamine shell. The reduction of Au^{3+} between the core/shell and $HAuCl_4$ solution resulted in numerous satellite structures of Au nanoparticles on the surface of the Fe_3O_4/polydopamine composite. The material exhibited excellent magnetic responsivity. It was also observed to be specific to the detection and removal of bovine hemoglobin. These properties confirm the potential use of these materials for protein removal in proteomic analyses.

7.3.6 Other Applications

Along with the above-mentioned applications, polymer magnetic nanomaterials have found applications in other advanced areas such as drug delivery [74-84], magnetic resonance (MR) imaging [85-94], tissue engineering [95-103], etc. Their specificity, responsivity towards magnetic environment as well as low toxicity are the most important characteristics driving their use in such applications. For instance, Lutz *et al.* [85] reported the synthesis of a well-defined copolymer poly(oligo(ethylene glycol) methacrylate-co-methacrylic acid) P(OEGMA-co-MAA) as a novel water-soluble biocompatible coating for superparamagnetic iron oxide nanoparticles. The authors reported that the diameter of the nanoparticles was controlled in the range 10-25 nm, which was dependent on the initial copolymer concentration. The magnetic nanomaterials were observed to have long term colloidal stability in physiological buffer. These materials could be studied in vivo by MR imaging. On injecting in rats, the nanoparticles were not observed to accumulate in the liver during the first 2 h, and exhibited maximum accumulation after 6 h, thus, indicating a prolonged circulation in bloodstream. Figure 7.9 presents the schematic of the preparation of copolymer as well as copolymer modified iron oxide nanoparticles.

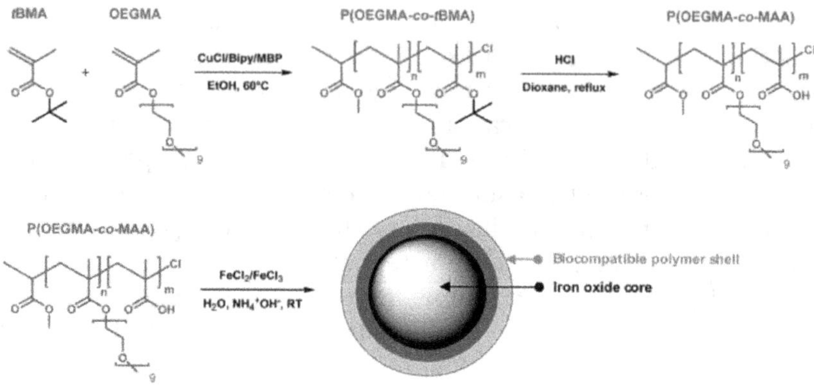

Figure 7.9 Schematic of the preparation of copolymer as well as copolymer modified iron oxide nanoparticles. Reproduced from Reference 85 with permission from American Chemical Society.

In another study, Berret *et al.* [89] also reported controlled clustering of superparamagnetic nanoparticles using block copolymers (Figure 7.10) [89]. The complexes were generated using 6.3 nm diameter superparamagnetic nanoparticles (maghemite γ-Fe_2O_3) and cationic-neutral copolymers such as poly(trimethylammonium ethylacrylate methyl sulfate)-b-poly-(acrylamide) with varying molecular weights of the blocks. The authors confirmed through magnetic resonance spin-echo measurements that the developed polymer-nanoparticle complexes exhibited a better contrast in magnetic resonance imaging as compared to individual nanoparticles.

Arora *et al.* [104] also reported the synthesis of cobalt ferrite doped graphene polyacrylamide cryogel nanocomposites synthesized by employing in situ-polymerization technique for bio-separation processes. In the formation of polyacrylamide based cryogel, divalent metal ions (such as Co^{2+} and Fe^{2+}) performed both as crosslinker and catalyst due to their ability of binding towards oxygen groups which led to crosslinking between graphene oxide and polyacrylamide skeleton, thus, enhancing the strength of the cryogel. The properties of cryogel such as pore dimensions, pore wall thickness, mechanical toughness and elasticity were affected by synthesis conditions, amount of solvent and precursors, temperature gradients, cooling rate and other conditions.

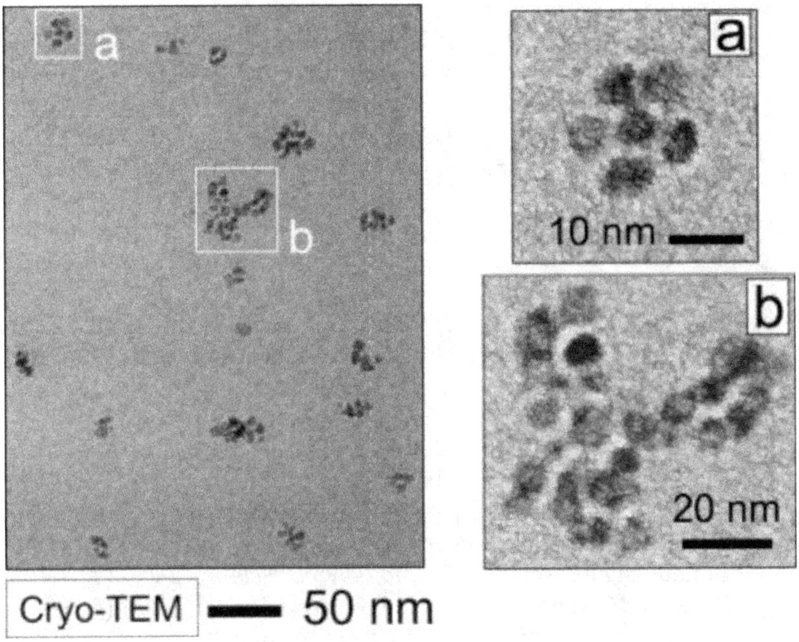

Figure 7.10 Cryo-transmission electron images of mixed aggregates obtained by complexation of PTEA(5K)-b-PAM(30K) and iron oxide nanoparticles. Reproduced from Reference 89 with permission from American Chemical Society.

7.4 Conclusions

Magnetic polymer nanocomposites (either nanoparticles embedded in polymer matrix or nanoparticles surface modified with polymer) are useful class of materials with functional characteristics. Polymer coating or grafting the nanoparticles to various polymer substrates avoids the usual agglomeration of nanoparticles. The magnetic polymer composites exhibit excellent improvement in properties as compared with individual components owing to the interactions between organic and inorganic frameworks. Specially, these nanomaterials have been utilized in various applications such as water purification, oil spill removal, tissue engineering, biocatalysis, biosensing, drug delivery, etc.

References

1. Smith, D. M., Baker Jr., J. R., and Simon, J. K. (2013) Applications of nanotechnology for immunology. *Nature Reviews Immunology,* **13**(8), 592-605.
2. Fischer, H. (2003) Polymer nanocomposites: from fundamental research to specific applications. *Materials Science and Engineering C,* **23**(6-8), 763-772.
3. Zhu, M., and Diao, G. (2011) Review on the progress in synthesis and application of magnetic carbon nanocomposites. *Nanoscale,* **3**, 2748-2767.
4. Liu, Y., Jiang, W., Wang, Y., Zhang, X. J., Song, D., and Li, F. S. (2009) Synthesis of Fe_3O_4/CNTs magnetic nanocomposites at the liquid–liquid interface using oleate as surfactant and reactant. *Journal of Magnetism and Magnetic Materials,* **321**(5), 408-412.
5. Behrens, S., and Appel, I. (2016) Magnetic nanocomposites. *Current Opinion in Biotechnology,* **39**, 89-96.
6. Kalia, S., Kango, S., Kumar, A., Haldorai, Y., Kumari, B., and Kumar, R. (2014) Magnetic polymer nanocomposites for environmental and biomedical applications. *Colloid Polymer Science,* **292**, 2025-2052.
7. Yoonessi, M., Gaier, J. R., Peck, J. A., and Meador, M. A. (2015) Controlled direction of electrical and mechanical properties in nickel tethered graphene polyimide nanocomposites using magnetic field. *Carbon,* **84,** 375-382.
8. Kainz, Q. M., and Reiser, O. (2014) Polymer- and Dendrimer-Coated Magnetic Nanoparticles as Versatile Supports for Catalysts, Scavengers, and Reagents. *Accounts of Chemical Research,* **47**(2), 667-677.
9. Fernandez, A. L., Cuellar, M. P., Herrera, J. M., Castillo, A. S., Pegalajar, M. D. C., Padilla, S. T., Colacio, E., and Vallvey L. F. C. (2014) Particle tuning and modulation of the magnetic/color synergy in Fe(II) spin crossover-polymer nanocomposites in a thermochromic sensor array. *Journal of Material Chemistry C,* **2**, 7292-7303.
10. Zhang, G., Li, Q., Gu, H., Jiang, S., Han, K., Gadinski, M. R., Haque, M. A., Zhang, Q., and Wang, Q. (2015) Ferroelectric polymer nanocomposites for room-temperature electrocaloric refrigeration. *Advanced Materials,* **27**, 1450-1454.
11. Gu, J., Jiang, W., Wang, F., Chen, M., Mao, J., and Xie, T. (2014) Facile removal of oils from water surfaces through highly hydrophobic and magnetic polymer nanocomposites. *Applied Surface Science,* **301,** 492-499.
12. Kalita, H., and Karak, N. (2013) Bio-based hyperbranched polyuret-

hane/Fe_3O_4 nanocomposites as shape memory materials. *Polymers for Advanced Technologies*, **24**, 819-823.

13. Baghayeri, M., Zare, E. N., and Lakouraj, M. M. (2014) Novel superparamagnetic PFu@Fe_3O_4 conductive nanocomposite as a suitable host for hemoglobin immobilization. *Sensors and Actuators B: Chemical*, **202**, 1200-1208.
14. Xiong, R., Lu, C., Wang, Y., Zhou, Z., and Zhang, X. (2013) Nanofibrillated cellulose as the support and reductant for the facile synthesis of Fe_3O_4/Ag nanocomposites with catalytic and antibacterial activity. *Journal of Material Chemistry A*, **1**, 14910-14918.
15. Jiang, X., Zhai, S., Jiang, X., Lu, G., and Huang, X. (2014) Synthesis of PAA-g-PNIPAM well-defined graft polymer by sequential RAFT and SET-LRP and its application in preparing size-controlled super-paramagnetic Fe_3O_4 nanoparticles as a stabilizer. *Polymer*, **55**(16), 3703-3712.
16. Sozeri, H., Kurtan, U., Topkaya, R., Baykal, A., and Toprak, M. S. (2013) Polyaniline (PANI)-$Co_{0.5}Mn_{0.5}Fe_2O_4$ nanocomposite: Synthesis, characterization and magnetic properties evaluation. *Ceramics International*, **39**(5), 5137-5143.
17. Ghanbari, D., Salavati-Niasari, M., and Ghasemi-Kooch, M. (2014) A sonochemical method for synthesis of Fe_3O_4 nanoparticles and thermal stable PVA-based magnetic nanocomposite. *Journal of Industrial and Engineering Chemistry*, **20**(6), 3970-3974.
18. Badruddoza, A. M., Bin Zakir Shawon, Z., Tay, W. J. D., Hidajat, K., and Uddin, M. S. (2013) Fe_3O_4/cyclodextrin polymer nanocomposites for selective heavy metals removal from industrial wastewater. *Carbohydrate Polymers*, **91**(1), 322-332.
19. Singh, J., Srivastava, M., Dutta, J., and Dutta, P. K. (2011) Preparation and properties of hybrid monodispersed magnetic α-Fe_2O_3 based chitosan nanocomposite film for industrial and biomedical applications. *International Journal of Biological Macromolecules*, **48**(1), 170-176.
20. Li, D., Luo, L., Pang, Z., Ding, L., Wang, Q., Ke, H., Huang, F., and Wei, Q. (2014) Novel phenolic biosensor based on a magnetic polydopamine-laccase-nickel nanoparticle loaded carbon nanofiber composite. *ACS Applied Materials and Interfaces*, **6**(7), 5144-5151.
21. Krupa, I., Cecen, V., Boudenne, A., Prokes, Jan., and Novak, I. (2013) The mechanical and adhesive properties of electrically and thermally conductive polymeric composites based on high density polyethylene filled with nickel powder. *Materials & Design*, **51**, 620-628.
22. Bora, C., Kalita, A., Das, D., Dolui, S. K. and Mukhopadhyay, P. K. (2014), Preparation of polyaniline/nickel oxide nanocomposites by liquid/liquid interfacial polymerization and evaluation of their electrical, electrochemical and magnetic properties. *Polymer Internatio-*

nal, **63**, 445-452.

23. Agarwal, S., Greiner, A., and Wendorff, J. H. (2013) Functional materials by electrospinning of polymers. *Progress in Polymer Science*, **38**(6), 963-991.
24. Sudagar, J., Lian, J., and Sha, W. (2013) Electroless nickel, alloy, composite and nano coatings - A critical review. *Journal of Alloys and Compounds*, **571**, 183-204.
25. He, S., Lu, C., Wang, G.-S., Wang, J.-W., Guo, H.-Y. and Guo, L. (2014), Synthesis and growth mechanism of white-fungus-like nickel sulfide microspheres, and their application in polymer composites with enhanced microwave-absorption properties. *ChemPlusChem*, **79**, 569-576.
26. Nandapure, B. I., Kondawar, S. B., Salunkhe, M. Y., and Nandapure, A. I. (2013) Magnetic and transport properties of conducting polyaniline/nickel oxide nanocomposites. *Advanced Materials Letters*, **4**(2), 134-140.
27. Zhai, W., Chen, Y., Ling, J., Wen, B., and Kim, Y.-W. (2014) Fabrication of lightweight, flexible polyetherimide/nickel composite foam with electromagnetic interference shielding effectiveness reaching 103 dB. *Journal of Cellular Plastics*, **50**(6), 537-550.
28. Romio, A. P., Rodrigues, H. H., Peres, A., Da Cas Viegas, A., Kobitskaya, E., Ziener, U., Landfester, K., Sayer, C. and Araujo, P. H. H. (2013) Encapsulation of magnetic nickel nanoparticles via inverse miniemulsion polymerization. *Journal of Applied Polymer Science*, **129**, 1426-1433.
29. Hui, N., Wang, S., Xie, H., Xu, S., Niu, S., and Luo, X. (2015) Nickel nanoparticles modified conducting polymer composite of reduced graphene oxide doped poly(3,4-ethylenedioxythiophene) for enhanced nonenzymatic glucose sensing. *Sensors and Actuators B: Chemical*, **221**(31), 606-613.
30. Yoonessi, M., Gaier, J. R., Peck, J. A., Meador, M. A. (2015) Controlled direction of electrical and mechanical properties in nickel tethered graphene polyimide nanocomposites using magnetic field. *Carbon*, **84**, 375-382.
31. Li, W., Qiu, T., Wang, L., Ren, S., Zhang, J., He, L., and Li, X. (2013) Preparation and electromagnetic properties of core/shell polystyrene@polypyrrole@Nickel composite microspheres. *ACS Applied Materials and Interfaces*, **5**(3), 883-891.
32. Yoonessi, M., Scheiman, D. A., Dittler, M., Peck, J. A., Ilavsky J., Gaier J. R., and Meador, M. A. (2013) High-temperature multifunctional magnetoactive nickel graphene polyimide nanocomposites. *Polymer*, **54**(11), 2776-2784.
33. Baykal, A., Günay, M., Toprak, M. S., and Sozeri, H. (2013) Effect of ionic liquids on the electrical and magnetic performance of polyaniline–nickel ferrite nanocomposite. *Materials Research Bulletin*, **48**,

378-382.

34. Boudenne, A., Mamunya, Y., Levchenko, V., Garnier, B., and Lebedev, E. (2015) Improvement of thermal and electrical properties of Silicone–Ni composites using magnetic field. *European Polymer Journal*, **63**, 11-19.
35. Malik, R., Sehdev, N., Lamba, S., Sharma, P., Makino, A., and Annapoorni, S. (2014) Magnetic memory effects in nickel ferrite/ polymer nanocomposites. *Applied Physics Letters*, **104**, 122407.
36. Zhu, L., Xie, D., Ma, J., Shao, J., and Shen, X. (2013) Fabrication of polydimethylsiloxane composites with nickel particles and nickel fibers and study of their magnetic properties. *Smart Materials and Structures*, **22** (4), 045015.
37. Amiri, S., and Shokrollahi, H. (2013) The role of cobalt ferrite magnetic nanoparticles in medical science. *Materials Science and Engineering C*, **33**(1), 1-8.
38. Wohlert, S., Tomkowicz, Z., Rams, M., Ebbinghaus, S. G., Fink, L., Schmidt, and M. U., Nather, C. (2014) Influence of the co-Ligand on the magnetic and relaxation properties of layered cobalt(II) thiocyanato coordination polymers. *Inorganic Chemistry*, **53**(16), 8298-8310.
39. Zeng, M.-H., Zheng Y., Tan, Y.-X., Zhang, W.-X., He, Y.-P., Kurmoo, M. (2014) Nanoporous cobalt(II) MOF exhibiting four magnetic ground states and changes in gas sorption upon post-synthetic modification. *Journal of American Chemical Society*, **136**(12), 4680-4688.
40. Pavia-Sanders, A., Zhang, S., Flores, J. A., Sanders, J. E., Raymond, J. E., and Wooley, K. L. (2013) Robust magnetic/polymer hybrid nanoparticles designed for crude oil entrapment and recovery in aqueous environments. *ACS Nano*, **7**(9), 7552-7561.
41. Khademolhosseini, F., and Chiao, M. (2013) Fabrication and patterning of magnetic polymer micropillar structures using a dry-nanoparticle embedding technique. *Journal of Microelectromechanical Systems*, **22**(1), 131-139.
42. Salunkhe, A. B., Khot, V. M., Thorat, N. D., Phadatare, M. R., Sathish, C. I., Dhawale, D. S. and Pawar, S. H. (2013) Polyvinyl alcohol functionalized cobalt ferrite nanoparticles for biomedical applications. *Applied Surface Science*, **264**, 598-604.
43. Zhu, Q., Tao, F., and Pan, Q. (2010) Fast and selective removal of oils from water surface via highly hydrophobic core–shell Fe_2O_3@C nanoparticles under magnetic field. *ACS Applied Materials and Interfaces*, **2**(11), 3141-3146.
44. Korhonen, J. T., Kettunen, M., Ras, R. H. A., and Ikkala, O. (2011) Hydrophobic nanocellulose aerogels as floating, sustainable, reusable, and recyclable oil absorbents. *ACS Applied Materials and Interfaces*, **3**(6), 1813-1816.

45. Chu, Y., and Pan, Q. (2012) Three-dimensionally macroporous Fe/C nanocomposites as highly selective oil-absorption materials. *ACS Applied Materials and Interfaces*, **4**(5), 2420-2425.
46. Kumar, A., Sharma, G., Naushad, M., and Thakur, S. (2015) SPION/β-cyclodextrin core–shell nanostructures for oil spill remediation and organic pollutant removal from waste water. *Chemical Engineering Journal*, **280,** 175-187.
47. Arora, N., Vengatesan, M. R., and Mittal, V. (2018) Resorcinol-formaldehyde cryogel nanocomposites for oil spill clean up. In: *Polymers in Oil and Gas Industry*, Mittal, V. (ed.), Central West Publishing, Australia, pp. 147-163.
48. Calcagnile, P., Fragouli, D., Bayer, I. S., Anyfantis, G. C., Martiradonna, L., Cozzoli, P. D., Cingolani, R., and Athanassios, A. (2012) Magnetically driven floating foams for the removal of oil contaminants from water. *ACS Nano*, **6**(6), 5413-5419.
49. Gu, J., Jiang, W., Wang, F., Chen, M., Mao, J., and Xie, T. (2014) Facile removal of oils from water surfaces through highly hydrophobic and magnetic polymer nanocomposites. *Applied Surface Science*, **301**, 492-499.
50. Thanikaivelan, P., Narayanan, N. T. Pradhan, B. K., and Ajayan, P. M. (2012) Collagen based magnetic nanocomposites for oil removal applications. *Scientific Reports*, **2**, 230.
51. Venkatanarasimhan, S., and Raghavachari, D. (2013) Epoxidized natural rubber–magnetite nanocomposites for oil spill recovery. *Journal of Material Chemistry A*, **1**, 868-876.
52. Parham, H., Zargar, B., and Shiralipour, R. (2012) Fast and efficient removal of mercury from water samples using magnetic iron oxide nanoparticles modified with 2-mercaptobenzothiazole. *Journal of Hazardous Materials*, **205-206**, 94-100.
53. Gao, H., Zhao, S., Cheng, X., Wang, X., and Zheng, L. (2013) Removal of anionic azo dyes from aqueous solution using magnetic polymer multi-wall carbon nanotube nanocomposite as adsorbent. *Chemical Engineering Journal*, **223**, 84-90.
54. Li, Y., Li, X., Chu, J., Dong, C., Qi, J., and Yuan, Y. (2010) Synthesis of core-shell magnetic molecular imprinted polymer by the surface RAFT polymerization for the fast and selective removal of endocrine disrupting chemicals from aqueous solutions. *Environmental Pollution*, **158**(6), 2317-2323.
55. Daraei, P., Madaeni, S. S., Ghaemi, N., Salehi, E., Khadivi, M. A., Moradian, R., and Astinchap, B. (2012) Novel polyethersulfone nanocomposite membrane prepared by PANI/Fe_3O_4 nanoparticles with enhanced performance for Cu(II) removal from water. *Journal of Membrane Science*, **415-416**, 250-259.
56. Bhaumik, M., Maity, A., Srinivasu, V. V., and Onyango, M. S. (2011) Enhanced removal of Cr(VI) from aqueous solution using polypyro-

ole/Fe_3O_4 magnetic nanocomposite. *Journal of Hazardous Materials*, **190**(1-3), 381-390.

57. Arora, N., Vengatesan, M. R., and Mittal, V. (2018) Cobalt ferrite doped graphene/phosphazene nanotubes (PZTs) - Resorcinol-Formaldehyde (R-F) cryogel nanocomposites: Synthesis and adsorption studies, *in preparation*.
58. Lin, M., Lu, D., Zhu, J., Yang, C., Zhang, Y. and Zheng, L. (2012) Magnetic enzyme nanogel (MENG): a universal synthetic route for biocatalysts. *Chemical Communications*, **48**, 3315-3317.
59. Qiu, G.-M., Zhu, B.-K. and Xu, Y.-Y. (2005) α-Amylase immobilized by Fe_3O_4/poly(styrene-co-maleic anhydride) magnetic composite microspheres: Preparation and characterization. *Journal of Applied Polymer Science*, **95**, 328-335.
60. Zhang, H., Zhong, X., Xu, J., and Chen, H. (2008) Fe_3O_4/polypyrrole/Au nanocomposites with core/shell/shell structure: synthesis, characterization, and their electrochemical properties. *Langmuir*, **24**(23), 13748-13752.
61. Kiasat, A. R., and Nazari, S. (2012) Magnetic nanoparticles grafted with β-cyclodextrin–polyurethane polymer as a novel nanomagnetic polymer brush catalyst for nucleophilic substitution reactions of benzyl halides in water. *Journal of Molecular Catalysis A: Chemical*, **365**, 80-86.
62. Lei, Z., Ren, N., Li, Y., Li, N., and Mu, B. (2009) Fe_3O_4/SiO_2-g-PSStNa polymer nanocomposite microspheres (PNCMs) from a surface-initiated atom transfer radical polymerization (SI-ATRP) approach for pectinase immobilization. *Journal of Agricultural and Food Chemistry*, **57**(4), 1544-1549.
63. Suri, K., Annapoorni, S., Sarkar, A. K., and Tandon, R. P. (2002) Gas and humidity sensors based on iron oxide–polypyrrole nanocomposites. *Sensors and Actuators B: Chemical*, **81**(2-3), 277-282.
64. Tandon, R. P., Tripathy, M. R., Arora, A. K., and Hotchandani, S. (2006) Gas and humidity response of iron oxide-polypyrrole nanocomposites. *Sensors and Actuators B: Chemical*, **114**(2), 768-773.
65. Taccola, S., Greco, F., Zucca, A., Innocenti, C., Fernandez, C. J., Campo, G., Sangregorio, C., Mazzolai, B., and Mattoli, V. (2013) Characterization of free-standing PEDOT:PSS/Iron oxide nanoparticle composite thin films and application as conformable humidity sensors. *ACS Applied Materials and Interfaces*, **5**(13), 6324-6332.
66. Sen, T., Shimpi, N. G., Mishra, S., and Sharma, R. (2014) Polyaniline/γ-Fe2O3 nanocomposite for room temperature LPG sensing. *Sensors and Actuators B: Chemical*, **190**, 120-126.
67. Wu, S., Wang, H., Tao, S., Wang, C., Zhang, L., Liu, Z., and Meng, C. (2011) Magnetic loading of tyrosinase-Fe_3O_4/mesoporous silica core/shell microspheres for high sensitive electrochemical biosensing. *Analytica Chimica Acta*, **686**(1-2), 81-86.

68. Oh, J. K., and Park, J. M. (2011) Iron oxide-based superparamagnetic polymeric nanomaterials: design, preparation, and biomedical application. *Progress in Polymer Science*, **36**(1), 168-189.
69. Kim, M. I., Ye, Y., Won, B. Y., Shin, S., Lee, J., and Park, H. G. (2011), A highly efficient electrochemical biosensing platform by employing conductive nanocomposite entrapping magnetic nanoparticles and oxidase in mesoporous carbon foam. *Advanced Functional Materials*, **21**, 2868-2875.
70. Wang, X., Wang, L., He, X., Zhang, Y., and Chen, L. (2009) A molecularly imprinted polymer-coated nanocomposite of magnetic nanoparticles for estrone recognition. *Talanta*, **78**(2), 327-332.
71. Baghayeri, M., Zare, E. N., and Lakouraj, M. M. (2014) A simple hydrogen peroxide biosensor based on a novel electro-magnetic poly(p-phenylenediamine)@Fe_3O_4 nanocomposite. *Biosensors and Bioelectronics*, **55**, 259-265.
72. Fredj, H. B., Helali, S., Esseghaier, C., Vonna, L., Vidal, L., and Abdelghani, A. (2008) Labeled magnetic nanoparticles assembly on polypyrrole film for biosensor applications. *Talanta*, **75**(3), 740-747.
73. Zhang, M., He, X., and Zhang, L. C. Y. (2010) Preparation of IDA-Cu functionalized core–satellite Fe_3O_4/polydopamine/Au magnetic nanocomposites and their application for depletion of abundant protein in bovine blood. *Journal of Materials Chemistry*, **20**, 10696-10704.
74. Singh, J., Srivastava, M., Dutta, J., and Dutta, P. K. (2011) Preparation and properties of hybrid monodispersed magnetic α-Fe_2O_3 based chitosan nanocomposite film for industrial and biomedical applications. *International Journal of Biological Macromolecules*, **48**(1), 170-176.
75. Perera, A. S., Zhang, S., Homer-Vanniasinkam, S., Coppens, M.-C., and Edirisinghe, M. (2018) Polymer-magnetic composite fibers for remote-controlled drug release. *ACS Applied Materials and Interfaces*, **10**(18), 15524-15531.
76. Xu, H., Cheng, L., Wang, C., Ma, X., Li, Y., and Liu, Z. (2011) Polymer encapsulated upconversion nanoparticle/iron oxide nanocomposites for multimodal imaging and magnetic targeted drug delivery. *Biomaterials*, **32**(35), 9364-9373.
77. Ma, X., Tao, H., Yang, K., Feng, L., Cheng, L., Shi, X., Li, Y., Guo, L., and Liu, Z. (2012) A functionalized graphene oxide-iron oxide nanocomposite for magnetically targeted drug delivery, photothermal therapy, and magnetic resonance imaging. *Nano Research*, **5**(3), 199-212.
78. Satarkar, N. S., and Hilt, J. Z. (2008) Magnetic hydrogel nanocomposites for remote controlled pulsatile drug release. *Journal of Controlled Release*, **130**(3), 246-251.

79. Kim, J., Lee, J. E., Lee, J., Yu, J. H., Kim B. C., An, K., Hwang, Y., Shin, C.-H., Park, J.-G., Kim, J., and Hyeon, T. (2006) Magnetic fluorescent delivery vehicle using uniform mesoporous silica spheres embedded with monodisperse magnetic and semiconductor nanocrystals. *Journal of American Chemical Society*, **128**(3), 688-689.
80. Zhao, W., Gu, J., Zhang, L., Chen, H., and Shi, J. (2005) Fabrication of uniform magnetic nanocomposite spheres with a magnetic core/mesoporous silica shell structure. *Journal of American Chemical Society*, **127**(25), 8916-8917.
81. Asmatulu, R., Zalich, M. A., Claus, R. O., and Riffle, J. S. (2005) Synthesis, characterization and targeting of biodegradable magnetic nanocomposite particles by external magnetic fields. *Journal of Magnetism and Magnetic Materials*, **292**, 108-119.
82. Urbina, M. C., Zinoveva, S., Miller, T., Sabliov, C. M., Monroe, W. T., and Kumar, C. S. S. R. (2008) Investigation of magnetic nanoparticle-polymer composites for multiple-controlled drug delivery. *Journal of Physical Chemistry C*, **112**, 11102-11108.
83. Brazel, C. S. (2009) Magnetothermally-responsive nanomaterials: combining magnetic nanostructures and thermally-sensitive polymers for triggered drug release. *Pharmaceutical Research*, **26**(3), 644-656.
84. Zhao, W., Chen, H., Li, Y., Li, L., Lang, M., and Shi, J. (2008), Uniform rattle-type hollow magnetic mesoporous spheres as drug delivery carriers and their sustained-release property. *Advanced Functional Materials*, **18**, 2780-2788.
85. Lutz, J.-F., Stiller, S., Hoth, A., Kaufner, L., Pison, U., and Cartier, R. (2006) One-pot synthesis of PEGylated ultrasmall iron-oxide nanoparticles and their in vivo evaluation as magnetic resonance imaging contrast agents. *Biomacromolecules*, **7**(11), 3132-3138.
86. Lee, J.-H., Jun, Y.-w., Yeon, S.-I., Shin, J.-S., and Cheon, J. (2006) Dual-mode nanoparticle probes for high-performance magnetic resonance and fluorescence imaging of neuroblastoma. *Angewandte Chemie*, **118**, 8340-8342.
87. Li, K., Ding, D., Huo, D., Pu, K.-Y., Thao, N. N. P., Hu, Y., Li, Z. and Liu, B. (2012) Conjugated polymer based nanoparticles as dual-modal probes for targeted in vivo fluorescence and magnetic resonance imaging. *Advanced Functional Materials*, **22**, 3107-3115.
88. Baldock, C., Burford, R. P., Billingham, N., Wagner, G. S., Patval, S., Badawi, R. D., and Keevil, S. F. (1998) Experimental procedure for the manufacture and calibration of polyacrylamide gel (PAG) for magnetic resonance imaging (MRI) radiation dosimetry, *Physics in Medicine and Biology*, **43**(3), 695-702.
89. Berret, J.-F., Schonbeck, N., Gazeau, F., Kharrat, D. E. Sandre, O., Vacher, A., and Airiau, M. (2006) Controlled clustering of superparamagnetic nanoparticles using block copolymers: design of new

contrast agents for magnetic resonance imaging. *Journal of American Chemical Society*, **128**(5), 1755-1761.

90. Lee, J. E., Lee, N., Kim, H., Kim, J., Choi, S. H., Kim, J. H., Kim, T., Song, I. C., Park, S. P., Moon, W. K., and Hyeon, T. (2010) Uniform mesoporous dye-doped silica nanoparticles decorated with multiple magnetite nanocrystals for simultaneous enhanced magnetic resonance imaging, fluorescence imaging, and drug delivery. *Journal of American Chemical Society*, **132**(2), 552-557.
91. Sitharaman, B., Van Der Zande, M., Ananta, J. S., Shi, X., Veltien, A., Walboomers, X. F., Wilson, L. J., Mikos, A. G., Heerschap, A., and Jansen, J. A. (2010) Magnetic resonance imaging studies on gadonanotube-reinforced biodegradable polymer nanocomposites. *Journal of Biomedical Materials Research*, **93A**, 1454-1462.
92. Jun, Y.-W., Huh, Y.-M., Choi, J.-S., Lee, J.-H., Song, H.-T., Sungjun, S., Yoon, S., Kim, K.-S., Shin, J.-S., Suh, J.-S., and Cheon, J. (2005) Nanoscale size effect of magnetic nanocrystals and their utilization for cancer diagnosis via magnetic resonance imaging. *Journal of American Chemical Society*, **127**(16), 5732-5733.
93. Ai, H., Flask, C., Weinberg, B., Shuai, X.-T., Pagel, M. D., Farrell, D., Duerk, J., and Gao, J. (2005) Magnetite-loaded polymeric micelles as ultrasensitive magnetic-resonance probes. *Advanced Materials*, **17**, 1949-1952.
94. Lu, J., Ma, S., Sun, J., Xia, C., Liu, C., Wang, Z., Zhao, X., Gao, F., Gong, Q., Song, B., Shuai, X., Ai, H., and Gu, Z. (2009) Manganese ferrite nanoparticle micellar nanocomposites as MRI contrast agent for liver imaging. *Biomaterials*, **30(**15), 2919-2928.
95. Armentano, I., Dottori, M., Fortunati, E., Mattioli, S., and Kenny, J. M. (2010) Biodegradable polymer matrix nanocomposites for tissue engineering: A review. *Polymer Degradation and Stability*, **95**(11), 2126-2146.
96. Gloria, A., Russo, T., D'Amora, U., Zeppetelli, S., D'Alessandro, T., Sandri, M., Banobre-Lopez, M., Pineiro-Redondo, Y., Uhlarz, M. Tampieri, A., Rivas, J., Herrmannsdorfer, T., Dediu, V. A., Ambrosio, L., and De Santis, R. (2013) Magnetic poly(ε-caprolactone)/iron-doped hydroxyapatite nanocomposite substrates for advanced bone tissue engineering. *Journal of the Royal Society Interface*, DOI: https://doi.org/10.1098/rsif.2012.0833.
97. Pankhurst, Q. A., Connolly, J., Jones, S. K., and Dobson, J. (2003) Applications of magnetic nanoparticles in biomedicine. *Journal of Physics D: Applied Physics*, **36**, R167.
98. De Santis, R., Gloria, A., Russo, T., D'Amora, U., Zeppetelli, S., Dionigi, C., Sytcheva, A., Herrmannsdorfer, T., Dediu, V. and Ambrosio, L. (2011) A basic approach toward the development of nanocomposite magnetic scaffolds for advanced bone tissue engineering. *Journal of Applied Polymer Science*, **122**, 3599-3605.

99. Zhang, Y., Venugopal, J. R., El-Turki, A., Ramakrishna, S. Su, B., and Lim, C. T. (2008) Electrospun biomimetic nanocomposite nanofibers of hydroxyapatite/chitosan for bone tissue engineering. *Biomaterials*, **29**(32), 4314-4322.
100. Gupta, A. K., Naregalkar, R. R., Vaidya, V. D., and Gupta, M. (2007) Recent advances on surface engineering of magnetic iron oxide nanoparticles and their biomedical applications. *Nanomedicine*, **2**(1), 23-39.
101. Meenach, S. A., Hilt, J. Z., and Anderson, K. W. (2010) Poly(ethylene glycol)-based magnetic hydrogel nanocomposites for hyperthermia cancer therapy. *Acta Biomaterialia*, **6**(3), 1039-1046.
102. Tan, S. T., Wendorff, J. H., Pietzonka, C., Jia, Z. H., and Wang, G. Q. (2005) Biocompatible and biodegradable polymer nanofibers displaying superparamagnetic properties. *Chemphyschem*, **6**(8), 1461-5.
103. Sureshkumar, M., Siswanto, D. Y., and Lee, C.-K. (2010) Magnetic antimicrobial nanocomposite based on bacterial cellulose and silver nanoparticles. *Journal of Materials Chemistry*, **20**, 6948-6955.
104. Arora, N., and Mittal, V. (2018) Cobalt ferrite doped graphene polyacrylamide cryogel nanocomposites: Synthesis and characterization studies, *in preparation*.

8

Inhibition of Corrosion by Polyetherimide-Graphene Nanocomposite Coatings for Long Term Protection of Carbon Steel

8.1 Introduction

Metallic materials exposed to aggressive environment during their application undergo deterioration and premature failure because of the processes of corrosion. Application of organic/polymeric coatings is the standard procedure to prevent the corrosive destruction of metal components. The barrier nature of organic/polymeric coatings prevents the corrosive ions and gases from reaching the metal surface. Recent studies indicate that addition of nanofillers enhance the barrier properties of the coatings [1-4]. Nanofillers improve the adhesion strength between the polymer and metal as well as the gas impermeability and mechanical strength of the polymer, thus, enhancing the corrosion resistance of the coating [2,3,5]. Graphene, one-atom thick two dimensional layers of sp^2 carbon atoms bonded together in a honey comb lattice, has been recently reported as an ideal filler for corrosion inhibition in reinforced polymer coatings because of its unique properties of exceptional chemical and thermal resistance, mechanical strength as well as high electrical and thermal conductivities [6-21]. Graphene prevents penetration of water and corrosive species like oxygen and chlorides due to barrier effect owing to its high aspect ratio and platelet structure which makes the diffusion pathway in the coating tortuous. Graphene produces electron depletion when in direct contact with the metal, leading to metal/metal oxide surface passivation through establishment of a Schottky barrier, thereby, providing protection of the metal against corrosion by an active-passive approach [6,22].

Several works have been reported on the corrosion inhibitive nature of graphene in polymer composite protective coatings [6-13,23,24]. Most of these demonstrated the diffusion barrier nature of

Tehsin Akhtar, Gisha E. Luckachan and Vikas Mittal, The Petroleum Institute (part of Khalifa University of Science and Technology), Abu Dhabi, UAE*
**Current address: Bletchington, Wellington County, Australia*

graphene, preventing direct interaction between underlying metal and corrosion promoting species. Recently, Chaudhry *et al.* [23] discussed anti-corrosion behavior of graphene incorporated self-cross-linked poly(vinyl butyral) (PVB) composites. Graphene platelets increased the diffusion path of oxygen and ionic species through the coating which improved corrosion resistance of the coating for a short period of time. For longer periods of time, corrosion promotion occurred due to induced porosity and electrochemical mechanisms related to excellent electrical conductivity of graphene. Figure 8.1 also demonstrates the potentiodynamic curves and optical microscopy images of PVB and graphene composites after 26 h of immersion. Similar mechanisms were reported for graphene/epoxy [11,12],

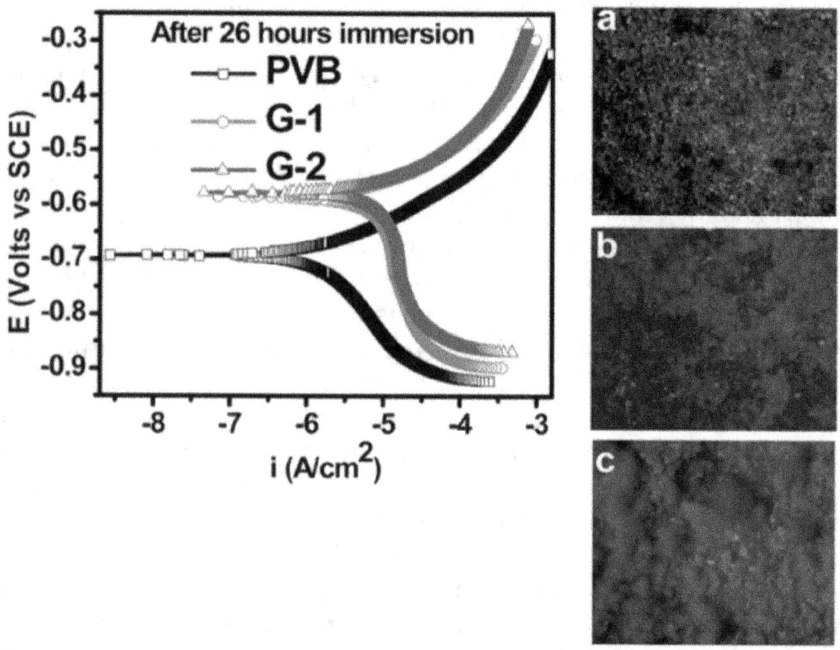

Figure 8.1 Potentiodynamic curve and optical microscopy images of (a) PVB, (b) G-1, (c) G-2 after 26 h of immersion. Reproduced form Reference 23 with permission from Royal Society of Chemistry.

polyurethane/graphene [7,13], polyimide/graphene [10], polyaniline/graphene [9] and poly(sodium styrene sulfonate)/graphene [25] composites in protective coatings. Some researchers also reported that graphene could act as cathode when in contact with metal due to its thermodynamic stability and increase the corrosion due to faster electron transfer owing to high conductivity, thereby, accelerating the metal dissolution and cathodic reduction when the graphene films are non-uniform and non-adherent [16,25,26]. This behavior restricted the protection of graphene grown by chemical vapor deposition (CVD) [26-29], electrophoretic deposition (EPD) [19] and other techniques [18,30] on the metal surface to short durations. Few works have reported the electrochemical response of underlying metal in presence of graphene layers [22,27,28]. Kirkland *et al.* [29] observed that the graphene affected metals differently. It primarily slowed the anodic dissolution reactions for Ni and the cathodic reduction reactions for Cu. Though graphene provided a barrier to metal dissolution of the metal substrate from the environment, however, the mechanism of protection imparted by the graphene layer needs to be elaborated. A quantitative description of p-n junction formed at the metal graphene interface was discussed by Sreevatsa *et al.* [31]. However, the authors used graphene in conjunction with polypyrrole (a conducting polymer) 'top coat' to form the p-n junction. Recently, Dumee *et al.* [22] described the reduced kinetics of corrosion reactions by graphene flakes grown on stainless steel resulted by the electron depletion layer formation which restricted the flow of electrons to the metal surface. Such a behavior of graphene in the protection mechanism was generally not reported in the polymer composite coatings [7-13]. If the electron depletion behavior of graphene could be implemented along with its diffusion barrier nature, the duration of protection by polymer composite coatings could be enhanced significantly. It could be attained by the good dispersion of graphene in the polymer matrix and by the strong interaction of graphene with the polymer chains and metal surface. Therefore, choice of polymer is important in obtaining a better performance of graphene in the polymer composite coatings.

It has been reported that good compatibility exists between polyetherimide (PEI) and graphene as these interact through π–π stacking which is non-covalent in nature and is formed by attractive forces between the adjacent aromatic rings of PEI and graphene [6]. Dennis *et al.* [6] reported such a behavior in graphene incorporated PEI coatings, though, the nature of coating in the aggressive environment as well as the protection mechanism of graphene need to be elaborated further. Therefore, a systematic study is required to further establish the fact that graphene reinforced coatings can serve as effective anticorrosion alternatives for long-term protection of metals.

The present work demonstrates a one-step procedure for the incorporation of graphene in polyetherimide coating formulations for effective corrosion protection of carbon steel substrates for long periods. Nature of coating in the aggressive environment as well as the mechanism of protection provided by the incorporated graphene platelets was studied by electrochemical analyses and microscopic techniques. Effect of graphene oxide reinforcement instead of graphene platelets in the corrosion protective nature of PEI composite coatings was also analyzed for comparison.

8.2 Experimental

8.2.1 Materials

Carbon steel of grade RST37-2 DIN 17100-80 was purchased from Qatar Steel Industries, Qatar. Its composition in weight% was C (0.125), Mn (0.519), Si (0.016), P (0.014), S (0.005), Al (0.034) and Fe (99.287). Carbon steel coupons of dimensions 5 cm x 2 cm x 1.8 cm were used as substrate. Polyetherimide with melt flow index 9 g/ 10 min (337 °C/6.6 kg) was purchased from Sigma-Aldrich. The graphene nano-platelets (N002-PDR, 98.48% purity) were purchased from Angstron Materials, Dayton, Ohio. N,N-dimethylacetamide (DMAc) solvent and hydrochloric acid (37%) were obtained from Merck and used as received.

8.2.2 Preparation of PEI-Graphene Coatings

Solutions of polyetherimide (15 wt%) were prepared by dissolving PEI pellets in DMAc in closed Pyrex bottles at 70 °C under magnetic stirring to produce homogeneous solutions. Graphene was added to DMAc and mixed under magnetic stirring for 10 min and was subsequently subjected to bath sonication for 6 h and overnight mixing to obtain a homogeneous dispersed graphene solution. This was followed by 2 h of bath sonication and shear mixing for 1 h. The graphene solution was added dropwise to polymer solution prepared separately and stirred vigorously for 30 min, shear mixed for 1 h followed by magnetic stirring overnight to obtain homogeneous solutions [21,32]. The solutions were allowed to stand for 10-15 min to remove air bubbles before dip coating of clean polished carbon steel coupons. Coating was carried out with an immersion speed of 50 mm/min, withdrawal speed of 150 mm/min and wetting time of 30

s. The concentration of graphene was fixed to 0.5, 1, 2 and 5 wt% of the polymer content in the solutions. The coatings were dried immediately in a stepwise manner starting at 80 °C for 30 min and subsequently at 120 °C, 150 °C, 180 °C, 210 °C, 240 °C for 15 min at each temperature and finally curing at 270 °C for 30 min. Graphene-PEI (Gr-PEI) films were also prepared on Teflon petri dishes and subjected to drying in the same manner. 0.5 and 5 wt% coatings with graphene oxide (GO) were also formulated in the same way as Gr-PEI coatings.

8.2.3 Characterization Techniques

The coating thickness was measured using PosiTector 6000 coating thickness gauge. The surface resistance of the coating was measured using the Prostat PRS-812 resistance meter set.

Transmission electron microscopy (TEM) was carried out for gaining insights about the dispersion state of graphene in polymer matrix (Philips CM 20, Germany). The copper grids were immersed for 20 s and quickly withdrawn from the Gr-PEI solutions and dried in the oven at 80 °C for 5 min to obtain a thin coating on copper grids for imaging. Scanning electron microscope (SEM) (FEI Quanta, FEG250, USA) at an accelerating voltage of 5 kV was used for observation of the metal surface underneath the coatings and energy dispersive X-ray spectroscopy (EDX) was employed to confirm the presence/absence of any corrosion products on the metal surface. To study the effect of graphene addition on the decomposition temperature of the coatings, thermogravimetric analysis (TGA) was carried out using the Discovery TGA equipment from TA instruments. The samples were heated from 25 °C to 700 °C at the rate of 10 °C/min under nitrogen flow and the changes in the weights were noted. The Fourier transform infrared spectroscopy of the coatings was performed using a Bruker VERTEX 70 FTIR spectrometer attached with a DRIFT accessory. IR acquisition was achieved by recording 120 scans at a resolution of 4 cm^{-1} in the frequency range of 370 cm^{-1} to 4000 cm^{-1} using OPUS software.

8.2.4 Immersion Tests

The coated coupons were immersed in 3.5 wt% sodium chloride solution and the corrosion was monitored by the appearance of corrosion products or rust with time at room temperature. With the help of optical microscopy (Digital Microscope KH-7700, Hirox Co. Ltd, USA), coated substrates were inspected closely for any signs of corrosion and local delamination. The coatings were scribed after immersion test completion to analyze the extent of undercoating corrosion.

8.2.5 Electrochemical Analysis

A three-electrode cell of 250 mL volume with a platinum gauze counter electrode and a saturated calomel reference electrode (SCE) with bridge tube was used to perform electrochemical tests on the flat coated coupons. The exposed working electrode area was 1 cm^2 and the coated coupon was kept in place with the help of an O ring and adjustable brass disc. All tests were carried out by connecting the corrosion cell to the BioLogic VMP-300 multipotentiostat (controlled by a computer running EC-Lab 10.40 software) in 3.5 wt% NaCl solution at room temperature. Ultra-low current cable connected to the potentiostat was used for the accurate measurement of the current. This option includes current ranges from 100 nA down to 100 pA with additional gains extending the current ranges to 10 pA and 1 pA. The resolution on the lowest range is 76 aA. The open circuit potential (OCP) was measured for 5 min in order to allow the potential to stabilize before the electrochemical impedance and potentiodynamic polarization test. The impedance measurements were performed at amplitude of 15 mV over frequency range from 200 kHz to 10 MHz. Linear polarization was carried out at potential range from -0.250 V to +0.250 V using a scan rate of 1.66 mV/s. Fits were performed by the associated software and corrosion rates were calculated using Faraday's equation for metal loss corrosion. The impedance test and linear polarization was conducted on three samples of each type to ensure reproducibility of the results.

8.3 Results and Discussion

Protective coatings of graphene-polyetherimide composites were prepared by dip coating on carbon steel substrate and subjected to a step wise heating from ambient conditions till 270 °C (as mentioned in the experimental part). Four different Gr-PEI coatings were obtained by changing the concentration of graphene as 0.5, 1, 2 and 5 wt% of PEI in the final coating formulation. Thickness of the obtained coatings was in the range of 20-30 μm.

Graphene in the coatings exhibited a sheet like morphology, as demonstrated in the TEM images shown in Figure 8.2, indicating a uniform distribution of graphene platelets in PEI matrix. The sheet

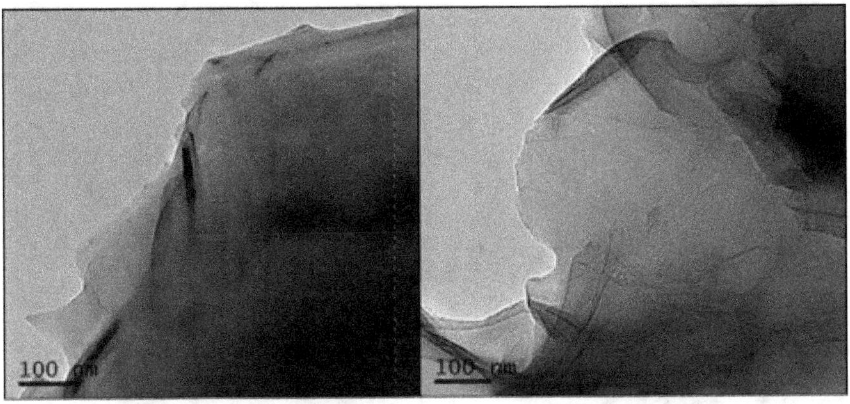

Figure 8.2 TEM images of Gr-PEI films: (left) 0.5 wt% Gr and (right) 5 wt% Gr in PEI.

like morphology added to the flexibility of the Gr-PEI coatings on steel. Further, the formability of the coated samples was also promoted. The sheet like morphology was due to π-π stacking between polyetherimide and graphene sheets [9,28]. Good interfacial compatibility was observed between PEI and graphene also due to similarities in their chemical structure. In addition, the graphene platelets were observed in Figure 8.3 to be effectively stabilized by PEI chains as the Gr-PEI formulations with different concentrations of graphene did not show phase separation even after 4 months of storage. As mentioned above, the uniform dispersion of graphene in PEI matrix was attributed to π-π interaction between aromatic moieties in the polymer and π conjugated graphene basal planes.

The incorporation of graphene also increased the thermal stability of the composite coatings in comparison to the pure PEI by 50 °C. Both 2 wt% and 5 wt% Gr-PEI samples exhibited a delayed decomposition temperature as compared to the pure polymer, as demonstrated in Figure 8.4.

The surface resistance of the coatings was recorded to analyze the changes in the conductivity of polymer coatings due to graphene addition, which can help in further understanding the corrosion behavior. The surface resistance of the pure PEI coating was observed to be high due to the insulating nature, as can be observed from Table 8.1.

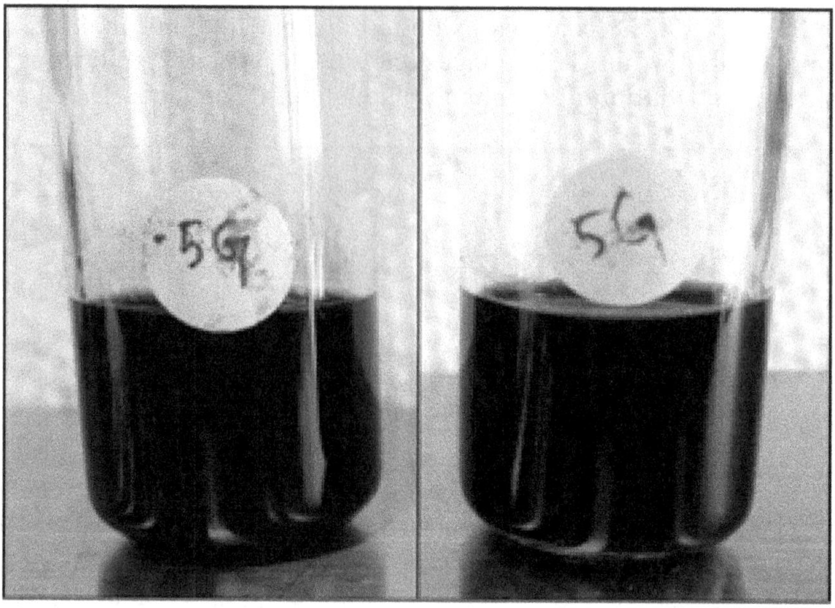

Figure 8.3 Gr-PEI formulations after 4 months of storage: (left) 0.5 wt% Gr and (right) 5 wt% Gr.

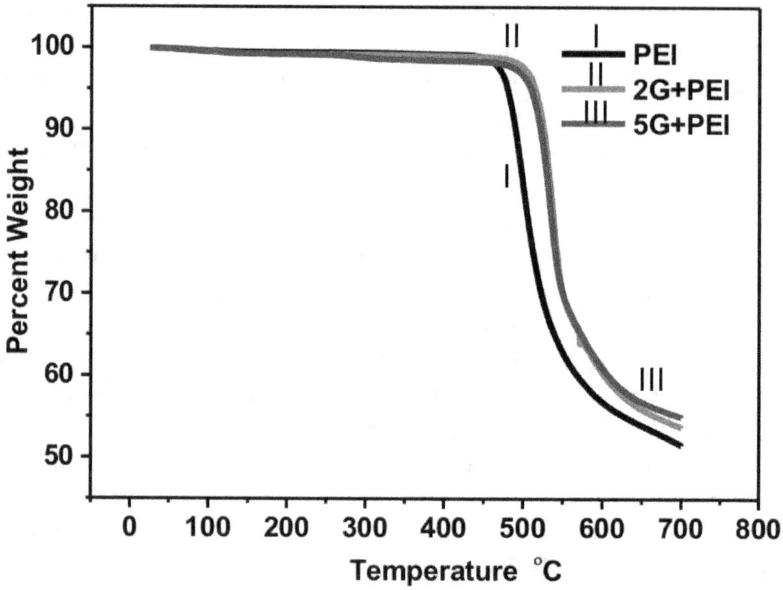

Figure 8.4 TGA thermograms of pure PEI and Gr-PEI films with 2 wt% and 5 wt% graphene.

Table 8.1 Surface conductivities of various coatings

Coating	Surface Resistance Ω	Surface Conductivity S/square
PEI	1.48×10^{11}	6.75×10^{-13}
0.5 wt% Gr-PEI	2.10×10^{11}	4.76×10^{-13}
1.0 wt% Gr-PEI	2.00×10^{11}	5.00×10^{-13}
2.0 wt% Gr- PEI	1.10×10^{4}	9.10×10^{-6}
5.0 wt% Gr-PEI	3.66×10^{2}	2.73×10^{-4}
0.5 wt% GO-PEI	1.10×10^{11}	9.10×10^{-13}
5.0 wt% GO-PEI	4.80×10^{5}	2.08×10^{-7}

A decrease in the resistance of coatings took place upon addition of graphene at concentrations higher than 1 wt%. A sharp reduction in resistance at 2 wt% loading of graphene nano-platelets as compared to 0.5 wt% and 1 wt% loading confirmed the formation of a percolation network structure throughout the polymer matrix [9,21]. At 5 wt% loading of graphene, the resistance declined rapidly to 3.66×10^2 Ω from 1.48×10^{11} Ω, exhibiting a decrease of 9 orders of magnitude in comparison with pure PEI. Meanwhile, the increased conductivity of the coating at 5 wt% graphene indicated a good platelet to platelet connection in forming a conducting path network in the polymer matrix [21,33]. The graphene sheets have high surface area to volume ratio and contact between the well-distributed sheets in the polymer matrix resulted in an increase in the surface conductivity of the coatings.

Protection of metal surface from corrosion by Gr-PEI composite coatings was studied by EIS analysis in 3.5 wt% NaCl solution. The 0.5 wt% and 1 wt% Gr-PEI coatings had similar corrosion performance. The Bode plots in Figures 8.5a and b indicated that the barrier properties of the 0.5 wt% and 1 wt% Gr-PEI coatings declined every week of immersion in NaCl solution. The initial impedance of the 0.5 wt% coating was very high ($10^{11.6}$ Ω.cm^2), however, a gradual decrease in the impedance modulus of the coating occurred and a loss of adhesion and pitting corrosion was observed during 6 weeks of immersion as seen in Figures 8.6a and b. The decline of impedance in 1 wt% Gr-PEI coating was faster than 0.5 wt% Gr-PEI coating and within a week's time of immersion, the coatings exhibited an impedance decline of 3 orders of magnitude, as can be seen from Figure 8.5b. In the subsequent weeks, the impedances decreased further and pitting was observed at several areas of the coating (Figures 8.6c and d). As a result, complete delamination of the coating was observed to

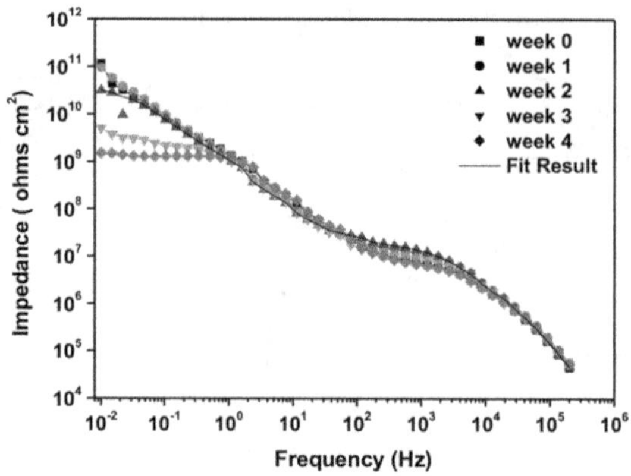

(a)

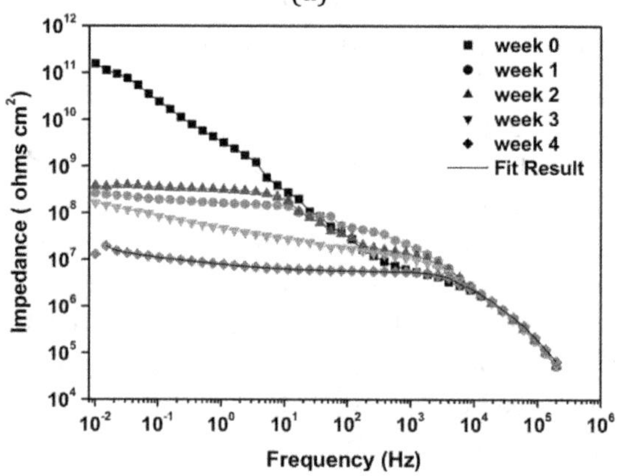

(b)

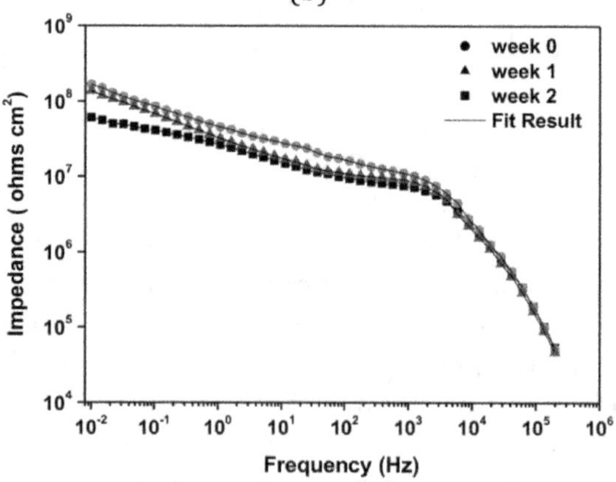

(c)

Figure 8.5 continued on next page

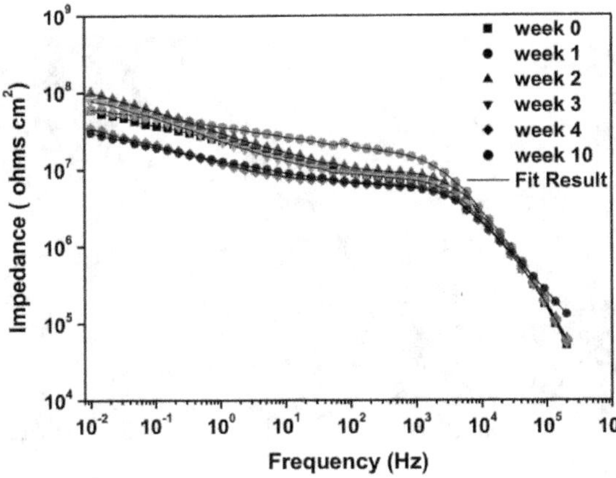

(d)

Figure 8.5 Bode plots of Gr-PEI coatings with graphene concentration of: (a) 0.5 wt% (b) 1 wt% (c) 2 wt% and (d) 5 wt%.

occur during fifth week of immersion. It indicated the diffusion of water into the coating and deterioration of barrier properties. Low frequency impedance decreased continuously to $10^{7.7}$ Ω with increase of graphene concentration to 2 wt% (Figure 8.5c). The coatings also did not achieve the desired level of pore blockage to prevent water from entering the coatings, as it delaminated before the beginning of third week of immersion (Figures 8.6e and f). Interestingly, further increase of graphene to 5 wt% did not decrease low frequency impedance from $10^{7.7}$ Ω and the value was retained over 10 weeks of exposure in 3.5 wt% NaCl solution (Figure 8.5d) without any visible corrosion products on the coating surface (Figures 8.6g and h). The constant coating impedance with immersion time exhibited the stability of adhesion and barrier protection of the 5 wt% Gr-PEI coatings. It can be explained that at low concentrations, the discontinuity of graphene particles created defects in the coatings leading to formation of pockets filled with electrolyte at the interface, and, thus, contributed to a localized corrosion on the metal surface as observed in the 0.5 wt% and 1 wt% graphene coatings (Figures 8.6b and d). At higher concentrations, the dispersed graphene particles, stabilized by the PEI matrix, formed sheet like structures with a good platelet to platelet contact (confirmed from the increased conductivity of the coating), which enhanced the impermeability of the coating because of the hydrophobic and diffusion barrier nature of the graphene platelets [10-12,27]. The metal surface was analyzed by scribing the 5 wt%

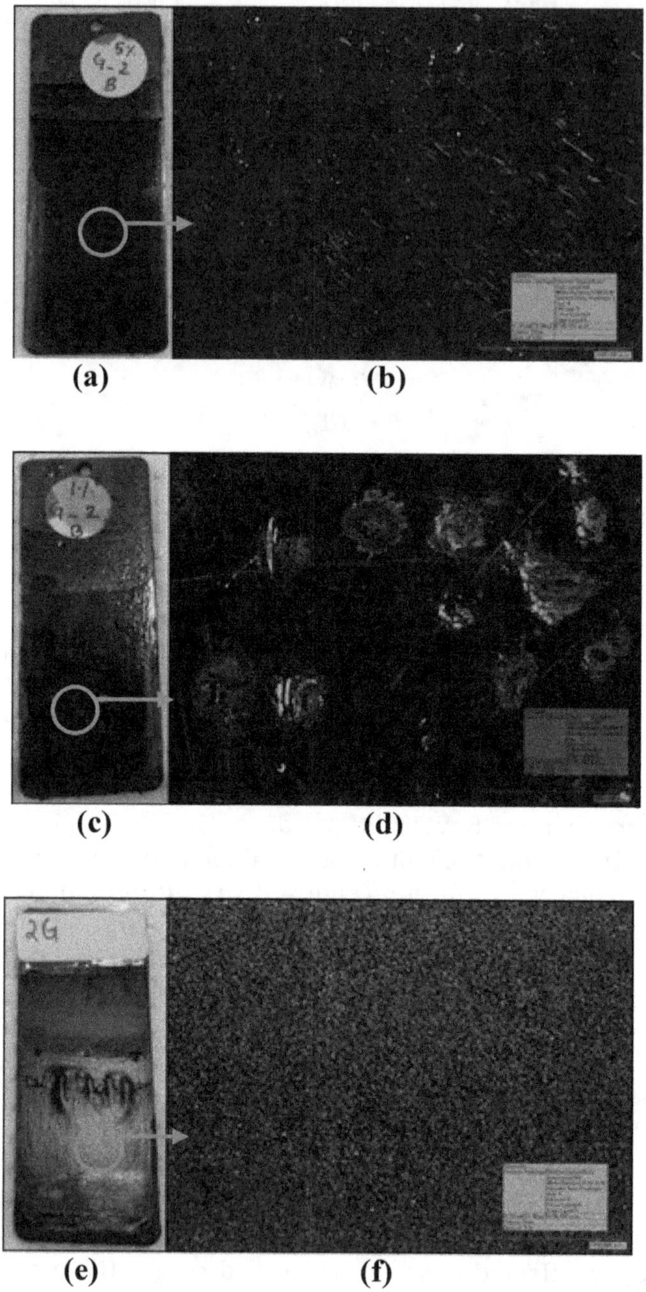

Figure 8.6 continued on next page

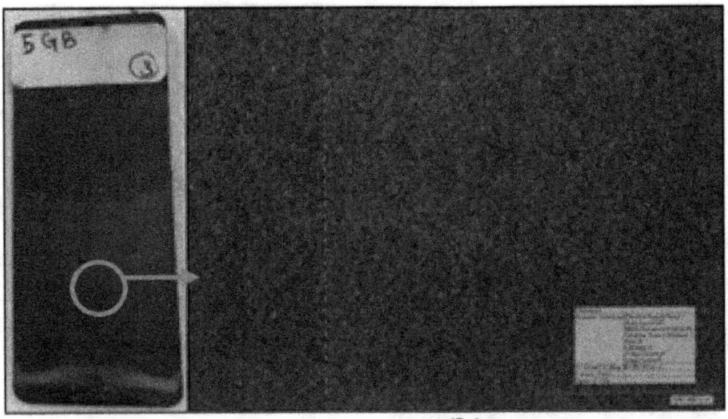

(g) (h)

Figure 8.6 Digital images of Gr-PEI composite coatings (a) 0.5 wt% Gr, (c) 1 wt% Gr, (e) 2 wt % Gr and (g) 5 wt % Gr after 6 weeks of immersion in 3.5 wt.% sodium chloride solution. (b), (d), (f) and (h) are optical images of the corresponding coatings marked with red circle.

Gr-PEI coating after 10 weeks of exposure to 3.5 wt% NaCl. The optical and scanning electron images of the metal surface at the scribed area did not exhibit any corrosion products (Figures 8.7a and b). It was confirmed further by EDX that observed signals corresponded to only metal ions (Figure 8.7c). In order to analyze the stability of adhesion, a scribed 5 wt% Gr-PEI coating was immersed in 3.5 wt% NaCl over 6 weeks. No coating disbonding was observed for the scribed coating exhibiting that it maintained an excellent adhesion (Figure 8.7d). The coating did not allow the entry of water from the scribed region and lateral diffusion, indicating the good interaction of graphene with PEI matrix and metal surface.

The EIS results were subsequently fitted using the equivalent electric circuit shown in Figure 8.8. R_s represents the solution resistance between the working electrode (coated sample) and the reference saturated calomel electrode. R_c is the coating/pore resistance and Q_c is the constant phase element related to the coating capacitance (C_c) as per the equation

$$C=Y_0\,(\omega_{max})^{\,n-1} \qquad (1)$$

where Y_0 is the magnitude of the constant phase element, ω_{max} is the frequency at which the imaginary impedance reaches a maximum for the respective time constant and 'n' is the exponential term of the CPE

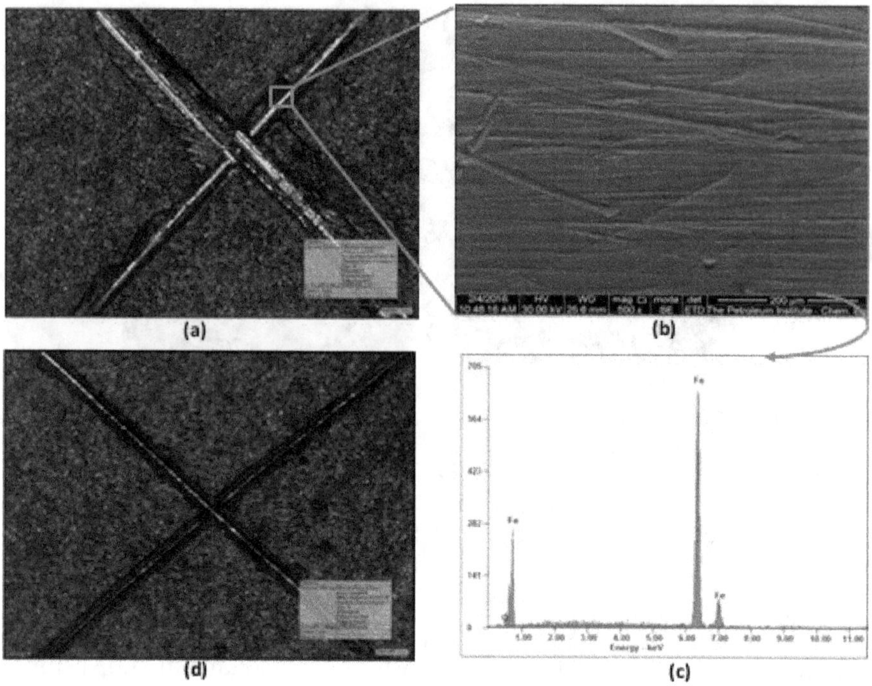

Figure 8.7 Optical image of scribed area on 5 wt% Gr-PEI coating (a). SEM image (b) and EDX (c) of the metal surface on the scribed area marked with red square. Optical image of 5 wt% Gr-PEI coating scribed and immersed for 6 weeks in 3.5 wt% sodium chloride solution.

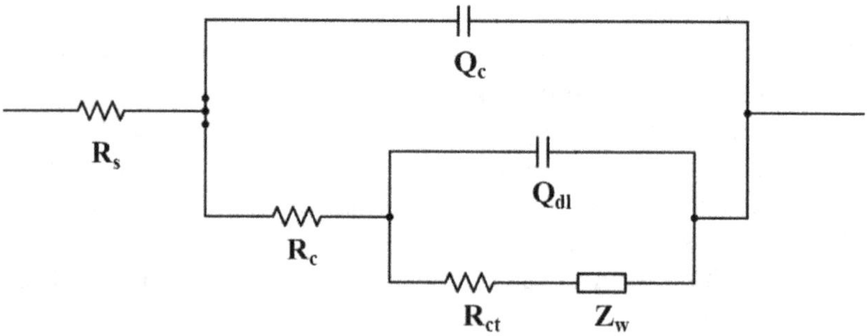

R_s - Solution resistance
R_c - Coating/pore resistance
R_{ct} - Charge transfer resistance
Q_c,Q_{dl} - Constant phase elements
Z_w - Warburg impedance

Figure 8.8 Electrical circuit used for the fitting.

which can vary between 1 for pure capacitance and 0 for a pure resistor [34,35]. Since constant 'n_1' was equal to 1 for all the coatings over the complete measurement time (Table 8.2), Q_c was considered as similar to pure capacitors and, hence, the Y_0 constant followed the same trend as the capacitance [35,36]. Capacitance of a coating (C_c) is

Table 8.2 Fitted parameters from Bode plots of Gr-PEI coatings

Coating	Week	R_{pore} (Ω cm^2)	Q_c		R_{ct} (Ω cm^2)	Q_{dl}		W (Ω s$^{-1/2}$)
			n_1	Y_0 (Ω^{-1}cm^2)		n_2	Y_0 (Ω^{-1}cm^2)	
0.5 wt%	0	6.04×10^6	1	1.04×10^{-11}	5.57×10^9	0.997	5.66×10^{-11}	5.17×10^{10}
	1	4.00×10^6	1	1.12×10^{-11}	3.06×10^9	0.913	1.97×10^{-10}	3.80×10^9
	2	1.99×10^6	1	8.21×10^{-11}	2.98×10^9	0.961	1.61×10^{-10}	9.24×10^7
	3	1.51×10^6	1	8.98×10^{-11}	1.22×10^9	0.953	2.08×10^{-10}	4.00×10^7
	4	1.09×10^6	1	9.37×10^{-11}	5.88×10^7	0.760	9.76×10^{-10}	4.71×10^6
1 wt%	0	6.42×10^6	1	9.67×10^{-12}	1.25×10^9	0.986	4.80×10^{-11}	8.48×10^{10}
	1	4.77×10^6	1	9.84×10^{-12}	1.14×10^8	0.903	1.38×10^{-10}	1.91×10^7
	2	1.90×10^6	1	9.78×10^{-12}	3.05×10^8	0.941	9.83×10^{-11}	1.42×10^7
	3	1.11×10^6	1	9.90×10^{-12}	2.66×10^8	0.379	1.19×10^{-8}	1.57×10^7
	4	9.52×10^5	1	1.05×10^{-11}	1.39×10^7	0.476	1.33×10^{-7}	3.41×10^5
2 wt%	0	7.63×10^6	1	9.43×10^{-12}	4.08×10^7	0.529	1.16×10^{-8}	3.25×10^6
	1	6.12×10^6	1	9.33×10^{-12}	1.76×10^7	0.407	1.06×10^{-8}	2.47×10^7
	2	4.20×10^6	1	1.06×10^{-11}	1.42×10^7	0.496	1.39×10^{-8}	1.3×10^7
5 wt%	0	7.21×10^6	1	9.93×10^{-12}	3.74×10^7	0.506	1.34×10^{-8}	-
	1	8.73×10^6	1	8.47×10^{-12}	3.74×10^7	0.513	1.14×10^{-8}	-
	2	8.63×10^6	1	8.35×10^{-12}	9.94×10^7	0.577	1.49×10^{-8}	-
	3	7.24×10^6	1	9.39×10^{-12}	8.71×10^7	0.589	1.95×10^{-8}	-
	4	7.17×10^6	1	9.11×10^{-12}	7.65×10^7	0.583	1.68×10^{-8}	-
	10	6.93×10^6	1	9.90×10^{-12}	6.90×10^7	0.618	1.66×10^{-8}	-

proportional to its dielectric constant and can be, therefore, attributed to the amount of water absorbed by the coating [36]. Therefore, the changes in Y_0 constant of Q_c with immersion time of the coatings (mentioned in Table 8.2) provides insights about the tendency of coatings to water and ionic species. For 0.5, 1 and 2 wt% Gr coatings, Y_0 constant increased significantly at the end of the measurement time. On the other hand, for 5 wt% Gr coating, the final value remained in the same order of 10^{-12} observed in the beginning of the immersion test, thus, exhibiting lower permeability of coatings at high graphene content. It indicated that the well dispersed graphene platelets provided diffusion barrier by enhancing both hydrophobicity of the coating and the tortuous paths of corrosive agents through the coating, thus, greatly reducing the corrosion tendency of underlying metal. It was also observed further in the R_{ct} data provided in

Table 8.2. R_{ct}, is attributed to the resistance to charge transfer processes on the metal surface. Addition of graphene particles diminished R_{ct} in the following order of 0.5>1>2>5 wt% graphene in PEI matrix. Except 5 wt% coating, all other coatings exhibited a large decrease in R_{ct} value during a 3-week immersion. However, a different behavior was observed for 5 wt% coating, where R_{ct} value increased after 2 weeks and the final R_{ct} value obtained after 10 weeks was higher than the initial value.

The behavior of Gr-PEI coatings was further analyzed by potentiodynamic measurements. Figures 8.9 a to c present the anodic and cathodic polarization curves of Gr-PEI coatings in 3.5 wt% NaCl solution. The corresponding E_{corr} and I_{corr} obtained by the fitting analysis are plotted in Figures 8.9 e to g. The polarization curves of 0.5 wt% Gr-PEI and 2 wt% Gr-PEI coatings in Figures 8.9 a and b showed the decline of coatings' barrier properties due to shifts of the curves from noble positive potentials to negative potentials with increase in immersion time. 2 wt% Gr-PEI coatings delaminated very fast during polarization testing and, therefore, accurate readings after 2 weeks of immersion could not be recorded. The polarization plots of 5 wt% Gr-PEI coatings in Figure 8.9 c showed an interesting behavior. A shift in the E_{corr} towards more noble or positive value (-117.728 mV) was recorded as compared with the initial value of -225.264 mV during immersion. After 4 weeks of immersion, the E_{corr} shifted further towards positive potential values (-53.960 mV). For comparison, the potentiodynamic curves of pure PEI coating were also plotted in Figure 8.9 d. Since E_{corr}, cathodic and anodic currents or slopes were absent in the initial weeks of immersion, I_{corr} of the pure PEI coating during this time was calculated directly from the Tafel plots without performing any fitting analysis [37]. In the beginning, E_{corr} value of pure PEI coating was higher than 5 wt% Gr-PEI coating. However, it decreased gradually, as expected from a barrier coating, to -185.88 mV after 4 weeks of immersion which was more negative than the E_{corr} value of 5 wt% Gr-PEI coating (-53.960 mV). Thus, all coatings, except with 5 wt% Gr, exhibited a shift of E_{corr} towards more negative potentials with time indicating that the protection provided by the coatings was purely barrier in nature, which could degrade over time. I_{corr} plotted in Figures 8.9 e to h showed a continuous increase with immersion time for pure PEI and 0.5 wt% Gr-PEI coatings. I_{corr} decreased continuously for 5 wt% Gr-PEI coatings, though, towards the end of the measurement it increased slightly, which can be accounted to the enhanced electrical conductivity of the coating due to the high graphene

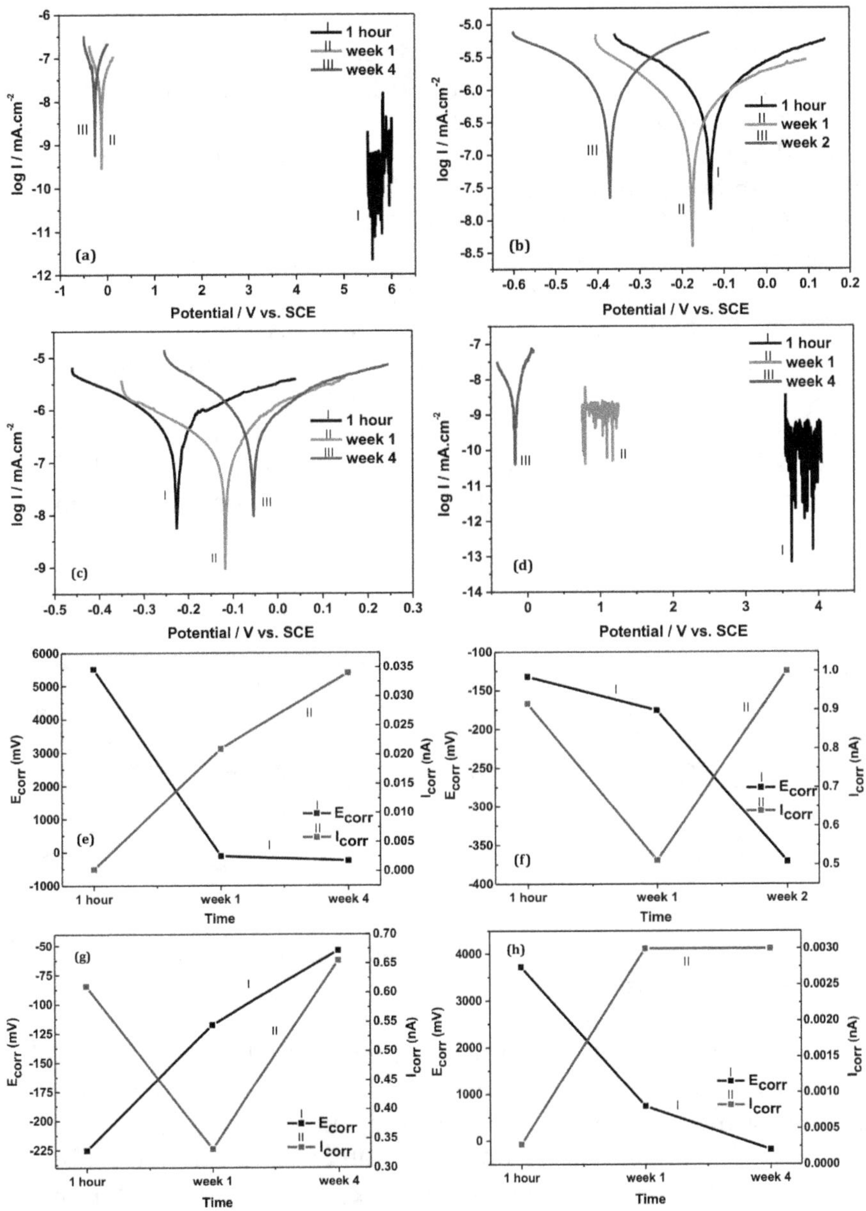

Figure 8.9 Tafel plots (a-d) and E_{corr} & I_{corr} vs. time of immersion (e-h) of Gr-PEI coatings with 0.5 wt% (a) & (e), 2 wt% (b) & (f), 5 wt% (c) & (g) graphene respectively. (d) & (h) represent the characteristics of pure PEI respectively.

content. The shift of the corrosion potential towards the positive scale only for 5 wt% coatings confirmed that graphene inhibited the corrosion of steel substrate at high concentrations when a substantial amount of graphene was in contact with the metal at the interface of the coating and metal. Similar shift was observed by Dennis *et al.* [6] in a Gr-PEI system with 2 wt% and 20 wt% graphene loadings. A corrosion rate of 5.52×10^{-4} mmpy and 8.46×10^{-4} mmpy respectively was reported for these coatings which is higher than that of 5 wt% Gr-PEI coating (3.9×10^{-6} mmpy) prepared in the current study.

A pictorial representation of the protection provided by the graphene in PEI matrix is presented in Figure 8.10. During the initial hours of immersion in NaCl solution, the 5 wt% Gr-PEI coating prevented water and oxygen permeation through the coating due its hydrophobic nature and platelet structure with high aspect ratio which increased the diffusion pathway of the corrosive species and inhibited the corrosion reaction by barrier effect. However, at prolonged

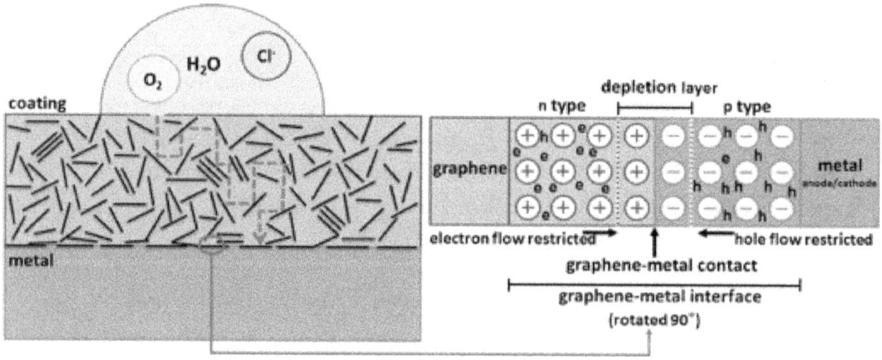

Figure 8.10 Pictorial representation of protection provided by graphene in 5 wt% Gr-PEI composite coating.

immersion, the protection exhibited by 5 wt% Gr-PEI coating was obtained by a different mechanism. Graphene promoted steel passivation at the interface of the steel surface and coating. This noble behavior of steel might have occurred due the formation of Schottky barrier at the steel-graphene junction [22]. The insert in Figure 8.10 shows the protection obtained by the band gap formation at the metal-graphene interface. The graphene is n-doped at the interface when contact with metal is established [38-41]. It is reported that adsorption of graphene on the metal includes initial charge transfer and hybridization [42-44]. Thus, the steel-graphene interface forms a p-n junction with metal surface as the p-type semiconductor and graphene as the n-type semiconductor (Figure 8.10). The depletion region or potential barrier does not allow movement of major charge carriers, i.e., electrons (e), from graphene to the metal surface and holes (h) from the metal surface to graphene. Electron depletion occurs at the metal substrate (anode) and no electrons are available for transfer to the cathode. This reduces the tendency of the metal to oxidize. The metal acts more nobly due to the passivation by the contact with graphene. In addition, the electron transfer from graphene coating to the metal (cathode) is prevented by the presence of potential barrier/depletion region. This also means that had there been electrons at the anode available for cathodic reaction, the movement of electrons from the anode through graphene to the cathode would not have been possible. Therefore, the cathodic reduction of oxygen is also arrested by the formation of p-n junction. Moreover, the repelling field created by the concentration of electrons in the p orbitals of graphene can repel negatively charged ions like chloride and improve anti-corrosion performance [45]. Such a protection was not reported in many graphene based coatings reported in the literature, where graphene provided protection for short time durations by increasing the tortuous path of diffusion through the coating. The passivation of steel by the n-doping of graphene at the metal interface provided protection for long time durations.

The surface area of the steel covered by graphene had an impact on the corrosion kinetics. The 0.5, 1 and 2 wt% coatings did not provide anodic protection of steel due to low concentration of graphene, as fuller coverage of the steel surface was not achieved. This exhibited that graphene coatings did not provide good anti-corrosion performance if these were not adsorbed on the metal at sufficiently high density [15,46]. Thus, any pathway leading to the unbound graphene can cause poor local adhesion as already observed for 0.5, 1, 2 wt% coatings. However, this was not the case for 5 wt% Gr coatings, which achieved excellent corrosion resistance for long periods because of the good contact between graphene and metal. It was achieved by the strong interaction of PEI with graphene and metal surface as observed from the FTIR spectra presented in Figure 8.11. Compared with the IR spectra of PEI pellet, the pure PEI coating exhibited two new signals at 1708 cm^{-1} and 1343 cm^{-1} corresponding respectively to changes in the imide carbonyl and C-N-C bonds (Figures 8.11 a & b) [47]. This observation indicated that the functional groups of imide ring in polyetherimide, C=O as well as C-N-C groups, interacted with

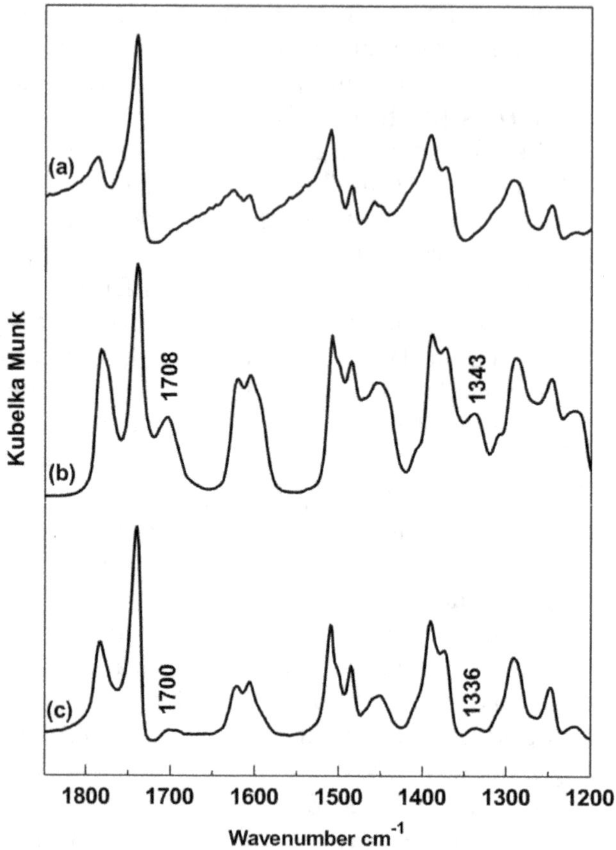

Figure 8.11 FTIR spectra of (a) pure PEI pellets, (b) pure PEI coating and (c) 5 wt% Gr-PEI composite coating.

the steel extensively to form strong adhesive bonds. These signals appeared in the IR spectra of 5 wt% Gr-PEI coating at low wavenumbers of 1700 cm^{-1} and 1336 cm^{-1} respectively. This might have occurred due to the interaction of PEI chains with metal surface as well as the graphene platelets. These strong adhesive forces stabilized the graphene platelets on the metal surface and, thus, enabled them to involve in the passivation process of steel by the formation of p-n junction. Therefore, strong binders like PEI are required as a matrix to maintain a good adhesion to the steel substrate which would help graphene particles to provide protection for long term.

Graphene oxide (GO) based coatings with filler concentrations of 0.5 and 5 wt% were also generated for comparing the performance with graphene based coatings. The Bode plots for these coatings are presented in Figure 8.12. It can be observed that the coatings with GO were unstable and exhibited decline in impedance values just after a week of immersion. Pitting corrosion and fast delamination occurred in 5 wt% GO-PEI coatings. The initial impedance of this coating was lower than that of 5 wt% Gr-PEI coating and further decreased significantly within a week of immersion. This demonstrated that GO did not protect the steel from corrosion and failed to provide an effective barrier against redox reactions. The corrosive electrolyte penetrated through the coatings and the coating performance deteriorated due to hydrophilic nature of GO as well as lower conductivity of the 5 wt% GO-PEI coating. For instance, the surface resistance of 5 wt% GO-PEI coating (4.80×10^5 Ω) was 3 orders higher than 5 wt% Gr-PEI coating (3.66×10^2 Ω), thus, indicating the lower conductivity of PEI matrix with GO incorporation.

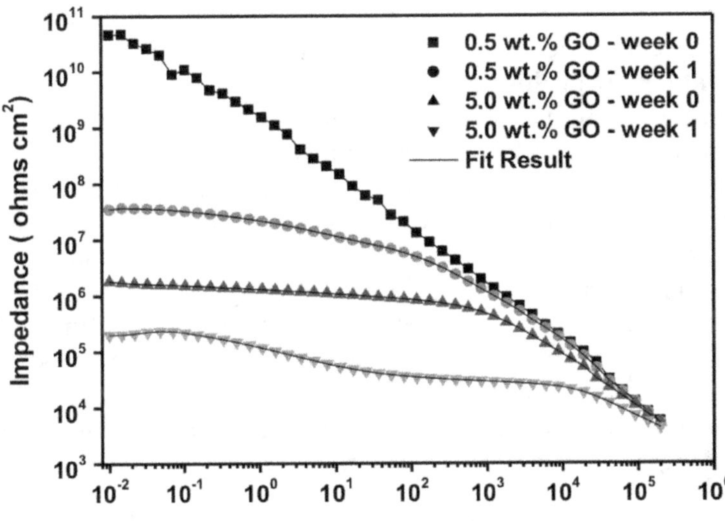

Figure 8.12 Bode plots of GO-PEI composite coatings.

8.4 Conclusions

The present work aimed to synthesize graphene incorporated stable polymer composite coatings for long term protection of carbon steel. Graphene (Gr) was incorporated in polyetherimide (PEI) matrix in different concentrations to investigate the anti-corrosion effect of graphene and nature of protection offered by Gr-PEI composite coatings. A maximum of 5 wt% graphene was successfully incorporated in the polyetherimide matrix, without sacrificing its native barrier property, by using a heating cycle starting from ambient temperature to 270 °C. Graphene loading of 5 wt% maintained good adhesion of the coating, while the coatings with other graphene concentrations (0.5, 1 and 2 wt%) failed due to their inability to stop the influx of water due to low concentration of graphene platelets as well as stresses which destroyed adhesive forces. The low frequency impedance modulus decreased continuously for 0.5, 1 and 2 wt% Gr-PEI coatings, whereas it remained constant at $10^{7.7}$ Ω over 10 weeks of exposure in 3.5 wt% NaCl solution without any visible corrosion products on the coating surface for 5 wt% Gr-PEI coating. During the initial hours of immersion, the 5 wt% Gr-PEI coating prevented water and oxygen permeation due to its hydrophobic nature and graphene platelets with high aspect ratio which increased the diffusion pathway of the corrosive species and inhibited the corrosion reaction by barrier effect. At prolonged immersion, the coating worked through electron depletion at the surface of steel by the formation of a p-n junction, which suppressed the metal's ability to donate electrons and oxidize. The potential barrier at the p-n junction also stopped electron transfer from the coating to the cathode. Thus, both anodic and cathodic reactions were stopped by the addition of graphene at high concentrations. It was also confirmed by an increased R_{ct} and a positively shifted E_{corr} of 5 wt% Gr-PEI composite coatings after a long period of 4-week immersion. Therefore, it can be suggested that graphene can provide a long-term protection when incorporated with suitable polymer matrices, thereby, generating protective barrier coatings with enhanced lifespan.

References

1. Behzadnasab, M., Mirabedini, S., and Esfandeh, M. (2013) Corrosion protection of steel by epoxy nanocomposite coatings containing various combinations of clay and nanoparticulate zirconia. *Corrosion Science*, **75**, 134-141.
2. Zhang, Y., Shao, Y., Zhang, T., Meng, G., and Wang, F. (2013) High corrosion protection of a polyaniline/organophilic montmorillonite coating for magnesium alloys. *Progress in Organic Coatings*, **76**, 804-811.
3. Piromruen, P., Kongparakul, S., and Prasassarakich, P. (2014) Synt-

hesis of polyaniline/montmorillonite nanocomposites with an enhanced anticorrosive performance. *Progress in Organic Coatings,* **77**, 691-700.

4. Radhakrishnan, S., Siju, C., Mahanta, D., Patil, S., and Madras, G. (2009) Conducting polyaniline–nano-TiO2 composites for smart corrosion resistant coatings. *Electrochimica Acta*, **54**, 1249-1254.
5. Bagherzadeh, M., and Mousavinejad, T. (2012) Preparation and investigation of anticorrosion properties of the water-based epoxy-clay nanocoating modified by Na+-MMT and Cloisite 30B. *Progress in Organic Coatings,* **74** 589-595.
6. Dennis, R. V., Viyannalage, L. T., Gaikwad, A. V., Rout, T. K., and Banerjee, S. (2013) Graphene nanocomposite coatings for protecting low-alloy steels from corrosion. *American Ceramic Society Bulletin*, **92**(5), 18-24.
7. Li, Y., Yang, Z., Qiu, H., Dai, Y., Zheng, Q., Li, J., and Yang, J. (2014) Self-aligned graphene as anticorrosive barrier in waterborne polyurethane composite coatings. *Journal of Materials Chemistry A,* **2,** 14139-14145.
8. Chang, K.-C., Hsu, M.-H., Lu, H.-I., Lai, M.-C., Liu, P.-J., Hsu, C.-H., Ji, W.-F., Chuang, T.-L., Wei, Y., and Yeh, J.-M. (2014) Room-temperature cured hydrophobic epoxy/graphene composites as corrosion inhibitor for cold-rolled steel. *Carbon,* **66,** 144-153.
9. Chang, C.-H., Huang, T.-C., Peng, C.-W., Yeh, T.-C., Lu, H.-I., Hung, W.-I., Weng, C.-J., Yang, T.-I., and Yeh, J.-M. (2012) Novel anticorrosion coatings prepared from polyaniline/graphene composites. *Carbon,* **50**(14), 5044-5051.
10. Chang, K., Hsu, C., Lu, H., Ji, W., Chang, C., Li, W., Chuang, T., Yeh, J., Liu, W., and Tsai, M. (2014) Advanced anticorrosive coatings prepared from electroactive polyimide/graphene nanocomposites with synergistic effects of redox catalytic capability and gas barrier properties. *Express Polymer Letters,* **8**(4), 243-255.
11. Liu, S., Gu, L., Zhao, H., Chen, J., Yu, H. (2016). Corrosion resistance of graphene-reinforced waterborne epoxy coatings. *Journal of Materials Science and Technology,* **32**(5), 425-431.
12. Zhang, Z., Zhang, W., Li, D., Sun, Y., Wang, Z., Hou, C., Chen, L., Cao, Y., and Liu, Y. (2015) Mechanical and anticorrosive properties of graphene/epoxy resin composites coating prepared by in-situ method. *International Journal of Molecular Ssciences,* **16**(1), 2239-2251.
13. Mo, M., Zhao, W., Chen, Z., Yu, Q., Zeng, Z., Wu, X., and Xue, Q. (2015) Excellent tribological and anti-corrosion performance of polyurethane composite coatings reinforced with functionalized graphene and graphene oxide nanosheets. *RSC Advances*, **5**, 56486-56497.
14. Sahu, S. C., Samantara, A. K., Seth, M., Parwaiz, S., Singh, B. P., Rath,

P. C., and Jena B. K. (2013) A facile electrochemical approach for development of highly corrosion protective coatings using graphene nanosheets. *Electrochemistry Communications*, **32**, 22-26.

15. Dong, Y., Liu, Q., and Zhou, Q. (2015) Time-dependent protection of ground and polished Cu using graphene film. *Corrosion Science*, **90**, 69-75.
16. Sun, W., Wang, L., Wu, T., Wang, M., Yang, Z., Pan, Y., and Liu, G. (2015) Inhibiting the corrosion-promotion activity of graphene. *Chemistry of Materials*, **27**(7), 2367-2373.
17. Miskovic-Stankovic, V., Jevremovic, I., Jung, I., and Rhee, K. (2014) Electrochemical study of corrosion behavior of graphene coatings on copper and aluminum in a chloride solution. *Carbon*, **75**, 335-344.
18. Liu, J., Hua, L., Li, S., and Yu, M. (2015) Graphene dip coatings: An effective anticorrosion barrier on aluminum. *Applied Surface Science*, **327**, 241-245.
19. Singh, B. P., Nayak, S., Nanda, K. K., Jena, B. K., Bhattacharjee, S., and Besra, L. (2013) The production of a corrosion resistant graphene reinforced composite coating on copper by electrophoretic deposition. *Carbon*, **61**, 47-56.
20. Ming, H., Wang, J., Zhang, Z., Wang, S., Han, E.-H., and Ke, W. (2014) Multilayer graphene: a potential anti-oxidation barrier in simulated primary water. *Journal of Materials Science and Technology*, **30**(11), 1084-1087.
21. Kumar, S., Sun, L., Caceres, S., Li, B., Wood, W., Perugini, A., Maguire, R., and Zhong, W. (2010) Dynamic synergy of graphitic nanoplatelets and multi-walled carbon nanotubes in polyetherimide nanocomposites. *Nanotechnology*, **21**(10), 105702.
22. Dumee, L. F., He, L., Wang, Z., Sheath, P., Xiong, J., Feng, C., Tan, M. Y., She, F., Duke, M., and Gray, S. (2015) Growth of nano-textured graphene coatings across highly porous stainless steel supports towards corrosion resistant coatings. *Carbon*, **87**, 395-408.
23. Chaudhry, A., Mittal, V., and Mishra, B. (2015) Inhibition and promotion of electrochemical reactions by graphene in organic coatings. *RSC Advances*, **5**, 80365-80368.
24. Mayavan, S., Siva, T., and Sathiyanarayanan, S. (2013) Graphene ink as a corrosion inhibiting blanket for iron in an aggressive chloride environment. *RSC Advances*, **3**(47), 24868-24871.
25. Hsieh, Y.-P., Hofmann, M., Chang, K.-W., Jhu, J. G., Li, Y.-Y., Chen, K. Y., Yang, C. C., Chang, W.-S., and Chen, L.-C. (2013) Complete corrosion inhibition through graphene defect passivation. *ACS Nano*, **8**(1), 443-448.
26. Schriver, M., Regan, W., Gannett, W. J., Zaniewski, A. M., Crommie, M. F., and Zettl, A. (2013) Graphene as a long-term metal oxidation barrier: worse than nothing. *ACS Nano*, **7**(7), 5763-5768.

27. Raman, R. S., Banerjee, P. C., Lobo, D. E., Gullapalli, H., Sumandasa, M., Kumar, A., Choudhary, L., Tkacz R., Ajayan P. M., and Majumder, M. (2012) Protecting copper from electrochemical degradation by graphene coating. *Carbon*, **50**(11), 4040-4045.
28. Prasai, D., Tuberquia, J. C., Harl, R. R., Jennings, G. K., and Bolotin, K. I. (2012) Graphene: corrosion-inhibiting coating. *ACS Nano*, **6**(2), 1102-1108.
29. Kirkland, N., Schiller, T., Medhekar, N., and Birbilis, N. (2012) Exploring graphene as a corrosion protection barrier. *Corrosion Science*, **56**, 1-4.
30. Nilsson, L., Andersen, M., Balog, R., Lægsgaard, E., Hofmann, P., Besenbacher, F., Hammer, B., Stensgaard, I., and Hornekær, L. (2012) Graphene coatings: probing the limits of the one atom thick protection layer. *ACS Nano*, **6**(11), 10258-10266.
31. Sreevatsa, S., Banerjee, A., and Haim, G. (2009) Graphene as a permeable ionic barrier. *ECS Transactions*, **19**(5), 259-264.
32. Longun J., and Iroh, J. (2012) Nano-graphene/polyimide composites with extremely high rubbery plateau modulus. *Carbon*, **50**(5), 1823-1832.
33. Luong, N. D., Hippi, U., Korhonen, J. T., Soininen, A. J., Ruokolainen, J., Johansson, L.-S., Nam, J.-D., and Seppala, J. (2011) Enhanced mechanical and electrical properties of polyimide film by graphene sheets via in situ polymerization. *Polymer*, **52**(23), 5237-5242.
34. Zheludkevich, M., Serra, R., Montemor, M., Yasakau, K., Salvado, I. M., and Ferreira, M. (2005) Nanostructured sol–gel coatings doped with cerium nitrate as pre-treatments for AA2024-T3: corrosion protection performance. *Electrochimica Acta*, **51**(2), 208-217.
35. Hsu, C. H., and Mansfeld, F. (2001) Technical Note: Concerning the conversion of the constant phase element parameter Y0 into a capacitance. *Corrosion*, **57**(9), 747-748.
36. Moreno, C., Hernandez, S., Santana, J., Gonzalez-Guzman, J., Souto, R., and Gonzalez, S. (2012) Characterization of water uptake by organic coatings used for the corrosion protection of steel as determined from capacitance measurements. *International Journal of Electrochemical Science*, **7**, 7390-7403
37. Rout, T. K., and Gaikwad, A. V. (2015) In-situ generation and application of nanocomposites on steel surface for anti-corrosion coating. *Progress in Organic Coatings*, **79**, 98-105.
38. Gong, C., Lee, G., Shan, B., Vogel, E. M., Wallace, R. M., and Cho, K. (2010) First-principles study of metal–graphene interfaces. *Journal of Applied Physics*, **108**, 123711.
39. Sevincli, H., Topsakal, M., Durgun, E., and Ciraci, S. (2008) Electronic and magnetic properties of 3d transition-metal atom adsorbed graphene and graphene nanoribbons. *Physical Review B*, **77**, 195434.

40. Enciu, D., Nemnes, G.A., and Ursu, I. (2014) Spintronic devices based on graphene nanoribbons with transition metal impurities. Towards space applications. *INCAS Bulletin*, **6**(1), 45.
41. Jaiswal, N. K., and Srivastava, P. (2013) Fe-doped armchair graphene nanoribbons for spintronic/interconnect applications. *IEEE Transactions on Nanotechnology*, **12**(5) 685-691.
42. Kim, S., Shin, D. H., Kim, C. O., Kang, S. S., Kim, J. M., Jang, C. W., Joo, S. S., Lee, J. S., Kim, J. H., and Choi, S.-H. (2013) Graphene p–n vertical tunneling diodes. *ACS Nano*, **7**(6), 5168-5174.
43. Batzill, M. (2012) The surface science of graphene: Metal interfaces, CVD synthesis, nanoribbons, chemical modifications, and defects. *Surface Science Reports*, **67**(3-4), 83-115.
44. Gadipelli, S., and Guo, Z. X. (2015) Graphene-based materials: synthesis and gas sorption, storage and separation. *Progress in Materials Science*, **69**, 1-60.
45. Berry, V. (2013) Impermeability of graphene and its applications. *Carbon*, **62**, 1-10.
46. Sun, M., Feng, J., Bu, Y., Wang, X., Duan, H., and Luo, C. (2015) Graphene coating bonded onto stainless steel wire as a solid-phase microextraction fiber *Talanta*, **134**, 200-205.
47. da Conceicao, T. F., Scharnagl, N., Dietzel, W., and Kainer, K. (2011) Corrosion protection of magnesium AZ31 alloy using poly(ether imide) [PEI] coatings prepared by the dip coating method: Influence of solvent and substrate pre-treatment. *Corrosion Science*, **53**(1), 338-346.

9

Eco-friendly Polylactic Acid-Silica Coatings for Corrosion Protection of Carbon Steel

9.1 Introduction

Corrosion protection using organic coatings has gained extensive research attention in the recent years, due to their potential to replace the coatings based on hexavalent chromium, which generally suffer from surface leaching. A large number of polymers, such as polyetherimide, polypyrrole, polyethylene, epoxy, polypropylene, polyurethane, etc., have been explored for the generation of anti-corrosion coatings. For instance, in a recent study, fabrication of stable self-healing organic coatings was reported by incorporating polypyrrole carbon black (PPyCB) composite pigment into a polyvinyl butyral (PVB) matrix [1]. The redox properties of PPy provided an anodic protection to the steel surface by inducing a passive film formation at the defect area, which, in turn, increased the amount of reduced PPy near the passivated areas. This passive film along with the hydrophobic nature of the reduced PPy limited the charge transfer reaction rate by acting as a diffusion barrier. Figure 9.1 also shows the cathodic and anodic polarization curves recorded for the PVB and PVB/PPyCB composite coatings (with different pigment concentration) after 1 day of exposure in 4 wt% sodium chloride solution.

Now a days, the interest for "green" or "bio" coatings has also increased tremendously because of the environmental as well as human health issues created by commonly used synthetic polymer coatings [2-3]. In addition to the environmental issues such as emission of volatile organic solvent and recycling or waste disposal of resins at the end of its usage, the trend for biopolymer coatings has also evolved from the realization that the fossil resources used for the conventional polymeric coatings are inherently finite. One of the possibilities to obtain the biopolymer coatings is to use the polymer film forming agents originated from bio-sources. This approach will cut

Tehsin Akhtar, Gisha E. Luckachan and Vikas Mittal, The Petroleum Institute (part of Khalifa University of Science and Technology), Abu Dhabi, UAE*
**Current address: Bletchington, Wellington County, Australia*

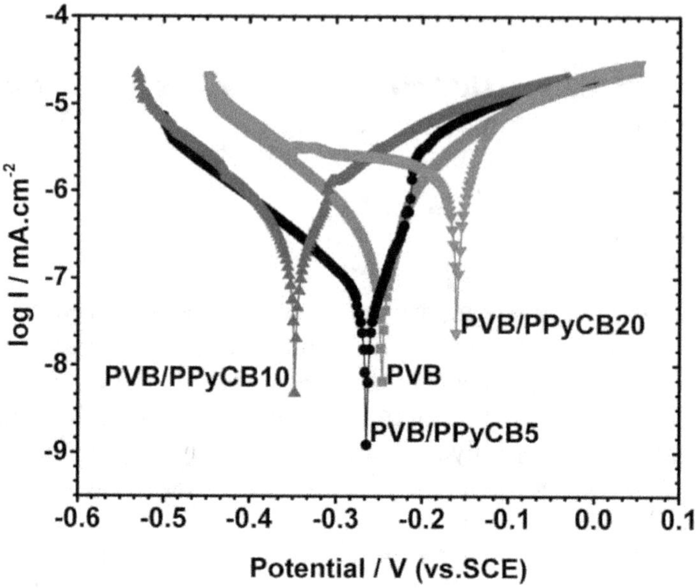

Figure 9.1 Polarization curves of PVB and PVB/PPyCB composite coatings after 1 d immersion in 4 wt% NaCl solution. Reproduced from Reference 1 with permission from Royal Society of Chemistry.

cut down the growing prices of raw materials, generate the environmentally friendly coatings, enhance the coating degradation after service and solve the waste disposal problems of resin materials. In a recent study, layer by layer addition of chitosan (Ch) and PVB was employed to generate multi-layer coatings on the carbon steel substrate [4]. Incorporation of glutaraldehyde in the chitosan layer in PVB_Ch_PVB coatings introduced a cross-linked structure, resulting in a physical barrier to water and ions reaching the metal substrate (Figure 9.2). Among the different bio-resources (cellulose, chitosan, starch, polylactide, vegetable oil, etc.) used commonly for the coating applications, poly(lactic acid) (PLA) has also gained recent interest due to its inherent properties of biodegradability, low toxicity, processability and excellent transparency [5-13].

Poly(lactic acid) is a biodegradable linear aliphatic thermoplastic polyester. It has excellent oil resistance and gas impermeability and offers easy processing. It has thermal and mechanical properties comparable to other conventional plastics [10,14,15]. These properties make PLA a potential alternative to the ubiquitous petroleum based materials for various applications in the areas such as medical,

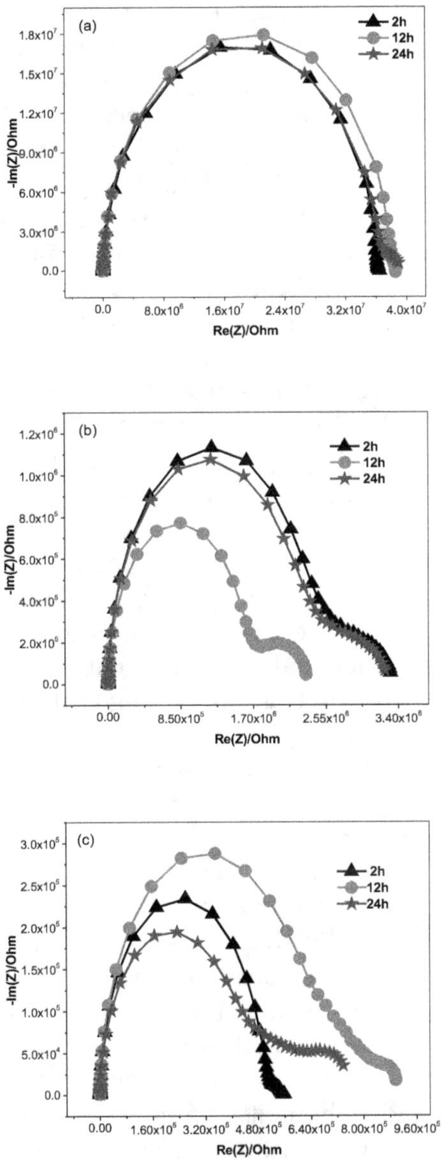

Figure 9.2 Nyquist plots of the coatings with different crosslinked content: PVB_Ch/1%Glu_PVB coating (a), PVB_Ch/5%Glu_PVB coating (b) and PVB_Ch/10%Glu_PVB coating (c) at different time of immersion (2 h, 12 h and 24 h) in 0.3 M salt solution. Reproduced from Reference 4 with permission from Springer.

packaging, textile, fiber, pharmaceutics, etc. [16-22]. PLA can also be modified to alter its physical and chemical properties to suit various applications [10,14,23-25]. Recently, PLA has attracted significant attention in the area of protective coatings. It has been used for the protection of exterior surfaces of buildings because of its hydrophobic nature and slow degradation rate [26]. Especially, historic monuments exposed to vigorous outdoor conditions have been reported to be effectively protected by PLA [27-31]. Svagan *et al.* [18] reported transparent films based on PLA and montmorillonite. To achieve this, layers of montmorillonite and chitosan were generated on PLA films by a layer by layer addition method. The hybrid films exhibited superior properties which reduced the oxygen permeation of PLA. Zeng *et al.* [27] studied the characteristics of microarc oxidation (MAO) and poly(L-lactic acid) (PLLA) composite coating, which was fabricated on Mg-1.21Li-1.12Ca-1Y alloy by dip-coating and freeze-drying. The MAO/PLLA composite coatings increased the corrosion resistance of alloy significantly.

PLA coatings have also been used in various studies to resist biocorrosion of magnesium and its alloys for temporary implant applications [20-22]. However, the protection was observed to be short lived and the coatings failed and disbonded within a week or even 48 hours of testing in saltwater, thus, indicating that the long-term corrosion performance was poor [32-34]. PLA did enhance the degradation resistance of the substrates, however, did not yield long lasting high impedance coatings due to poor adhesion. In some studies, low thickness of coatings was also a factor contributing to the inability of PLA to provide long-term barrier properties to prevent corrosion.

Overall, to the best of our knowledge, the use of PLA as corrosion resistant coating for metallic structures exposed to corrosive environmental conditions has been scantly explored. Therefore, the current study aims to further explore the potential of PLA as a strong coating material for the corrosion protection of metal substrates. PLA produces flat, smooth, uniform and nonporous coatings on metal surfaces. Coupled with hydrophobicity, a moderate glass transition temperature of 60-70 °C, high melting temperature of 170-190 °C and chemical inertness, PLA can be optimized for use as an efficient coating material for anti-corrosion applications [1,10,26]. It is known that strong coating adhesion with the substrate, in addition to restriction of diffusion of corrosive agents through the coating, is of prime importance for good anti-corrosion performance [19,20]. In this context, PLA due to its high polarity has the ability to exhibit enhanced adhesion to the metallic substrates and produce hydrophobic flexible films with low water and oxygen permeability [15,26,35]. In addition, as mentioned above, the performance of PLA can also be further enhanced by blending it with other biopolymers to achieve higher degree of flexibility and to combine the useful properties of different polymers. Structural modification of PLA to achieve chain extension and hyperbranching can also lead to further enhancement in its property profiles, especially anti-corrosion performance. In addition, many research studies have also reported the modification of PLA properties by the incorporation of nanoparticles [36-38]. Among various nanoparticle types, SiO_2 has been widely used for the modification of polymer properties because of its high surface area, high density of silanol groups (Si-OH) on the surface and small particle size. Keeping this in perspective, adhesion promotion of the PLA coatings was explored in the current study by fabricating silica coatings as primer layer by dip-coating in silica sols prepared by acidic hydrolysis of tetraethyl orthosilicate (TEOS) [39,40]. The objective of depositing sol-gel layer was to increase the adhesion between PLA and the steel substrate and promote the compatibility between them. All coatings were subjected to corrosion analysis by electrochemical methods as well as through visual inspections.

9.2 Experimental

9.2.1 Materials

Carbon steel of grade RST37-2 DIN 17100-80, purchased from Qatar Steel Industries, Qatar, was used in the study. Its composition in weight % was C (0.125), Mn (0.519), Si (0.016), P (0.014), S (0.005), Al (0.034) and Fe (99.287). Carbon steel coupons of dimensions 5 cm x 2 cm x 1.8 cm were used for coating purposes. Two different polylactide grades (named as PLA1 and PLA2) were procured from Nature Works. To study the effect of PLA modification on performance, the PLA samples were chain extended using a multifunctional epoxide using reactive extrusion in a compounder and were denoted at PLA1-CE and PLA2-CE. Tetrahydrofuran (THF) of synthesis grade was obtained from Merck and was used as received. Hydrochloric acid (37%) was supplied by Merck while ethyl alcohol (99.5%) and tetraethyl orthosilicate (99.9%) were supplied by Sigma Aldrich. The two-pack polyamide cured epoxy primer was purchased locally from National Paints.

9.2.2 Substrate Preparation

Concentrated hydrochloric acid was used to pickle the as-received carbon steel coupons to dissolve mill scale. The coupons were subsequently rinsed with excess distilled water to remove acid residue. It was followed by quick drying to avoid flash rusting as well as polishing with three grades of sand paper (60, 150 and 180 grits) to remove the top layer of pickling compounds, oxides and hydroxides so as to obtain a smooth surface. The polished coupons were rinsed with acetone and distilled water to remove metal and grit dust, dried by wiping and heating in the oven at 80 °C before use.

9.2.3 Preparation of PLA Solutions, Silica Sol and Coatings

PLA samples were dried under vacuum at 50 °C overnight to remove any adsorbed surface moisture and were subsequently dissolved in THF at 40 °C under magnetic stirring to produce homogeneous 15 wt% PLA solutions. The cleaned steel coupons were then dip coated in the polymer solutions using an immersion and withdrawal speed of 100 mm/minute with a wetting time of 30 s to allow the polymer solution to penetrate into the substrate's surface irregularities. The coatings were dried at 50 °C as the starting temperature, with stepwise increase in temperature of 30 °C/15 min to 190 °C to prevent coating blistering and defect formation due to rapid solvent evaporation. This way, smooth and visually defect-free coatings of PLA1, PLA2, PLA1-CE and PLA2-CE were obtained. Free standing PLA films were also prepared by casting the polymer solutions on Teflon petri dishes and drying under the same conditions as the coatings.

TEOS (9 mL) was subjected to hydrolysis in a mixture of 36.5 mL ethanol, 1.5 mL water and 1.8 mL HCl (0.01N) by stirring at 35 °C for 24 h as described elsewhere [41]. Acidic hydrolysis of TEOS produced pure silica sol in which the cleaned coupons were then dip coated to produce 1-2 μm thick coatings. The coatings were dried overnight at room temperature and subsequently heated at 40 °C for 2 h, at 60 °C for 1 h and at 100 °C for another hour to ensure evaporation of ethanol and water. The silica coated coupons were then coated with PLA solutions (PLA1, PLA2, PLA1-CE and PLA2-CE) and subjected to same drying conditions as the pure PLA coatings till 190 °C to produce PLA-silica coatings. The final thickness of the coatings ranged from 30 to 40 μm. The coating thickness was measured using PosiTector 6000 coating thickness gage. For preparing the epoxy coatings, the base and hardener were mixed in a ratio of 4:1 and kept for 10-15 min before use in accordance to the manufacturer's instructions. The epoxy solution was then dip coated on the carbon steel substrates, which were subsequently kept for 24 h in ambient conditions to obtain defect-free smooth epoxy coatings.

9.2.4 Salt Spray Tests

The coated coupons were subjected to continuous salt spray in the Ascott salt spray chamber. 3.5 wt% sodium chloride (NaCl) solution was sprayed at the rate of 5 mL/80cm^2/h at 25 °C. The corrosion was monitored by the appearance of corrosion products or rust and progress of coating delamination with time. The coatings were also analyzed for their protective behavior using electrochemical impedance spectroscopy (EIS) and Tafel every week as described later.

9.2.5 UV Weathering Test

Two sets of coated coupons were subjected to 4 cycles of weathering tests. Samples were exposed to 0.67 W/m^2 ultraviolet (UV) irradiance for 8 h followed by 4 h of water condensation at 25 °C and 50 °C (ASTM D4587 method) in the Biuged UV weathering cabinet. EIS of the coatings was carried out after every cycle to observe the degradation in corrosion performance due to UV weathering.

9.2.6 Electrochemical Analysis

A three-electrode cell of volume 250 mL with a platinum gauze counter electrode and a saturated calomel reference electrode (SCE) with bridge tube was used to perform electrochemical tests on the flat coated coupons. The exposed working electrode area was 1 cm^2 and the coated coupon was kept in place with the help of an O ring and adjustable brass disc. All tests were carried out by connecting the corrosion cell to the BioLogic VMP-300 multi-potentiostat (controlled by a computer installed with EC-Lab 10.40 software) in 3.5 wt% NaCl solution at room temperature. Ultra-low current cable connected to the potentiostat was used for the accurate measurement of the current. This option included current ranges from 100 nA down to 100 pA with additional gains extending the current ranges to 10 pA and 1 pA. The resolution on the lowest range is 76 aA. The open circuit voltage (OCV) was measured for 10 min to allow the potential to stabilize before the electrochemical impedance and potentiodynamic polarization test. The impedance measurements were performed at amplitude of 10 mV over frequency range from 200 kHz to 10 mHz. Linear polarization was carried out at potential range from -0.250 V to +0.250 V using a scan rate of 1.66 mV/s. The impedance test and linear polarization was conducted on three samples of each type to ensure reproducibility of the results.

9.2.7 FTIR Spectroscopy

The Fourier transform infrared (FTIR) spectroscopy of the coatings and films was performed using a Bruker VERTEX 70 FTIR spectrometer attached with a DRIFT accessory. IR acquisition was achieved by recording 120 scans at a resolution of 4 cm^{-1} in the frequency range of 370 cm^{-1} to 4000 cm^{-1} using the OPUS software. The necessary sample conditioning was carried out before the analysis so as to remove any adsorbed moisture on the surface of the samples.

9.2.8 Scanning Electron Microscopy

The metal surface after EIS analysis was observed using energy dispersive X-ray spectroscopy (EDAX) and scanning electron microscopy (SEM) (FEI Quanta, FEG250, USA) at accelerating voltages of 10-15 kV by removing the coatings with a sharp blade.

9.3 Results and Discussion

In the current study, strong and corrosion resistant PLA coatings were prepared on carbon steel surface by a layer by layer approach, where the first layer consisted of silica coating, followed by a top layer of PLA coating. Silica coating, applied as first layer by sol gel coating method, enhanced the adhesion between the metallic substrate and the PLA polymer. Different coatings were prepared by using PLA1 and PLA2 polymers, as well as chain extended PLA1-CE and PLA2-CE.

Figure 9.3 shows the DRIFT spectra of PLA1 film in comparison with PLA1 coated and PLA1-silica coated steel. The peak observed at 1748 cm^{-1} in the PLA film is assigned to the C=O stretching and the band at 1451 cm^{-1} corresponds to CH_3 group. The two peaks at 1384 cm^{-1} and 1360 cm^{-1} are produced due to the deformation of CH groups. The stretching of ester groups produces signals at 1271 cm^{-1},

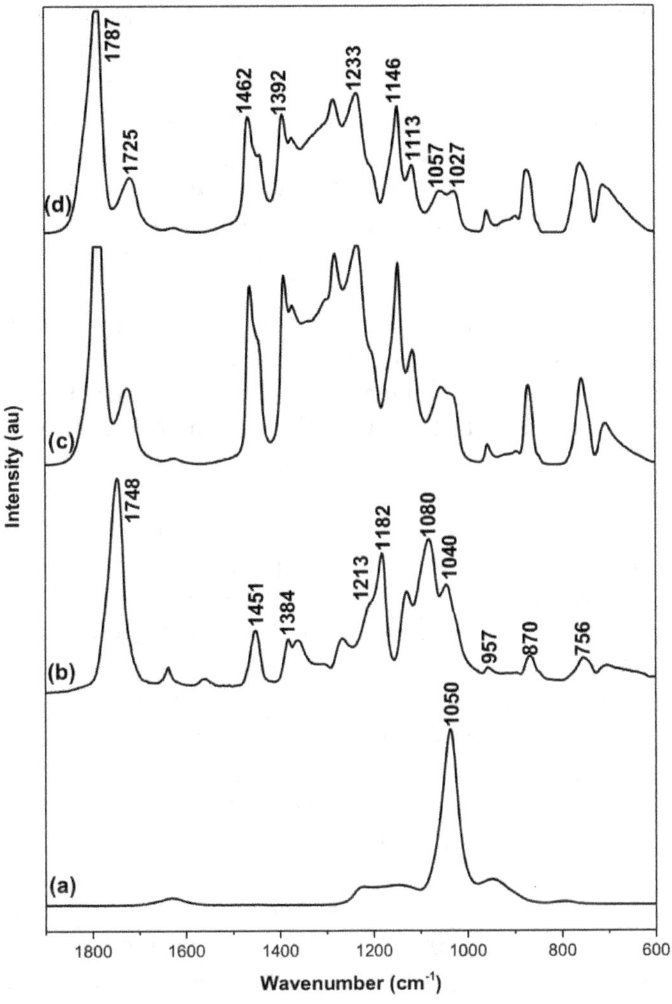

Figure 9.3 FTIR spectra of (a) silica, (b) PLA1 film, (c) PLA1 coating on carbon steel (CS) and (d) PLA1-silica coating on metal surface.

1213 cm^{-1} and 1182 cm^{-1}. The signals at 1130 cm^{-1}, 1081 cm^{-1} and 1043 cm^{-1} arise due to the C-O-C asymmetric mode. Helical backbone vibrations with the CH_3 rocking modes produce a signal at 957 cm^{-1}. The peaks at 870 cm^{-1} and 756 cm^{-1} are due to the crystalline and amorphous phases of PLA [42-44]. However, for the PLA coated steel, the shift of C-O-C stretching modes to 1282 cm^{-1}, 1233 cm^{-1} and 1201 cm^{-1} and asymmetric bands to 1146 cm^{-1}, 1113 cm^{-1}, 1057 cm^{-1} and 1026 cm^{-1} was observed due to the involvement of the PLA backbone in bonding with steel. The shifting of carbonyl peak to 1787cm^{-1} occurred by the melt adhesion of PLA onto the steel surface, thus, the strong adhesion prevented the free vibration of PLA carbonyl groups. The high temperature treatment may have caused a slight degradation of the PLA chains, as it resulted in the generation of new peak at 1725 cm^{-1} corresponding to carboxylic carbonyl groups. Slight changes appeared in the CH_3 vibrations which shifted to 1462 cm^{-1} and 1443 cm^{-1}, while the peaks due to CH deformation moved to 1392 cm^{-1} and 1373 cm^{-1}. All coatings of different PLA grades showed the same spectral features. IR spectra of PLA1-silica coating shown in Figure 9.3d also exhibited the same spectral features as the PLA1 coating. Thus, no significant changes were noted in IR for the PLA-silica coatings in comparison to the pure PLA coatings. This was probably due to the thickness of the top layer (PLA coating), due to which the interaction between the silica and PLA could not be distinguished from the strong PLA signals. The Si-O-Si bond in silica coating on steel showed strong vibrations at 1050 cm^{-1} [45]. This peak was observed in the IR spectra of PLA1-silica coating as merged within the broad vibrations at 1100 cm^{-1} to 1000 cm^{-1}. The two well distinguished doublets at 1057 cm^{-1} and 1027 cm^{-1} might have been formed from the C-O-C and Si-O-Si backbone vibrations respectively. However, it has to be noted that even though similar spectral features were observed in all PLA and PLA-silica coatings, however, the anti-corrosion performance of PLA-silica coatings is expected to be significantly affected by the incorporation of a first layer of thin silica coating on the metal surface. The subsequent sections provide more insights in the electrochemical measurements of these coatings as a function of PLA chain extension, silica coating as well as UV weathering.

As mentioned earlier, the protective nature of PLA-silica coatings was studied by electrochemical impedance measurements in 3.5 wt% sodium chloride solution. For comparison, the EIS analysis was also conducted on PLA, PLA-CE as well as silica coatings. Figure 9.4 presents the EIS results of PLA1 and silica coatings as Bode and phase plots. The initial impedance of 10.6 Ωcm^2 shows the protectiveness of PLA1 coating. Continuous immersion in salt solution decreased the impedance to 7.6 Ωcm^2 indicating the loss of adhesion. After 1 week, an undercoating corrosion was observed to spread throughout. It was due to the lack of strong adhesive bonding with the substrate, which led to the lateral penetration of water under the coatings, thus, resulting in uniform corrosion by disbonding. Similar results were also observed for PLA2 coating as well. The PLA1-CE and PLA2-CE coatings

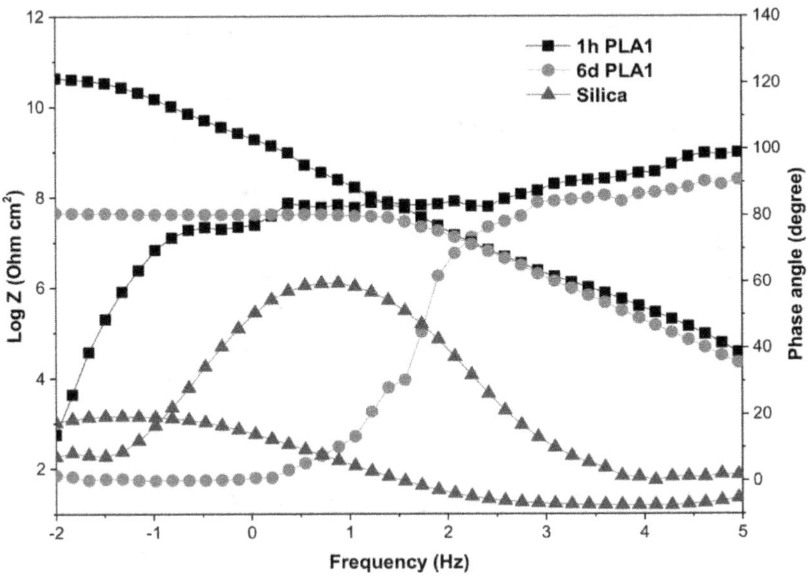

Figure 9.4 EIS Bode and phase plots of PLA1 and silica coating.

lasted without corrosion or delamination for at least 10 days. The Bode and phase plots of the sol-gel film in Figure 9.4 clearly suggested that no additional barrier was provided by the silica coatings alone. On the other hand, the delamination was prevented by the base layer of silica coating which accounted for strong adhesion of PLA chains with metal surface (Figure 9.5 and 9.6), thus, enhancing the barrier and protective nature of PLA coatings. This was evidenced from the high low frequency impedance (log Z), in the range of 11 Ω cm^2 to 12 Ω cm^2, for all the PLA-silica coatings (Figures 9.5 and 9.6) resulting from the silica layer in between the metal and PLA coating.

The low frequency impedance (log Z) of PLA1-silica and PLA2-silica coatings declined with increased salt spraying time that reached respectively to 7 Ωcm^2 and 6 Ωcm^2 by 7 weeks (Figure 9.5). The phase plots recorded in the beginning of the tests demonstrated the initial capacitative behavior of coatings. The plateau in the phase angle at high frequency region, corresponding to the barrier nature of the coating, remained close to 90° during the complete measurement time. This meant that the coatings had no defects and possessed a very high value of coating resistance. In addition, it also indicated that no part of the coating underwent any disbonding to give rise to the Faradaic process. However, after 3 weeks of immersion, the coatings

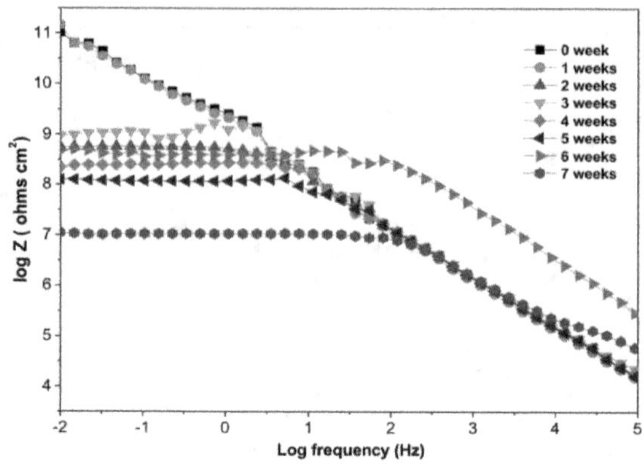

(a)

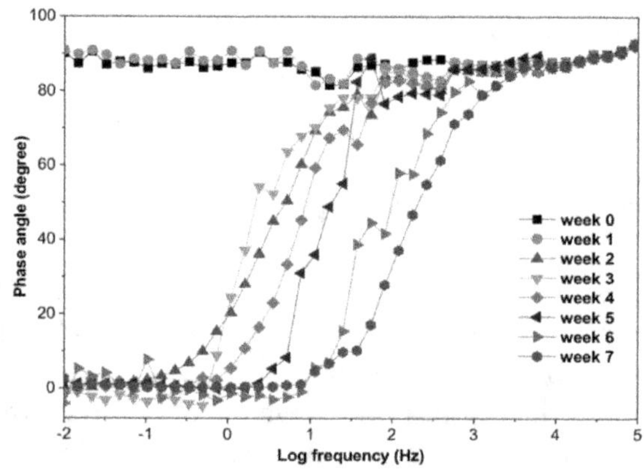

(b)

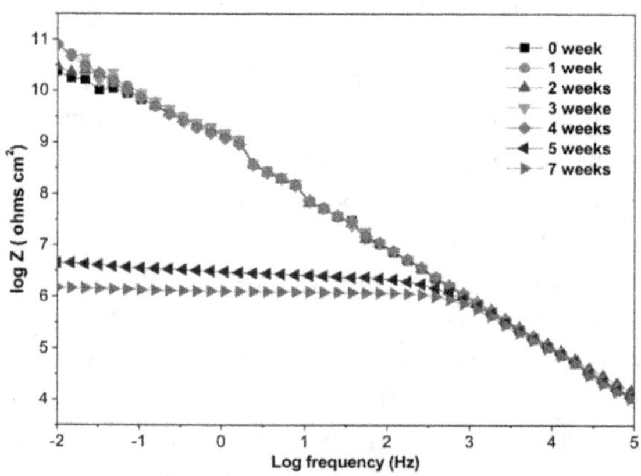

(c)

Figure 9.5 continued on next page

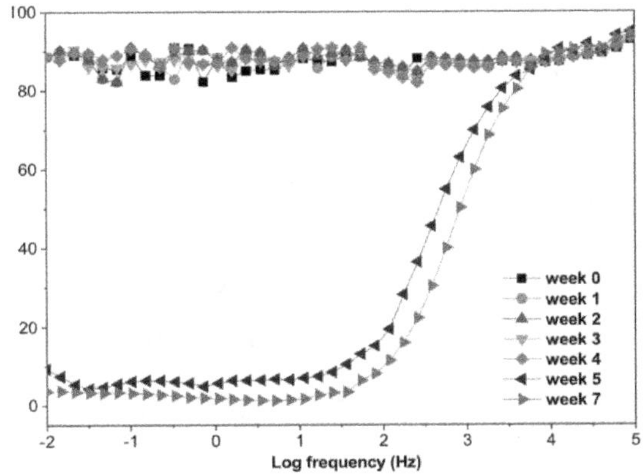

(d)

Figure 9.5 EIS Bode and phase plots of (a,b) PLA1-silica coating and (c,d) PLA2-silica coating at different time of immersion in 3.5 wt% sodium chloride solution.

exhibited deviation from the capacitative behavior and the coating resistance declined due to influx of water in the coatings which led to the contact of the steel with corrosive electrolyte, thereby, resulting in corrosion initiation. However, the modified PLA coatings, i.e. PLA1-CE-silica and PLA2-CE-silica coatings, maintained the initial high impedance values in the order of 10^{11} Ωcm^2 to 10^{10} Ωcm^2 over an experimental duration of 24 weeks (Figure 9.6). It can be clearly seen in the Bode plots recorded both in the beginning and at the end of the test that the behavior was similar and no drop of the impedance values took place during the entire observation period. The phase angle continued to be 90° for more than half of testing duration which indicated that the coatings were capacitative and highly protective in nature. This indicated that the chain extended samples had much superior performance as compared to the unmodified PLA samples. The chain extension led to molecular enhancement of chain length as well as branches, which generally increases viscosity and other physical properties, thereby, also positively affecting the anti-corrosion performance. Thus, PLA modification is a useful strategy to enhance its application areas.

Since epoxy coatings are widely used for the corrosion protection of metal substrates, a comparison study of PLA-silica coating was conducted with an epoxy coating of 33 μm thickness. Figure 9.7 demonstrates the EIS plots of epoxy coating in 3.5 wt% NaCl solution at different exposure durations. It can be seen that the low frequency

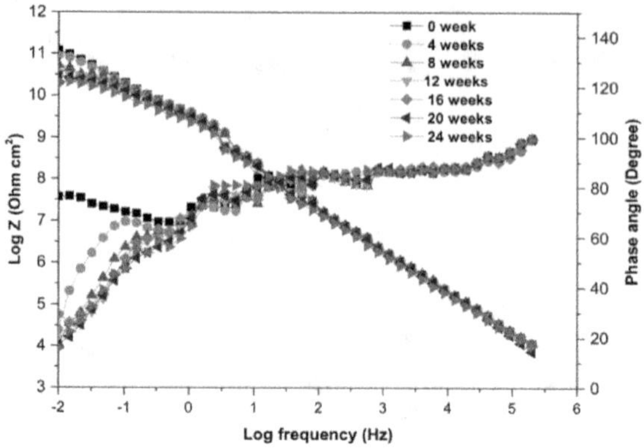

(a)

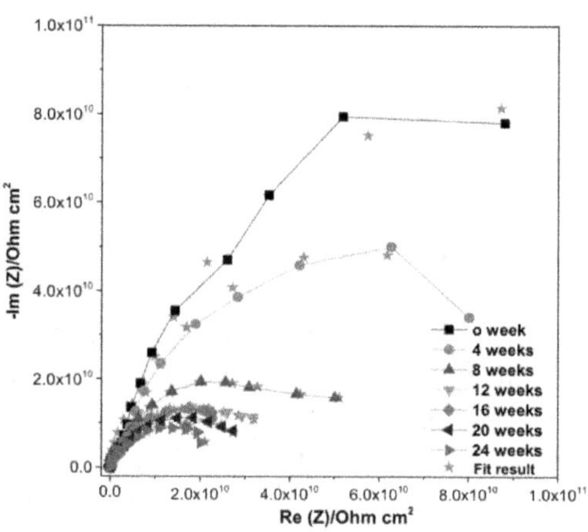

(b)

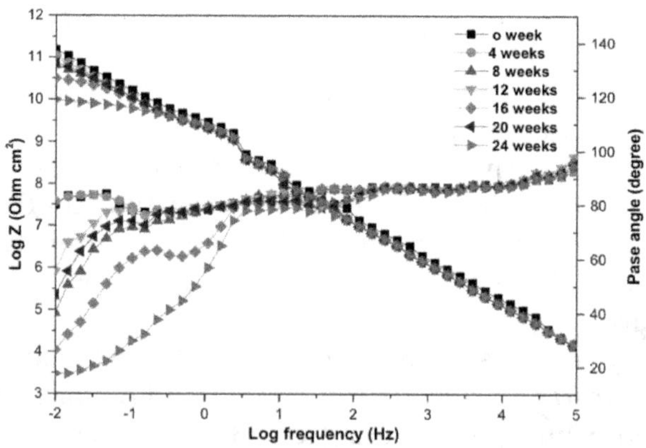

(c)

Figure 9.6 continued on next page

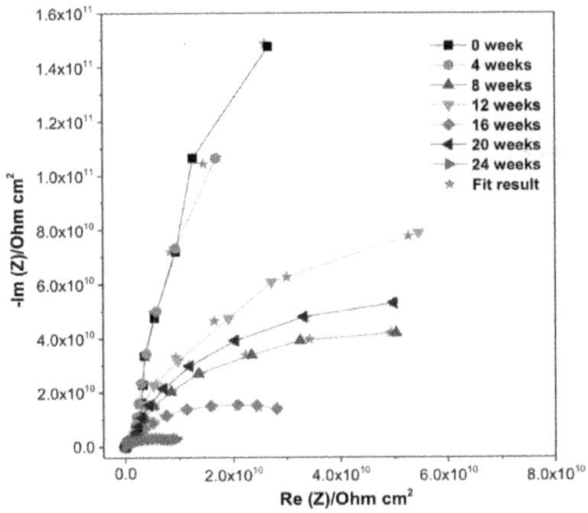

(d)

Figure 9.6 EIS plots of PLA1-CE-silica coating (a) Bode and phase plots, (b) Nyquist plots and PLA2-CE-silica coating (c) Bode and phase plots, (d) Nyquist plots at different time of immersion in 3.5 wt% sodium chloride solution.

impedance (log Z) of the coating decreased every week and reached 7.5 Ωcm^2 at the end of 4 weeks of testing. It is obvious in Nyquist plots where the diameter of the semicircle decreased continuously, thus, showing its tendency to deteriorate with time of exposure. It indicates that PLA-silica coatings protected metal surface from corrosive species more effectively than epoxy coating. Therefore, it can be suggested that such strong and eco-friendly coatings would be a beneficial alternative to the commercial barrier coatings for the advanced protective performance of metal substrates exposed to aggressive environments.

To study further the behavior of PLA-silica coatings in corrosive environment, Nyquist plots of PLA1-CE-silica and PLA2-CE-silica coatings were fitted with different equivalent circuits and the best fitted one is shown in Figure 9.8. The symbol R_s represents the solution resistance of the bulk electrolyte between the reference electrode and working electrode, R_1 and C_1 represent the resistance and capacitance of defect-free PLA top coating. R_2 and C_2 represent the resistance and capacitance of first layer of silica coating filled with PLA chains at the boundaries. R_{ct} is the charge transfer resistance and C_{dl} is the double layer capacitance [46]. R_1 of both PLA1-CE-silica and PLA2-CE-silica coatings remained in the range of 10^9 Ωcm^2 even after 24 weeks of analysis with slight decrease in the magnitude, whereas

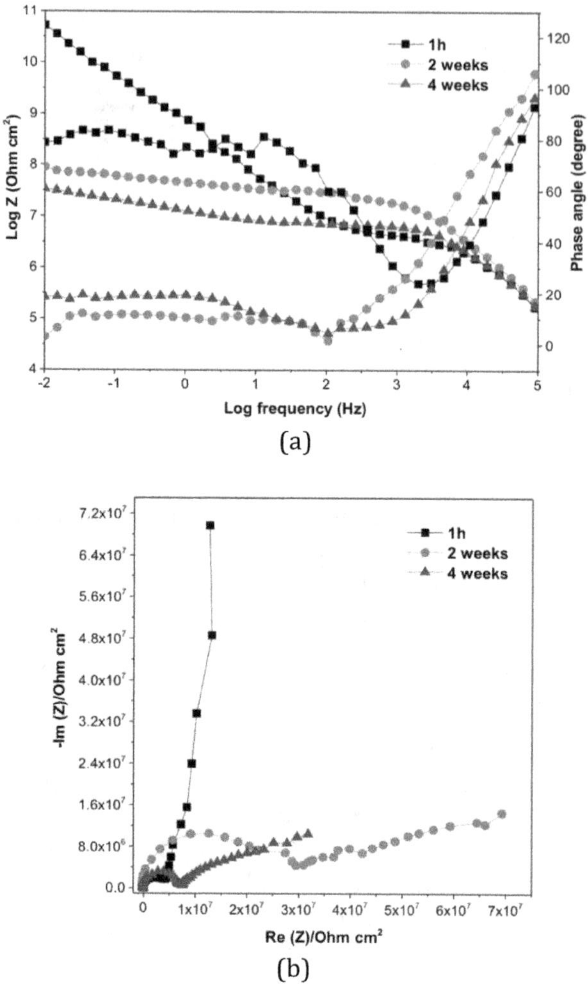

Figure 9.7 EIS Bode and phase plots (a) and Nyquist plots (b) of epoxy coating in 3.5 wt% sodium chloride solution.

R_2 and R_{ct} decreased from 10^{12} Ωcm^2 to 10^9 Ωcm^2. Since capacitance of a coating (C_c) is proportional to its dielectric constant, this value directly attributes to the amount of water absorbed by the coating [47,48]. Therefore, the changes in capacitance of different layers shown in Tables 9.1 and 9.2 clearly indicates the tendency of the coatings to water and ionic species. Both C_2 and C_{dl} showed a similar trend, as these values increased with analysis time. However, C_2 of both PLA1-CE-silica and PLA2-CE-silica coatings increased slightly at 4 weeks and then remained constant over 16 weeks, which indicated that the water molecules in the coatings were distributed uniformly without further ingress down to the metal surface. C_1 also increased slightly during the initial weeks and then remained almost constant over the complete analysis time. Such a behavior indicating efficient protection occurred by the strong adhesion of PLA1-CE and PLA2-CE with the metal surface. Since PLA1-CE and PLA2-CE were generated by the chain extension modifications, these may contain more interactive sites which can be attributed to an extended interaction or adhesion with metal surface and silica particles as compared with the pure PLA1 and PLA2 coatings.

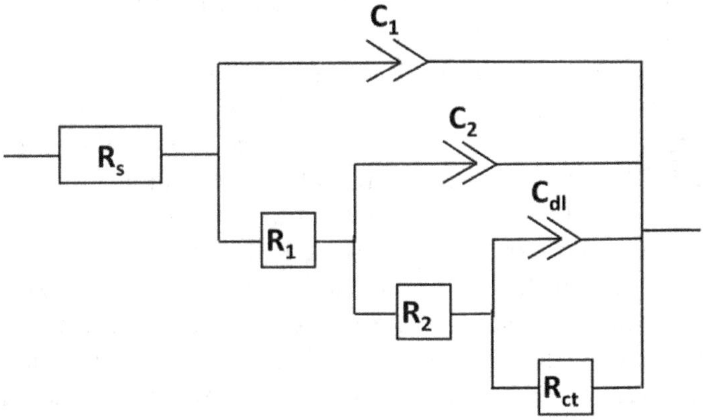

Figure 9.8 Equivalent circuit used for the fitting of Nyquist plots of PLA1-CE-silica and PLA2-CE-silica coatings.

Table 9.1 EIS fitted data of PLA1-CE-silica coatings in 3.5 wt% NaCl

	R_1 (Ω cm^2)	C_1 (F cm^2)	R_2 (Ω cm^2)	C_2 (F cm^2)	R_{ct} (Ω cm^2)	C_{dl} (F cm^2)
Initial	10.52e9	37.43e-12	16.64e12	46.02e-12	1.125e12	22.7e-12
4 weeks	9.18e9	37.09e-12	13.36e12	60.07e-12	0.799e12	55.36e-12
8 weeks	8.93e9	38.84e-12	0.13e12	62.1e-12	2.25e12	22.83e-12
12 weeks	8.92e9	39.26e-12	45.44e9	62.61e-12	95.07e9	10.54e-12
16 weeks	8.80e9	43.4e-12	56.94e9	62.8e-12	80.62e9	0.84e-9
20 weeks	8.71e9	44.67e-12	69.17e9	0.115e-9	60.46e9	1.01e-9
24 weeks	7.65e9	45.77e-12	10.07e9	1.26e-9	34.78e9	0.46e-9

It is widely proven that the introduction of small amounts of nanostructured materials may represent an effective way to enhance the mechanical performance of thermoplastics [49-51]. Inorganic nanoparticles have often been added to polymeric matrices to improve their toughness and strength, to increase their thermal stability and to enhance the barrier properties of the pristine polymer. However, a good care is needed in the optimization of the concentration of nanoparticles in the coating formulation, otherwise it may negatively affect the performance of the coating by creating more defect areas. The conventional plastic films were usually mixed with metal oxides like silicon oxide, aluminum oxide, silicon nitride, etc., to improve their barrier properties, however, these systems are inadequate due to well-known coating imperfections such as pinholes, cracks and grain boundaries that diminish their performance as barrier coatings. In the present work, instead of adding silicon oxide into the coating matrix, a fine layer of silica was fabricated on the metal surface before the application of PLA layer to create multi-layered hybrid composite structure. High temperature curing at 190 °C enabled PLA to fill the cavities in between the silica particles in the first layer of the hybrid coating and formed a strong and defect-free uniform top layer. The hybrid formation contributed to a significantly enhanced performance of PLA coating in the corrosive environments. Silica particles

Table 9.2 EIS fitted data of PLA2-CE-silica coatings in 3.5 wt% sodium chloride solution

	R_1 (Ω cm^2)	C_1 (F cm^2)	R_2 (Ω cm^2)	C_2 (F cm^2)	R_{ct} (Ω cm^2)	C_{dl} (F cm^2)
Initial	8.683e9	45.45e-12	0.88e12	54.62e-12	0.49e12	19.33e-12
4 weeks	7.16e9	65.44e-12	0.166e12	81.13e-12	0.608e12	12.63e-12
8 weeks	5.214e9	60.13e-12	0.165e12	86.91e-12	3.06e12	0.31e-12
12 weeks	5.274e9	62.15e-12	0.108e12	84.75e-12	56.69e9	63.28e-12
16 weeks	5.699e9	62.91e-12	96.05e9	90.63e-12	18.47e9	70.33e-12
20 weeks	5.998e9	62.41e-12	26.56e9	0.117e-9	16.88e9	1.93e-9
24 weeks	4.40e9	54.97e-12	4.17e9	0.362e-9	4.33e9	4.68e-9

not only promoted the strong adhesion of PLA chains with the metal surface, but also created a long tortuous path for oxygen and moisture diffusion through the coating. It was attributed in PLA1-CE-silica and PLA2-CE-silica coatings with a constant C_1 and C_2 values in the order of 10^{-12} Fcm^2 over 16 weeks in sodium chloride solution. In addition, the hydrophilic nature and high surface area of silica enabled it to bind with water molecules and, thus, reduced the chances of further ingress of moisture down to the metal surface. Thus, the overall permeation of corrosive species through the PLA coating was minimized by the presence of a thin silica base layer which, in turn, attributed to an improved corrosion resistance of the hybrid coating.

After the EIS analysis, PLA1-silica and PLA1-CE-silica coatings were also examined visually for any signs of corrosion. No damage or delamination was observed in any coating, but a discoloration was observed to occur on PLA1-silica coating as shown in Figures 9.9a and b. Therefore, the metal surface was analyzed with SEM and EDX after removing the coatings. The patches of corrosion products were clearly observed on the PLA1-silica coated metal surface after 7 weeks of EIS. The EDX of the metal surface exhibited the presence of

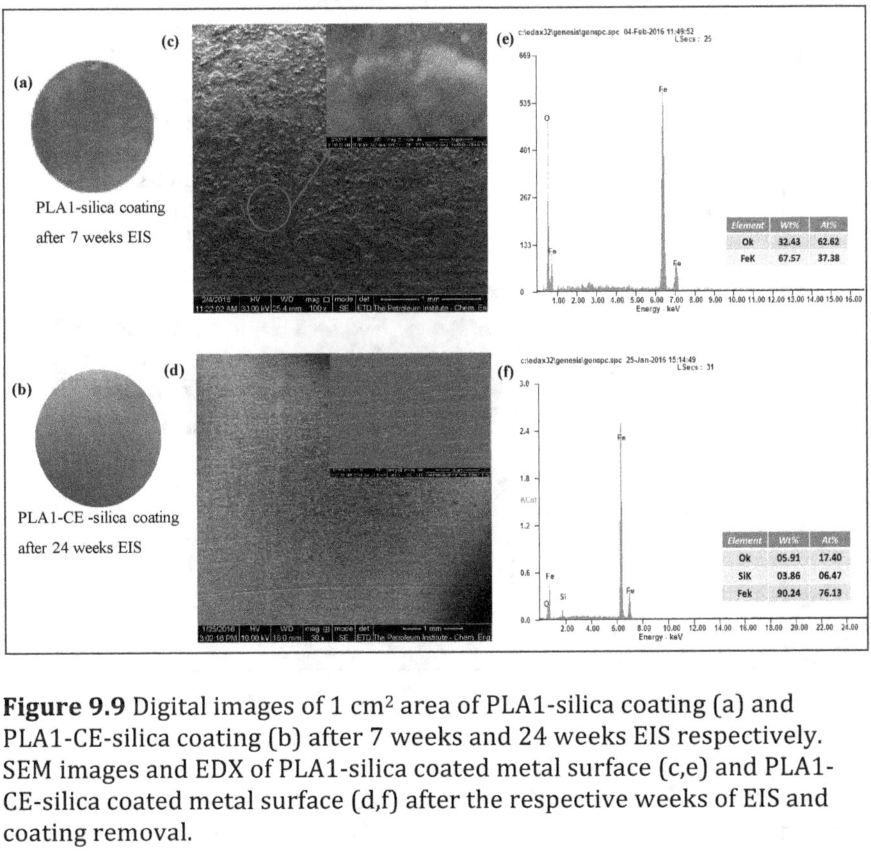

Figure 9.9 Digital images of 1 cm^2 area of PLA1-silica coating (a) and PLA1-CE-silica coating (b) after 7 weeks and 24 weeks EIS respectively. SEM images and EDX of PLA1-silica coated metal surface (c,e) and PLA1-CE-silica coated metal surface (d,f) after the respective weeks of EIS and coating removal.

elemental oxygen confirming the formation of iron oxide corrosion products beneath the PLA1-silica coating (Figures 9.9c and e). Such patches of corrosion products were not observed on the PLA1-CE-silica coated metal surface in SEM image and its EDX analysis exhibited a very small amount of elemental oxygen (Figures 9.9d and f). It is obvious from the images that modified PLA chains formed strong coatings which provided even better protection than that of conventional epoxy coatings. In addition, these coatings presented additional functionality due to their completely environmental friendly nature, as well as easy availability.

PLA coatings were further studied using potentiodynamic measurements and the corresponding Tafel plots and Tafel data of PLA1-silica coating and PLA1-CE-silica coatings are presented in Figure 9.10 and Table 9.3 respectively. Initially, the Tafel plots obtained by linear polarization of both samples didn't display well defined anodic and cathodic currents. This indicated the strong suppression of corrosion activity and restriction of penetration of water. Since E_{corr},

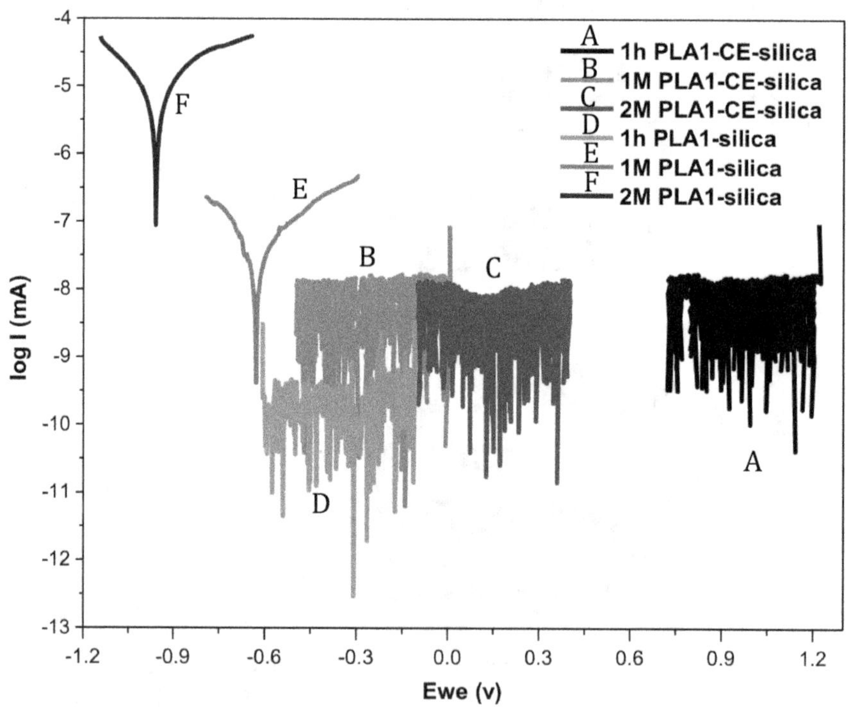

Figure 9.10 Tafel plots of PLA1-silica and PLA1-CE-silica coatings after 1 h, 1 month and 2 month exposure in 3.5 wt% sodium chloride solution.

Table 9.3 Tafel data of PLA-silica coatings

	PLA1-silica		PLA1-CE-silica	
	E_{corr} (V)	I_{corr} (μA)	E_{corr} (V)	I_{corr} (μA)
1 hour	-	5.997e-8	-	4.38e-7
1 month	-0.629	0.81e-5	-	3.98e-7
2 months	-0.962	0.019	-	1.27e-6

cathodic and anodic currents or slopes were absent in the initial weeks of immersion, I_{corr} of the coating during this time was calculated directly from the Tafel plots without performing any fitting analysis [52]. The enhanced protection of PLA1-CE-silica coating is obvious from the position on more positive side of Tafel plots as compared to PLA1-silica coating during the entire analysis time. In addition, E_{corr}, cathodic and anodic current slopes were observed to appear for to PLA1-silica coating at 1 month of immersion time in 3.5 wt% sodium chloride solution, whereas for PLA1-CE-silica coating, the initial behavior was observed to persist over the 2 months of analysis. I_{corr} of PLA1-silica coating increased to 0.019 μA after an immersion period of 2 months, whereas this value was still very low for PLA1-CE-silica at 1.27e-6 μA, thus, showing the strong protection provided by the chain extended PLA-silica coating. Such improvement in the performance was attributed to the strong bonding generated by silica particles, thus, promoting adhesion between the polymer and the steel surface as well as positive effect of the treatment of coated samples at 190 °C, which resulted in uniform, defect-free, and glossy coatings.

PLA coatings were further subjected to UV weathering analysis in order to analyze its effect on the anti-corrosion performance of the coatings. Actual environmental conditions, where the barrier coatings come in contact with UV during their service time, were mimicked in the UV weathering analysis. Four cycles of 8 h UV exposure and 4 h water condensation were used and EIS was performed after each cycle of exposure (Figure 9.11). PLA1-CE-silica and PLA2-CE-silica coatings retained a high low frequency impedance (log Z) of 11 Ωcm^2 even after four cycles of UV exposure. The plateau in the phase angle plot remained between 80° to 90° during the complete measurement time. These results clearly indicated the barrier protection of carbon steel surface by modified PLA-silica coatings and it did not

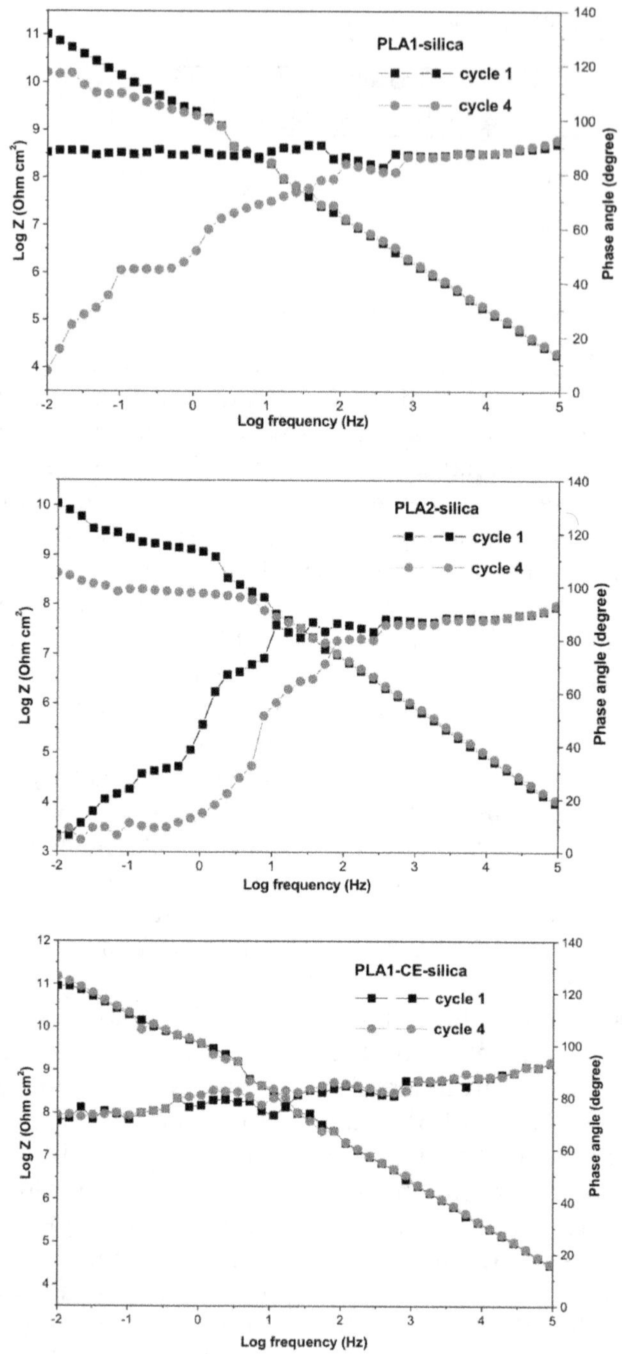

Figure 9.11 continued on next page

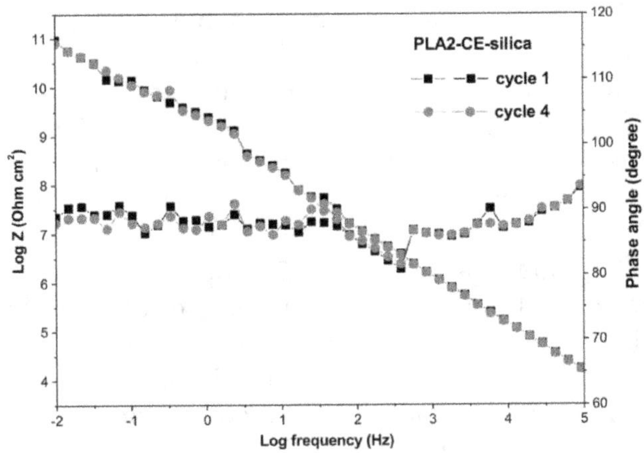

Figure 9.11 EIS Bode and phase plots of PLA1-silica, PLA2-silica, PLA1-CE-silica and PLA2-CE-silica coatings after 1st and 4th cycle of UV and weathering exposure.

degrade or deteriorate by the exposure of high dose of UV radiation, in contrast with other traditional polymer coatings which generally degrade when exposed to UV. However, for PLA1-silica and PLA2-silica coatings, a slow degradation of coating was apparent due to the observed slow decrease in the low frequency impedance values (log Z). The phase angle also deviated from 90° after four cycles of UV weathering. These observed results, overall, confirm that PLA has a strong potential to provide effective and long term protection to metal substrates in natural as well as corrosive environmental conditions. Specifically, the structural modification of PLA opens the opportunities to tune the anti-corrosion performance of the coatings even further. Synergistic combination of PLA with filler also leads to significant enhancement in its performance, which is even better than widely used epoxy based coatings. In addition, the bio-coatings are completely environmentally friendly and are, thus, good alternative to the existing barrier coatings based on non-biodegradable polymeric materials.

9.4 Conclusions

Eco-friendly and strong PLA coatings with thickness in the range of 30-40 μm were fabricated on carbon steel by a layer by layer approach of first layer of silica followed by a PLA over-coating. The silica particles improved the adhesion strength of PLA chains with metal surface. The high temperature curing enabled PLA chains to completely fill the cavities between the silica domains which further enhanced the adhesion strength and, thus, contributed to a better corrosion protection of PLA-silica coatings, as compared with pure PLA coatings. The chain extended PLA coatings, i.e. PLA1-CE-silica and PLA2-CE-silica coatings, exhibited a very high low frequency impedance (log Z) in the range of 10 Ωcm^2 - 11 Ωcm^2. These values were retained in the same range even after 24-week immersion in 3.5 wt% sodium chloride solution, which confirmed the strong ability of modified PLA-silica hybrid coatings in providing protection to metal substrates from corrosive species. Silica particles not only promoted the strong adhesion of PLA chains with the metal surface, but also created a long tortuous path for oxygen and moisture diffusion through the coating. It was attributed in PLA1-CE-silica and PLA2-CE-silica coatings to a constant C_2 values in the order of 10^{-12} Fcm^2 for over 16 weeks in sodium chloride solution. In addition, the hydrophilic nature and high surface area of silica enabled the silica particles to bind with water molecules, thus, inhibiting its further ingress down to the metal surface. The SEM images and EDX of the metal surface analyzed after the coating removal confirmed the protective nature of the modified PLA-silica coatings, which prevented the formation of any type of corrosion products on the metal surface. In addition, the exposure of the modified PLA-silica coatings to UV weathering did not deteriorate their protective nature and exhibited an effective anti-corrosion performance even after 4 cycles of high dose of UV irradiation. The performance of these coatings was also significantly better than commercially used epoxy coatings.

The results conclude that the PLA coatings, especially generated with structurally modified PLAs, can provide strong protection to metal substrates in natural as well as corrosive environmental conditions. The results also confirmed that specific hybridization of the filler with the polymer is necessary to obtain the synergistic effect. In the current study, layer by layer hybridization concept could be successfully demonstrated to provide a significantly improved anti-corrosion performance as compared to traditional composites where the filler is simply embedded into the polymer matrix. Thus, the obtained bio-coatings exhibit much superior performance along with environmental friendliness and have strong potential of commercial use by replacing the traditionally used polymeric coating systems. Also, more research studies should be channeled in this direction to fully explore various opportunities available with such bio-systems.

References

1. Niratiwongkorn, T., Luckachan, G.E., and Mittal, V. (2016) Self-healing protective coatings of polyvinyl butyral/polypyrrole-carbon black composite on carbon steel. *RSC Advances,* **6**, 43237-43249.
2. Chaudhari, A., Kuwar, A., Mahulikar, P., Hundiwale, D., Kulkarni, R., and Gite, V. (2014) Development of anticorrosive two pack polyurethane coatings based on modified fatty amide of Azadirachta indica Juss oil cured at room temperature - a sustainable resource. *RSC Advances,* **4**, 17866-17872.
3. Rajput, S. D., Mahulikar, P. P., and Gite, V. V. (2014) Biobased dimer fatty acid containing two pack polyurethane for wood finished coatings. *Progress in Organic Coatings,* **77**, 38-46.
4. Luckachan, G. E., and Mittal, V. (2015) Anti-corrosion behavior of layer by layer coatings of cross-linked chitosan and poly(vinyl butyral) on carbon steel. *Cellulose,* **22**, 3275-3290.
5. Gandini, A., and Belgacem, M. N. (2002) Recent contributions to the preparation of polymers derived from renewable resources. *Journal of Polymers and Environment,* **10**, 105-114.
6. Derksen, J. T. P., Cuperus, F. P., and Kolster, P. (1996) Renewable resources in coatings technology: a review. *Progress in Organic Coatings,* **27**, 45-53.
7. Williams, C. K., and Hillmyer, M. A. (2008) Polymers from renewable resources: a perspective for a special issue of polymer reviews. *Polymer Reviews,* **48**, 1-10.
8. Lasprilla, A. J. R., Martinez, G. A. R., Lunelli, B. H., Jardini, A. L., and Filho, R. M. (2012) Poly-lactic acid synthesis for application in biomedical devices - A review. *Biotechnology Advances,* **30**, 321-328.
9. Nampoothiri, K. M., Nair, N. R., and John, R. P. (2010) An overview of the recent developments in polylactide (PLA) research. *Bioresource Technology,* **101**, 8493-8501.
10. Gupta, B., Revagadea, N., and Hilborn, J. (2007) Poly(lactic acid) fiber: an overview. *Progress in Polymer Science,* **32**, 455-482.
11. Mittal, V., Akhtar, T., and Matsko, N. (2015) Mechanical, thermal, rheological and morphological properties of binary and ternary blends of PLA, TPS and PCL. *Macromolecular Materials and Engineering,* **300**, 423-435.
12. Rhim, J. W., Hong, S. I., and Ha, C. S. (2009) Tensile, water vapor barrier and antimicrobial properties of PLA/nanoclay composite films. *LWT - Food Science and Technology,* **42**, 612-617.
13. Xu, L., and Yamamoto, A. (2012) Characteristics and cytocompatibility of biodegradable polymer film on magnesium by spin coating. *Colloids and Surfaces B: Biointerfaces,* **93**, 67-74.
14. Carrasco, F., Pagesb, P., Gamez-Perez, J., Santana, O. O., and Maspoch, M. L. (2010) Kinetics of the thermal decomposition of processed po-

ly(lactic acid). *Polymer Degradation and Stability*, **95**, 2508-2514.

15. Wang, S., Cui, W., and Bei, J. (2005) Bulk and surface modifications of polylactide. *Analytical and Bioanalytical Chemistry*, **381**, 547-556.
16. Lim, L. T., Auras, R., and Rubino, M. (2008) Processing technologies for poly(lactic acid). *Progress in Polymer Science*, **33**, 820-852.
17. Carrasco, F., Pagesb, P., Gamez-Perez, J., Santana, O. O., and Maspoch, M. L. (2010) Processing of poly(lactic acid): Characterization of chemical structure, thermal stability and mechanical properties. *Polymer Degradation and Stability*, **95**, 116-125.
18. Svagan, A. J., Akesson, A., Cardenas, M., Bulut, S., Knudsen, J. C., Risbo, J., and Plackett, D. (2012) Transparent films based on PLA and montmorillonite with tunable oxygen barrier properties. *Biomacromolecules.* **13**(2), 397-405.
19. Abdal-hay, A., Hwang, M. G., and Lim, J. K. (2012) In vitro bioactivity of titanium implants coated with bicomponent hybrid biodegradable polymers. *Journal of Sol-Gel Science and Technology*, **64**, 756-764.
20. Alabbasi, A., Liyanaarachchi, S., and Kannan, M. B. (2012) Polylactic acid coating on a biodegradable magnesium alloy: An in vitro degradation study by electrochemical impedance spectroscopy. *Thin Solid Films*, **520**, 6841-6844.
21. Shi, P., Niua, B., Eb, S., Chen Y., and Li, Q. (2015) Preparation and characterization of PLA coating and PLA/MAO composite coatings on AZ31 magnesium alloy for improvement of corrosion resistance. *Surface & Coatings Technology*, **262**, 26-32.
22. Abdal-hay, A., Barakat, N. A. M., and Lim, J. K. (2013) Hydroxyapatite-doped poly(lactic acid) porous film coating for enhanced bioactivity and corrosion behavior of AZ31 Mg alloy for orthopedic applications. *Ceramics International*, **39**, 183-195.
23. Perumal, G., Pappuru, S., Chakraborty, D., Nandkumar, A. M., Chand, D. K., and Doble, M. (2017) Synthesis and characterization of curcumin loaded PLA -Hyperbranched polyglycerol electrospun blend for wound dressing applications. *Materials Science and Engineering: C*, **76**, 1196-1204.
24. Ocak, Y., Aysun, S., Funda, T., and Hasan B. (2009) Protection of marble surfaces by using biodegradable polymers as coating agent. *Progress in Organic Coatings*, **66**, 213-220.
25. Frediani, M., Rosi, L., Camaiti, M., Berti, D., Mariotti, A., Comucci, A., Vannucci, C., and Malesci, I. (2010) Polylactide/perfluoropolyether block copolymers: Potential candidates for protective and surface modifiers. *Macromolecular Chemistry and Physics*, **211**, 988-995.
26. Rasal, R. M., Janorkarc, A. V., and Hirt, D. E. (2010) Poly(lactic acid) modifications. *Progress in Polymer Science* **35**, 338-356.
27. Zeng, R. C., Cui, L. Y., Jiang, K., Liu, R., Zhao, B. D., and Zheng, Y. F. (2016) In vitro corrosion and cytocompatibility of a microarc oxida-

tion coating and poly(L-lactic acid) composite coating on Mg-1Li-1Ca alloy for orthopedic implants. *ACS Applied Materials & Interfaces,* **8**(15), 10014-10028.

28. Ocak, Y., Sofuoglu, A., Tihminlioglu, F., and Boke, H. (2015) Sustainable bio-nano composite coatings for the protection of marble surfaces. *Journal of Cultural Heritage,* **16**(3), 299-306.
29. Giuntoli, G., Rosi, L., Frediani, M., Sacchi, B., and Frediani, P. (2012) Fluoro-functionalized PLA polymers as potential water-repellent coating materials for protection of stone. *Journal of Applied Polymer Science,* **125**, 3125-3133.
30. Giuntoli, G., Frediani, M., Pedna, A., Rosi, L., and Frediani, P. (2012) New perspectives for the application of PLA in cultural heritage. In: *Polylactic Acid: Synthesis, Properties and Applications,* Nova Science Publishers, USA, pp.161-189.
31. Pedna, A., Pinho, L., Frediani, P., and Mosquera, M. J. (2016) Obtaining SiO_2-fluorinated PLA bionanocomposites with application as reversible and highly-hydrophobic coatings of buildings. *Progress in Organic Coatings,* **90**, 91-100.
32. Kannan, M. B., and Liyanaarachchi, S. (2013) Hybrid coating on a magnesium alloy for minimizing the localized degradation for load-bearing biodegradable mini-implant applications. *Materials Chemistry and Physics,* **142**, 350-354.
33. Singer, F., Distler, T., and Virtanen, S. (2014) Long-term corrosion behavior of poly-L-lactic acid coated magnesium in dulbecco's modified eagle medium at body temperature. *International Journal of Electrochemical Science,* **9**, 7965-7976.
34. Shi, P., Niua, B., Eb, S., Chen, Y., and Li, Q. (2015) Preparation and characterization of PLA coating and PLA/MAO composite coatings on AZ31 magnesium alloy for improvement of corrosion resistance. *Surface & Coatings Technology,* **262**, 26-32.
35. Rhim, J. W. (2013) Preparation and characterization of vacuum sputter silver coated PLA Film. *LWT - Food Science and Technology,* **54**, 477-484.
36. Lee, B. K., Yun, Y., and Park, K. (2016) PLA micro- and nano-particles. *Advanced Drug Delivery Reviews,* **107,** 176-191.
37. Yalcinkaya, E. E., Puglia, D., Fortunati, E., Bertoglio, F., Bruni, G., Visai, L., and Kenny, J.M. (2017) Cellulose nanocrystals as templates for cetyltrimethylammonium bromide mediated synthesis of Ag nanoparticles and their novel use in PLA films. *Carbohydrate Polymers,* **157,** 1557-1567.
38. Zhang, H., Hortal, M., Beneyto, M. J., Rosa, E., Lledo, M. L., and Lorente, I. (2017). ZnO-PLA nanocomposite coated paper for antimicrobial packaging application. *LWT - Food Science and Technology,* **78,** 250-257.
39. Tsai, P. A., Chiu, W. M., Lin, C. E., and Wu, J. H. (2013) Fabrication and

characterization of PLA/sio$_2$/Al_2O_3 composites prepared by sol-gel process. *Polymer-Plastics Technology and Engineering*, **52**, 1488-1495.

40. Luckachan, G. E., and Mittal, V. (2016) Self-healing anticorrosion coatings for gas pipelines and storage tanks. *Corrosion Science and Technology*, **15** (5), 209-216.
41. Mahltig, B., and Bottcher, H. (2003) Modified silica sol coatings for water-repellent textiles. *Journal of Sol-Gel Science and Technology*, **27**, 43-52.
42. Wu, F., Lan X., Ji, D., Liu, Z., Yang, W., and Yang, M. (2013) Grafting polymerization of polylactic acid on the surface of nano-SiO2 and properties of PLA/PLA-grafted-SiO_2 nanocomposites. *Journal of Applied Polymer Science*, **129**(5), 3019-3027.
43. Mittal. V., Akhtar, T., Luckachan, G. E., and Matsko, N. (2015) PLA, TPS and PCL binary and ternary blends: structural characterization and time-dependent morphological changes. *Colloid and Polymer Science*, **293**, 573-585.
44. Kister, G., Cassanas, G., and Vert, M. (1998) Effects of morphology, conformation and configuration on the IR and Raman spectra of various poly(lactic acid)s. *Polymer*, **39**(2), 267-273.
45. Lin, R. H., Lin, C. W., Lee, A. C., Chen, Y. H., and Yen, F. S. (2007) Highly crosslinked nanocomposites of aromatic dicyanates/SiO_2 via the sol-gel method. *Journal of Applied Polymer Science*, **103**, 1356-1366.
46. Akid, R., Gobara, M., and Wang, H. (2011) Corrosion protection performance of novel hybrid polyaniline/sol–gel coatings on an aluminium 2024 alloy in neutral, alkaline and acidic solutions. *Electrochimica Acta*, **56**, 2483-2492.
47. Latnikova, A., Grigoriev, D., Schenderlein, M., Mohwald, H., and Shchukin, D. (2012) A new approach towards "active" self-healing coatings: exploitation of microgels. *Soft Matter*, **8**, 10837-10844.
48. Akhtar, T. (2015) *Polyetherimide Coatings for Corrosion Protection*, MSc Thesis, The Petroleum Institute, UAE.
49. Murariu, M., Doumbia, A., Bonnaud, L., Dechief, A. L., Paint, Y., Ferreira, M., Campagne, C., Devaux, E., and Dubois, P. (2011) High-performance polylactide/ZnO nanocomposites designed for films and fibers with special end-use properties. *Biomacromolecules*, **12**, 1762-1771.
50. Mittal, V., Chaudhry, A. U., and Luckachan, G. E. (2014) Biopolymer-thermally reduced graphene nanocomposites: Structural characterization and properties. *Materials Chemistry and Physics*, **147**(1-2), 319-332.
51. Thellen, C., Orroth, C., Froio, D., Ziegler, D., Lucciarini, J., Farrell, R., Ann D'Souza, N., and Ratto, J. A. (2005) Influence of montmorillonite layered silicate on plasticized poly(L-lactide) blown films. *Polymer*, **46**(25), 11716-11727.

52. Rout, T. K., and Gaikwad, A. V. (2015) In-situ generation and application of nanocomposites on steel surface for anti-corrosion coating. *Progress in Organic Coatings*, **79**, 98-105.

Index

A

B

C

D

E

F

G

H

I

L

M

N

O

P

R

S

T

U

W

www.ingramcontent.com/pod-product-compliance
Lightning Source LLC
LaVergne TN
LVHW011527200926
848484LV00001B/181
* 9 7 8 1 9 2 5 8 2 3 0 4 2 *